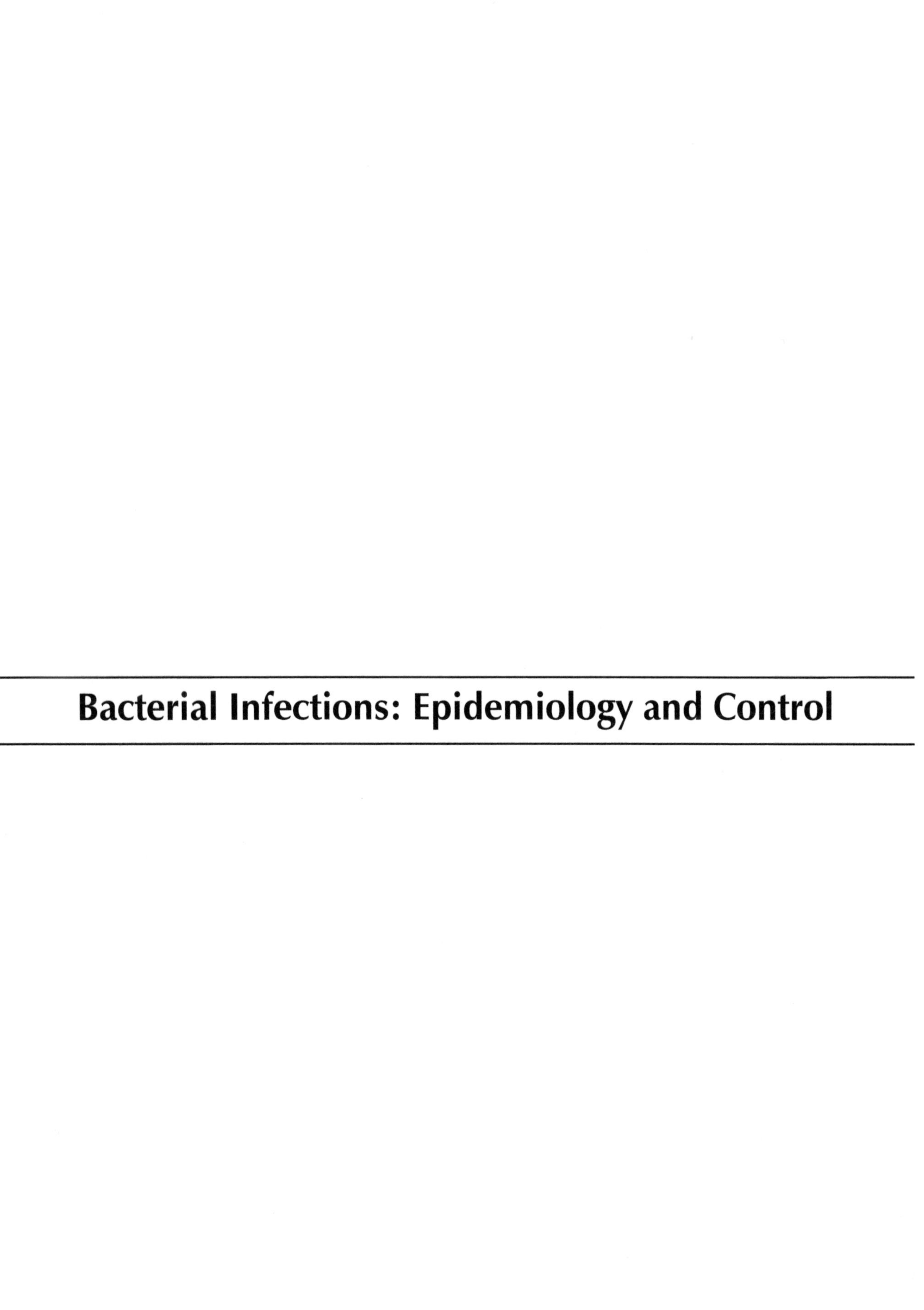

# Bacterial Infections: Epidemiology and Control

# Bacterial Infections: Epidemiology and Control

Editor: Colin Hunt

**Cataloging-in-Publication Data**

Bacterial infections : epidemiology and control / edited by Colin Hunt.
p. cm.
Includes bibliographical references and index.
ISBN 978-1-63927-592-2
1. Bacterial diseases. 2. Bacterial diseases--Epidemiology. 3. Bacterial diseases--Prevention.
4. Communicable diseases. 5. Medical bacteriology. 6. Pathogenic bacteria. I. Hunt, Colin.
RA644.B32 B33 2023
614.57--dc23



American Medical Publishers,
41 Flatbush Avenue,
1st Floor, New York,
NY 11217, USA

ISBN 978-1-63927-592-2 (Hardback)

# Preface

Bacteria are single cell microbes, which have the capability to survive without any support, both inside and outside the human body. Several bacteria can be found in the human body, such as the bacteria present in the gut, which are helpful in the digestion of food. However, some bacteria can result in infections and can affect the skin, throat, bowel, lungs and other body parts. Majority of the infections are mild but in certain cases infection can be severe. Urinary tract infection, strep throat, ear infection, and cough are some examples of bacterial infections. They are caused due to bacteria entering the human body through cuts and openings in the skin, which multiply in number, thereby causing a reaction. Antibiotics are effective in treating majority of the infections, which kill and stop the bacteria from multiplying. Prevention of bacterial infections can be done by covering mouth while coughing and sneezing, washing hands and not sharing utensils. This book includes some of the vital pieces of work being conducted across the world, on various topics related to bacterial infections. Also included herein is a detailed explanation of the epidemiology and control of these infections. This book will serve as a reference to a broad spectrum of readers.

The information shared in this book is based on empirical researches made by veterans in this field of study. The elaborative information provided in this book will help the readers further their scope of knowledge leading to advancements in this field.

Finally, I would like to thank my fellow researchers who gave constructive feedback and my family members who supported me at every step of my research.

**Editor**

# Preface

# 1

# Host–Receptor Post-Translational Modifications Refine Staphylococcal Leukocidin Cytotoxicity

**Angelino T. Tromp [1], Michiel Van Gent [1,2], Joris P. Jansen [1], Lisette M. Scheepmaker [1], Anneroos Velthuizen [1], Carla J.C. De Haas [1], Kok P.M. Van Kessel [1], Bart W. Bardoel [1], Michael Boettcher [3,4], Michael T. McManus [3], Jos A.G. Van Strijp [1], Robert Jan Lebbink [1], Pieter-Jan A. Haas [1] and András N. Spaan [1,5,*]**

1 Department of Medical Microbiology, University Medical Center Utrecht, 3584 CX Utrecht, The Netherlands; A.T.Tromp-3@umcutrecht.nl (A.T.T.); mvangent@uchicago.edu (M.V.G.); joris_jansen24@hotmail.com (J.P.J.); L.M.Scheepmaker-2@umcutrecht.nl (L.M.S.); A.Velthuizen@umcutrecht.nl (A.V.); C.J.C.deHaas@umcutrecht.nl (C.J.C.D.H.); K.Kessel@umcutrecht.nl (K.P.M.V.K.); B.W.Bardoel-2@umcutrecht.nl (B.W.B.); J.vanStrijp@umcutrecht.nl (J.A.G.V.S.); R.J.Lebbink-2@umcutrecht.nl (R.J.L.); P.J.A.Haas@umcutrecht.nl (P.-J.A.H.)

2 Department of Microbiology, University of Chicago, Chicago, IL 60637, USA

3 Department of Microbiology and Immunology, UCSF Diabetes Center, Keck Center for Noncoding RNA, University of California, San Francisco, San Francisco, CA 94143, USA; michael.boettcher@medizin.uni-halle.de (M.B.); Michael.McManus@ucsf.edu (M.T.M.)

4 Medical Faculty, Martin Luther University Halle-Wittenberg, 06120 Halle (Saale), Germany

5 St. Giles Laboratory of Human Genetics of Infectious Diseases, Rockefeller Branch, The Rockefeller University, New York, NY 10065, USA

* Correspondence: a.n.spaan@umcutrecht.nl

**Abstract:** Staphylococcal bi-component pore-forming toxins, also known as leukocidins, target and lyse human phagocytes in a receptor-dependent manner. S-components of the leukocidins Panton-Valentine leukocidin (PVL), $\gamma$-haemolysin AB (HlgAB) and CB (HlgCB), and leukocidin ED (LukED) specifically employ receptors that belong to the class of G-protein coupled receptors (GPCRs). Although these receptors share a common structural architecture, little is known about the conserved characteristics of the interaction between leukocidins and GPCRs. In this study, we investigated host cellular pathways contributing to susceptibility towards *S. aureus* leukocidin cytotoxicity. We performed a genome-wide CRISPR/Cas9 library screen for toxin-resistance in U937 cells sensitized to leukocidins by ectopic expression of different GPCRs. Our screen identifies post-translational modification (PTM) pathways involved in the sulfation and sialylation of the leukocidin-receptors. Subsequent validation experiments show differences in the impact of PTM moieties on leukocidin toxicity, highlighting an additional layer of refinement and divergence in the staphylococcal host-pathogen interface. Leukocidin receptors may serve as targets for anti-staphylococcal interventions and understanding toxin-receptor interactions will facilitate the development of innovative therapeutics. Variations in the genes encoding PTM pathways could provide insight into observed differences in susceptibility of humans to infections with *S. aureus*.

**Keywords:** *Staphylococcus aureus*; bi-component pore-forming toxins; leukocidins; receptors; G-protein coupled receptors; post-translational modifications

**Key Contribution:** By taking advantage of leukocidin-specific receptor-engagement, this study identifies host cellular post-translational modification pathways that refine GPCR-mediated susceptibility of human phagocytes to the staphylococcal leukocidins.

## 1. Introduction

*Staphylococcus aureus* is a Gram-positive bacterium that colonizes the skin and anterior nares of 20%–30% of the general human population [1]. *S. aureus* also causes a variety of diseases, ranging from superficial skin and soft tissue infections to severe invasive infections with a poor prognosis and high mortality [2]. Upon infection, *S. aureus* is faced with the host humoral and cellular innate immune response [3]. *S. aureus*, in return, secretes an arsenal of virulence factors to circumvent host defenses and to avoid killing by phagocytes [4]. An important group of *S. aureus* virulence factors, the leukocidins, specifically target and lyse host phagocytes [5,6]. *S. aureus* leukocidins are bi-component beta-barrel pore-forming toxins [6]. Human *S. aureus* isolates secrete up to five leukocidins: Panton-Valentine leukocidin (PVL), γ-haemolysin AB (HlgAB) and CB (HlgCB), leukocidin ED (LukED) and leukocidin AB (LukAB, also knowns as LukGH) [6]. Based on chromatography elution profiles, the two individual leukocidin subunits are designated S- (slow migrating) or F- (fast migrating) components [5].

Proteinaceous targets have been identified for all *S. aureus* leukocidins. The S-component of the leukocidins, with the exception of LukAB, target specific G-protein coupled receptors (GPCRs) expressed on the surface of host cells [5]. The C5a anaphylatoxin chemotactic receptor 1 (C5aR1, also known as CD88) and C5a anaphylatoxin chemotactic receptor 2 (C5aR2, also known as C5L2) were identified as targets for PVL and HlgCB [7,8]. LukED targets leukocytes via CC-chemokine receptor 5 (CCR5), as well as CXC chemokine receptor 1 (CXCR1) and CXC chemokine receptor 2 (CXCR2) [9,10]. HlgAB targets CXCR1, CXCR2 and CC-chemokine receptor 2 (CCR2) [8]. In addition, HlgAB and LukED both target the Duffy antigen receptor for chemokines (DARC, also known as ACKR1), an atypical chemokine receptor expressed on erythrocytes [11]. Although these receptors share a seven-transmembrane spanning structural architecture common to all GPCRs, little is known about the conserved or divergent characteristics of the interaction between leukocidins and their respective GPCR host-counterparts. The apparent redundancy of the leukocidins in terms of overlapping receptors and host target cell populations remains enigmatic. Furthermore, additional molecular determinants of the host target cell involved in leukocidin-receptor interactions are incompletely understood.

In this study, we applied a genome-wide CRISPR/Cas9 library screen to identify host factors involved in PVL- and HlgCB-mediated cytotoxicity. We identify post-translational modification (PTM) pathways that refine GPCR-mediated susceptibility of human phagocytes to leukocidins. Sulfation-mediated receptor-employment serves as a major and conserved feature for C5aR1-interacting leukocidins. In contrast, sialylation rather than sulfation is a major PTM motif facilitating cytotoxicity of CXCR2-targeting leukocidins. These findings further substantiate the complexity underlying the divergent interaction between *S. aureus* bi-component pore-forming toxins and their target cells.

## 2. Results

### *2.1. PTM Pathways Affect Susceptibility to PVL and HlgCB Cytotoxicity.*

To identify host factors involved in PVL- and HlgCB-mediated susceptibility of human phagocytes, a genome-wide CRISPR/Cas9 library screen for both PVL- and HlgCB resistance was set up in human U937 promyelocytic cells. Cells were sensitized to PVL- and HlgCB mediated pore-formation by overexpressing C5aR1 (U937-C5aR1), followed by the introduction of a human codon-optimized nuclear-localized *S. pyogenes* cas9 gene (U937-C5aR1-SpCas9). Host factors involved in PVL and HlgCB toxicity were detected via the introduction of a genome-wide sgRNA library coupled to deep sequencing, allowing for the identification of genes inactivated in cells surviving toxin treatment. *C5AR1*, encoding the LukS-PV and HlgC receptor C5aR1, was identified as a top hit in both the HlgCB- and PVL-resistance screen, validating the screening method (Figure 1 and Tables S1 and S2). As recently reported, *PTPRC*, encoding the Luk-F-PV specific F-component receptor CD45, was identified in the PVL-, but not HlgCB-screen [12]. Consistent with screenings using other pore-forming toxins, the gene encoding the alpha-hemolysin (Hla) determinant sphingomyelin synthase 1 (*SGMS1*) [13] was also identified for both PVL and HlgCB (Figure 1).

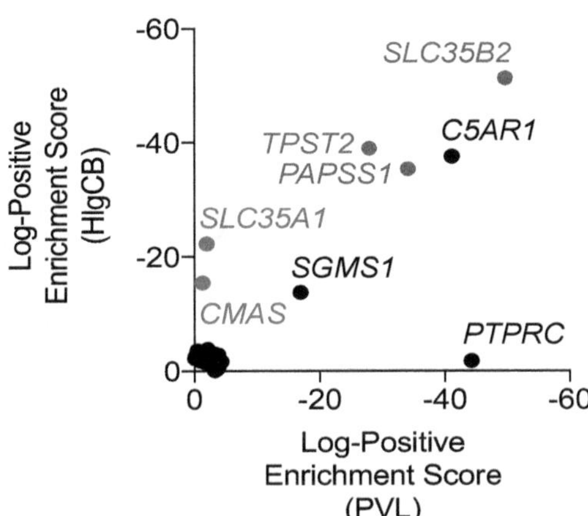


**Figure 1.** Genome-wide CRISPR/Cas9 library screen reveals post-translational modification pathways involved in PVL and HlgCB toxicity. Cellular factors involved in PVL- and HlgCB-mediated cytotoxicity as identified by the introduction of a genome-wide sgRNA library in U937-C5aR1-SpCas9 cells coupled to deep sequencing. Visualized are the most significantly enriched genes after PVL and HlgCB challenge as calculated by the MaGeCK 'positive enrichment score'. See Tables S1 and S2 for the full list of genes. Grey: genes encoding known host cellular factors; blue: genes belonging to the tyrosine sulfation pathway; red: genes involved in the sialylation pathway.

Unexpectedly, the screenings for both PVL and HlgCB-resistance revealed enrichment for the genes encoding the Solute Carrier Family 35 Member B2 (*SLC35B2*), 3′-Phosphoadenosine 5′-Phosphosulfate Synthase 1 (*PAPSS1*) and Tyrosylprotein Sulfotransferase 2 (*TPST2*) (Figure 1). In addition, the screening for HlgCB-resistance was enriched for the genes encoding the Solute Carrier Family 35 Member A1 (*SLC35A1*) and Cytidine Monophosphate N-Acetylneuraminic Acid Synthetase (*CMAS*) (Figure 1). SLC35b2, PAPSS1 and TPST2 are key components of the protein sulfation pathway in which tyrosine residues are enzymatically decorated with a sulfate group [14]. SLC35a1 and CMAS are part of the sialylation pathway, a modification process resulting in the attachment of sialic acids to other molecules [15]. As our screenings with PVL and HlgCB in C5aR1-expressing cells identified multiple genes in two major PTM pathways, we subsequently assessed the contribution of sulfation and sialylation to *S. aureus* leukocidin susceptibility.

### *2.2. Sulfation of C5aR1 Facilitates both PVL and HlgCB Cytotoxicity.*

To validate the involvement of *SLC35B2*, *PAPSS1* and *TPST2* in PVL and HlgCB cytotoxicity, single knock-out cells were generated in U937-C5aR1-SpCas9 cells. Single knock-out cells where incubated with different antibodies to assess the expression of specific targets and evaluated by flow cytometry [12]. Individually knocking-out *SLC35B2* or *PAPSS1*, and to an extent *TPST2*, resulted in a decrease of overall cell surface tyrosine sulfation, as determined using an anti-sulfotyrosine antibody (Figure 2a). Consistent results were obtained in single knock-out U937-SpCas9 cells (Figure S1). Lack of tyrosine sulfation did not affect the overall levels of C5aR1 expression, as detected with a sulfation-independent binding anti-C5aR1 antibody (clone S5/1, Figure 2a). Subsequently, $C5aR1^+SLC35b2^-$, $C5aR1^+PAPSS1^-$ and $C5aR1^+TPST2^-$ cells were challenged with PVL and HlgCB. Consistent with the screening results, $C5aR1^+SLC35b2^-$, $C5aR1^+PAPSS1^-$ and $C5aR1^+TPST2^-$ cells showed resistance to pore formation induced by both PVL and HlgCB (Figure 2b). These data identify sulfation as a conserved posttranslational modification for the interaction of both PVL and HlgCB with their target cells.

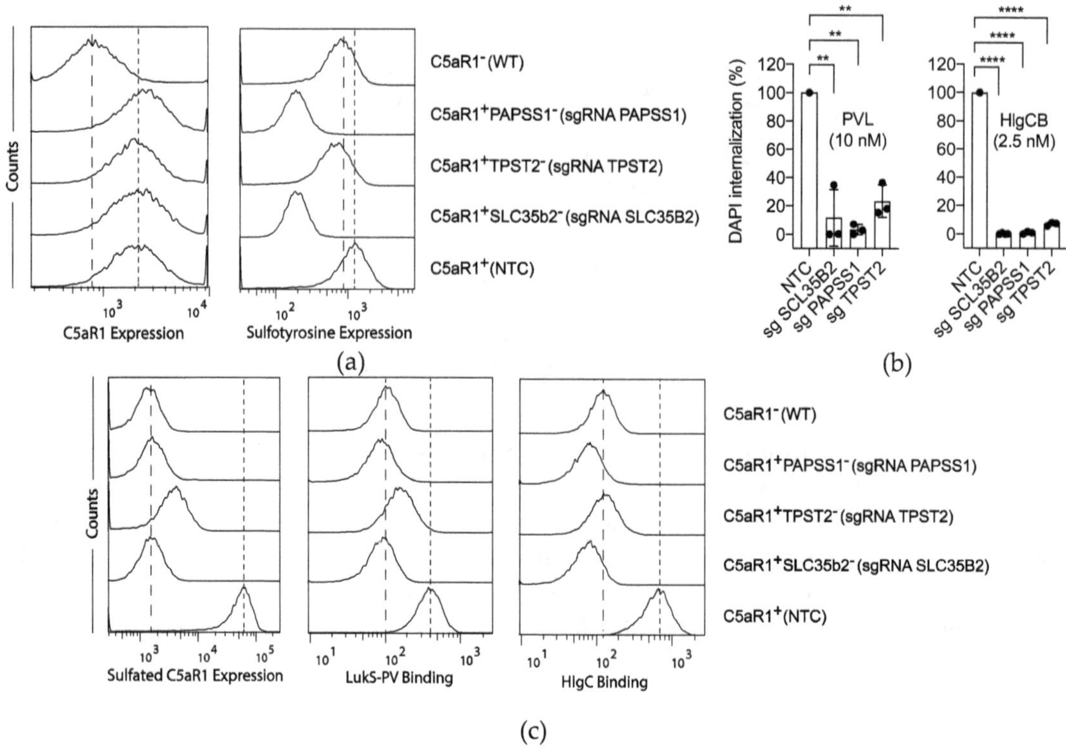


**Figure 2.** Sulfation of C5aR1 facilitates PVL and HlgCB cytotoxicity. (**a**) Anti-C5aR1 (clone S5/1) and anti-sulfotyrosine antibodies were used to assess the overall expression of C5aR1 and total sulfotyrosine in U937-C5aR1-SpCas9 cell lines transduced with sgRNA for *PAPSS1* ($C5aR1^+$ $PAPSS1^-$), *TPST2* ($C5aR1^+$ $TPST2^-$), *SLC35B2* ($C5aR1^+$ $SLC35b2^-$), non-targeting control sgRNA (NTC, $C5aR1^+$), and U937-SpCas9 (WT, $C5aR1^-$) cells. Antibody binding was determined by a fluorescent secondary antibody and the fluorescence measured and analyzed by flow cytometry. Dashed line: expression in U937-SpCas9 (WT, $C5aR1^-$) cells; dotted line: C5aR1 expression in NTC ($C5aR1^+$) U937 cells. Histograms depict representative examples of two independently repeated experiments. (**b**) Validation of the sulfation-pathway hits after genome-wide CRISPR/Cas9 screen for PVL and HlgCB resistance in U937-C5aR1-SpCas9 cells. As a readout for cell permeability, internalization of DAPI was tested at 30 min post-toxin treatment on a monochromator-based microplate reader and expressed in relation to U937-C5aR1-SpCas9 cells transduced with an NTC sgRNA. Mean and s.d. are shown, with $n = 3$. Statistical significance was calculated using ANOVA analysis of variance with Bonferroni posttest correction for multiple comparison. Statistical significance is displayed as ** for $p < 0.01$ and **** for $p < 0.0001$. (**c**) Expression of sulfated C5aR1 in, and binding of polyhistidine-tagged LukS-PV and HlgC to, U937-C5aR1-SpCas9 cell lines transduced with sgRNA for *PAPSS1* ($C5aR1^+$ $PAPSS1^-$), *TPST2* ($C5aR1^+$ $TPST2^-$), *SLC35B2* ($C5aR1^+$ $SLC35b2^-$), non-targeting control sgRNA (NTC, $C5aR1^+$), and U937-SpCas9 (WT, $C5aR1^-$) cells. Anti-C5aR1 (clone 347214) antibodies were used to assess the expression of sulfated C5aR1. Antibody binding was determined by a fluorescent secondary antibody. Binding of LukS-PV and HlgC was detected with anti-His-FITC antibodies. Fluorescence was measured and analyzed by flow cytometry. Dashed line: expression on and binding to U937-SpCas9 (WT, $C5aR1^-$) cells; dotted line: expression on and binding to NTC ($C5aR1^+$) U937 cells. Histograms depict representative examples of two independently repeated experiments.

Binding of LukS-PV, the S-component of PVL, to a synthetic N-terminal C5aR1-peptide was previously shown to be mediated by sulfation of two peptide-tyrosine residues [7]. We hypothesized that knocking-out *SLC35B2, PAPSS1* or *TPST2* in U937-C5aR1 cells reduces C5aR1-sulfation and subsequent binding of the S-components LukS-PV and HlgC. During the evaluation of multiple anti-C5aR1 antibodies, we observed a sulfation-dependent binding of the anti-human C5aR1 antibody clone 347214. While the overall level of C5aR1 expression was not affected (Figure 2a), knocking-out *SLC35B2* or *PAPSS1* indeed reduced C5aR1-sulfation as assessed with this sulfation-dependent binding anti-C5aR1 antibody (Figure 2c). In addition, although knocking-out *TPST2* only slightly affected overall tyrosine sulfation (Figure 2a), C5aR1-sulfation was reduced in C5aR1$^+$TPST2$^-$ cells (Figure 2c). Next, we assessed the binding of LukS-PV and HlgC to these cells. As expected, binding of LukS-PV and HlgC was impaired in C5aR1$^+$SLC35b2$^-$, C5aR1$^+$PAPSS1$^-$ and C5aR1$^+$TPST2$^-$ cells (Figure 2c). Thus, these results show that cellular susceptibility towards both PVL and HlgCB is mediated by a conserved sulfation-dependent binding of LukS-PV and HlgC to C5aR1.

### 2.3. *The role of C5aR1 Sialylation in HlgCB- and PVL-Induced Cytotoxicity.*

In addition to genes involved in tyrosine sulfation, our screening for HlgCB resistance identified genes involved in the sialylation pathway. To confirm these findings, *SLC35A1* and *CMAS* single knock-out cells were generated in U937-C5aR1-SpCas9 cells. The level of sialylated LewisX (CD15s) expression was used as a readout to assess overall cellular sialylation. C5aR1$^+$SLC35a1$^-$ and C5aR1$^+$CMAS$^-$ cells showed no binding of anti-Sialyl-LewisX antibodies (Figure 3a), confirming the lack of sialic acid modifications in these cells. Consistent results were obtained in single knock-out U937-SpCas9 cells (Figure S1). The overall C5aR1-expression was mildly impaired in the mutant U937-C5aR1-SpCas9 cell lines (Figure 3a).

Subsequent toxin challenge confirmed a reduced susceptibility to pore formation induced by HlgCB in C5aR1$^+$SLC35A1$^-$ and C5aR1$^+$CMAS$^-$ cells (Figure 3b). As *SLC35A1* was also identified in our PVL-resistance screen (yet with a lower enrichment score, Figure 1 and Tables S1 and S2), we challenged C5aR1$^+$SLC35a1$^-$ and C5aR1$^+$CMAS$^-$ cells with PVL as well. Knocking-out *SLC35A1* or *CMAS* however reduced cell susceptibility towards PVL only mildly and in a statistically insignificant manner (Figure 3b). Thus, these data identify sialylation of C5aR1 as a PTM necessary for the interaction with HlgCB and, to a lesser extent, PVL.

As with the sulfation pathway, we hypothesized that reduced susceptibility of C5aR1$^+$SLC35A1$^-$ and C5aR1$^+$CMAS$^-$ cells to HlgCB and PVL is due to an impaired interaction of the toxin's S-component with C5aR1. To test this, we determined the binding of HlgC and LukS-PV to C5aR1$^+$SLC35a1$^-$ and C5aR1$^+$CMAS$^-$ cells. Although binding of HlgC to C5aR1$^+$SLC35a1$^-$ and C5aR1$^+$CMAS$^-$ cells was indeed reduced (Figure 3c), no differences were detected in the binding of LukS-PV (Figure 3c).

Taken together, these results show that sialylation of C5aR1 is a molecular determinant mediating susceptibly to HlgCB more than PVL. For HlgCB, sialylation drives refinement of cytotoxicity by facilitating the binding of the S-component, combined with a modulation of receptor expression levels. For PVL, the limited impact of sialylation on cellular susceptibility is likely mediated by receptor surface expression levels solely.

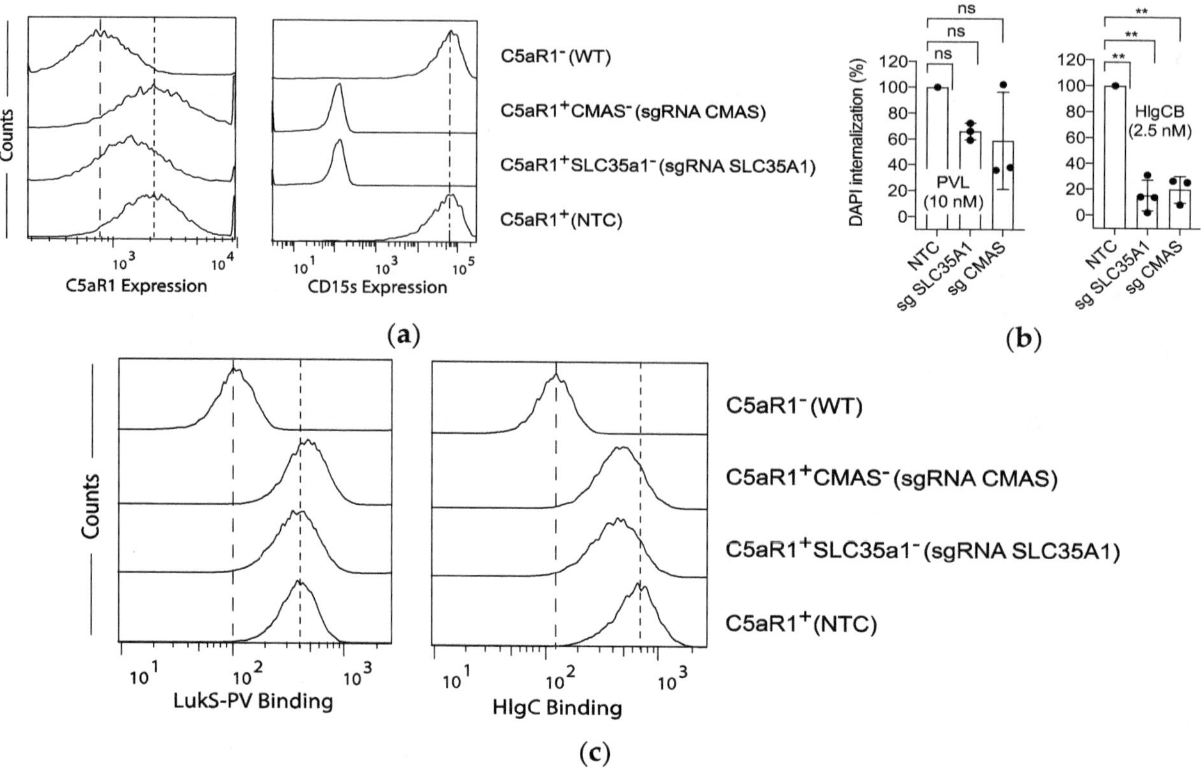


**Figure 3.** The role of sialylation in HlgCB- and PVL-induced cytotoxicity. (**a**) Anti-C5aR1 and anti-CD15s antibodies were used to assess the expression of C5aR1 and CD15s on U937-C5aR1-SpCas9 cell lines transduced with sgRNA for *CMAS* (C5aR1$^+$ CMAS$^-$), *SLC35A1* (C5aR1$^+$ SLC35a1$^-$), non-targeting control sgRNA (NTC, C5aR1$^+$), and U937-SpCas9 (WT, C5aR1$^-$) cells. Antibody binding was determined by a fluorescent secondary antibody and the fluorescence measured and analyzed by flow cytometry. Dashed line: expression in U937-SpCas9 (WT, C5aR1$^-$) cells; dotted line: C5aR1 expression in NTC (C5aR1$^+$) U937 cells. Histograms depict representative examples of two independently repeated experiments. (**b**) Validation of the sialylation-pathway hits after genome-wide CRISPR/Cas9 screen for PVL and HlgCB resistance in U937-C5aR1-SpCas9 cells. As a readout for cell permeability, internalization of DAPI was tested at 30 min post-toxin treatment on a monochromator-based microplate reader and expressed in relation to U937-C5aR1-SpCas9 cells transduced with an NTC sgRNA. Mean and s.d. are shown, with $n = 3$. Statistical significance was calculated using ANOVA analysis of variance with Bonferroni posttest correction for multiple comparison. Statistical significance is displayed as ** for $p < 0.01$ and NS for not significant. (**c**) Binding of polyhistidine-tagged LukS-PV and HlgC to U937-C5aR1-SpCas9 cell lines transduced with sgRNA for *CMAS* (C5aR1$^+$ CMAS$^-$), *SLC35A1* (C5aR1$^+$ SLC35a1$^-$), non-targeting control sgRNA (NTC, C5aR1$^+$) and wild type U937 (C5aR1$^-$). Cells were subsequently incubated with anti-His-FITC antibodies and the fluorescence measured and analyzed by flow cytometry. Dashed line: binding to U937-SpCas9 (WT, C5aR1$^-$) cells; dotted line: binding to NTC (C5aR1$^+$) U937 cells. Histograms depict representative examples of two independently repeated experiments.

### *2.4. Sulfation and to a Lesser Extent Sialylation Refine Susceptibility to PVL and HlgCB.*

To further assess the extent of the role of *SLC35B2, PAPSS1, TPST2, SLC35A1* and *CMAS* in cellular susceptibility to PVL and HlgCB, mutant cells were incubated with different concentrations of PVL or HlgCB. Pore formation was defined as the time and concentration dependent collective DAPI-internalization (Figure S2). The area under the curve was calculated and subsequently related to U937-C5aR1-SpCas9 cells transduced with a non-targeting control sgRNA [12].

As previously reported, the expression of C5aR1 was essential for sensitization of cells towards PVL and HlgCB cytotoxicity [7,16]. Results obtained for the individual genes in the sulfation pathway (*SLC35B2, PAPSS1,* and *TPST2*) were similar, as were those for the genes in the sialylation pathway (*SLC35A1* and *CMAS*). The absence of cellular sulfation resulted in a ~5-fold increase in the half-maximum effective concentration (EC50) for PVL and a ~15-fold increase in the EC50 for HlgCB (Figure 4a and Table 1). The increase in the EC50 in cells lacking sialylation was limited, with a ~2.5-fold shift for HlgCB and a statistically insignificant trend towards a ~1.5-fold shift for PVL (Figure 4b and Table 1).

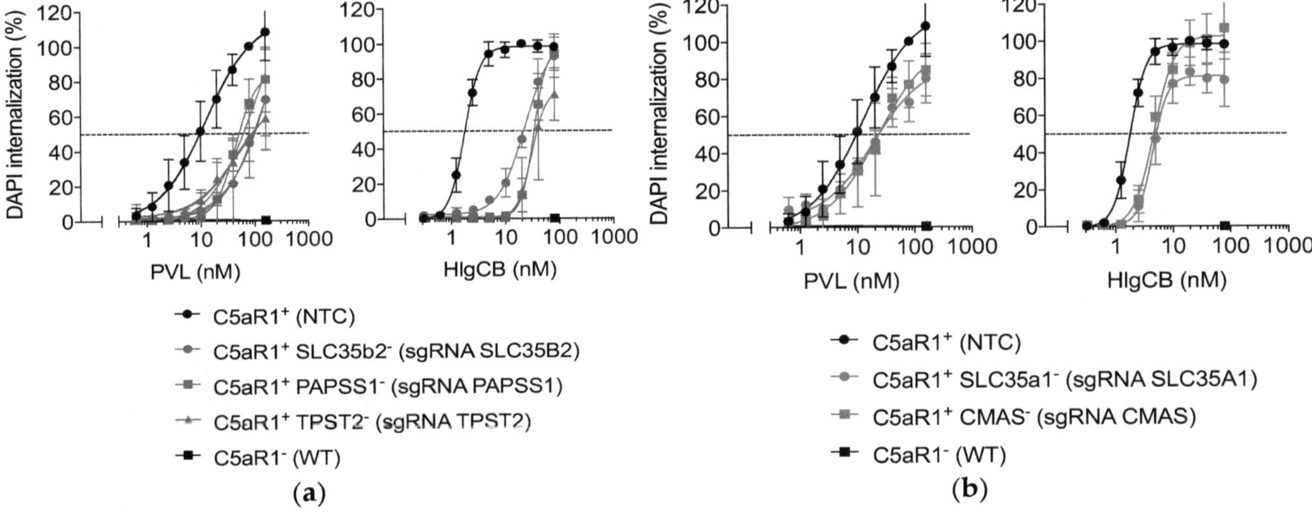


**Figure 4.** Sulfation and sialylation of C5aR1 refine susceptibility to PVL and HlgCB. Susceptibility of U937-C5aR1-SpCas9 cell lines transduced with sgRNA for (**a**) the sulfation pathway genes *SLC35B2* ($C5aR1^+$ $SLC35b2^-$), *PAPSS1* ($C5aR1^+$ $PAPSS1^-$), *TPST2* ($C5aR1^+$ $TPST2^-$), and (**b**) the sialylation pathway genes *SLC35A1* ($C5aR1^+$ $SLC35a1^-$), *CMAS* ($C5aR1^+$ $CMAS^-$), a non-targeting control sgRNA (NTC, $C5aR1^+$), and U937-SpCas9 (WT, $C5aR1^-$) cells to PVL and HlgCB. As a readout for cell permeability, internalization of DAPI was measured during 30 min post-toxin treatment on a monochromator-based microplate reader and expressed in relation to the area under the curve for NTC sgRNA transduced U937-C5aR1-SpCas9 cells at 80 nM PVL or 20 nM HlgCB.

**Table 1.** Sulfation and sialylation of C5aR1 refine susceptibility to PVL and HlgCB. Half-maximum effective concentrations (EC50s) of U937-C5aR1-SpCas9 cells transduced with a non-targeting control sgRNA (NTC, $C5aR1^+$) and the EC50 and fold increase in U937-C5aR1-SpCas9 cell lines transduced with the sgRNA *SLC35B2* ($C5aR1^+$ $SLC35b2^-$), *PAPSS1* ($C5aR1^+$ $PAPSS1^-$), *TPST2* ($C5aR1^+$ $TPST2^-$), *SLC35A1* ($C5aR1^+$ $SLC35a1^-$) and *CMAS* ($C5aR1^+$ $CMAS^-$) after exposure to PVL or HlgCB. EC50 values were calculated using four-parametric non-linear regression analyses. Fold increased EC50 values are expressed in relation to the EC50 of the NTC, with their corresponding 95% confidence interval.

| sgRNA | PVL | | | HlgCB | | |
|---|---|---|---|---|---|---|
| | EC50 (nM) | 95% Conf. Interval (nM) | Fold Increased EC50 | EC50 (nM) | 95% Conf. Interval (nM) | Fold Increased EC50 |
| NTC | 12.9 | 9.1–21.5 | | 1.8 | 1.7–1.9 | |
| *SLC35B2* | 123.1 | 38.1–infin | 9.5 | 30.3 | 23.1–50.6 | 16.8 |
| *PAPSS1* | 44.6 | 35.7–63.4 | 3.5 | 33.3 | 30.6–36.3 | 18.5 |
| *TPST2* | 42.5 | 22.2–infin | 3.3 | 23.1 | 18.6–34.3 | 12.8 |
| *SLC35A1* | 18.5 | 11.5–67.4 | 1.4 | 4.5 | 3.9–5.1 | 2.5 |
| *CMAS* | 19.4 | 11.3–79.6 | 1.5 | 4.7 | 4.0–5.6 | 2.6 |

These findings show that, although sialylation and sulfation of C5aR1 are not essential for PVL and HlgCB cytotoxicity at high toxin concentrations, these PTM pathways further refine toxin susceptibility of human phagocytic cells. For the C5aR1-targeting leukocidins PVL and HlgCB, the impact of sulfation is more extensive when compared to sialylation.

### *2.5. Sialyation and to a Lesser Extent Sulfation Refine Susceptibility to LukED and HlgAB.*

We hypothesized that PTMs of host receptors not only play a role in the interaction of C5aR1-interacting toxins, but also in the interaction of other leukocidins with their GPCR-targets. To study the role of sulfation and sialylation in LukED and HlgAB receptor interactions, we engineered specific gene deletants in U937 cells ectopically expressing their shared cell surface target CXCR2 (U937-CXCR2-SpCas9). As representatives of the sialylation and sulfation pathways, *SLC35B2* and *SLC35A1* were selected for gene editing. As with C5aR1$^+$SLC35a1$^-$ cells, expression of CXCR2 on CXCR2$^+$SLC35a1$^-$ cells was mildly impaired while CXCR2 expression on CXCR2$^+$SLC35b2$^-$ cells was unaffected (Figure 5a). CXCR2$^+$SLC35b2$^-$ cells lacked detectable levels of sulfation while CXCR2$^+$SLC35a1$^-$ cells were devoid of sialylation (Figure 5a).

Next, CXCR2$^+$SLC35a1$^-$ and CXCR2$^+$SLC35b2$^-$ cells were challenged with a concentration range of HlgAB or LukED. In line with previous reports, CXCR2-expression was essential for sensitization of cells towards both HlgAB and LukED cytotoxicity (Figure 5b) [8,9]. Opposite to our observations with C5aR1, CXCR2$^+$SLC35a1$^-$ cells were ~4 and ~5 fold less sensitive to both HlgAB and LukED, respectively (Figure 5b and Table 2) while CXCR2$^+$SLC35b2$^-$ cells showed only a minor trend towards a decreased sensitivity for HlgAB and LukED (Figure 5b and Table 2). Thus, for the CXCR1-targeting leukocidins HlgAB and LukED, the impact of sialylation is more extensive when compared to sulfation.

**Table 2.** Sulfation and sialylation of CXCR2 refine susceptibility to LukED and HlgAB. Half-maximum effective concentrations (EC50s) of U937-CXCR2-SpCas9 cells transduced with a non-targeting control sgRNA (NTC, CXCR2$^+$) and the EC50 and fold increase in U937-CXCR2-SpCas9 cell lines transduced with the sgRNA *SLC35B2* (CXCR2$^+$ SLC35b2$^-$) and *SLC35A1* (CXCR2$^+$ SLC35a1$^-$) after exposure to HlgAB or LukED. EC50 values were calculated using four-parametric non-linear regression analyses. Fold increased EC50 values are expressed in relation to the EC50 of the NTC, with their corresponding 95% confidence interval.

| **sgRNA** | **HlgAB** | | | **LukED** | | |
|---|---|---|---|---|---|---|
| | **EC50 (nM)** | **95% Conf. Interval (nM)** | **Fold Increased EC50** | **EC50 (nM)** | **95% Conf. Interval (nM)** | **Fold Increased EC50** |
| NTC | 0.8 | 0.7–0.9 | | 4.4 | 3.4–5.7 | |
| *SLC35B2* | 1.1 | 0.7–1.7 | 1.4 | 11.8 | 10.1–14.0 | 2.7 |
| *SLC35A1* | 2.9 | 2.1–5.1 | 3.6 | 22.7 | 19.8–27.3 | 5.2 |

As with C5aR1, we subsequently investigated if the observed PTM-dependent resistance in CXCR2-expressing cells could be correlated to a reduced binding of HlgA or LukE, the respective S-components of HlgAB and LukED. Reminiscent of C5aR1-expressing cells and LukS-PV and HlgC, binding of HlgA and LukE was reduced in CXCR2$^+$SLC35b2$^-$ cells. Although CXCR2$^+$SLC35a1$^-$ cells showed a reduced susceptibility to pore formation when challenged with HlgAB or LukED, only a reduced binding of LukE, but not HlgA, could be detected in CXCR2$^+$SLC35a1$^-$ cells (Figure 5c). This apparent discrepancy may be attributed to the facts that 1) the overall signals for HlgA and LukE-binding in CXCR2$^+$ cells are low (Figure 5c), and 2) different strategies were applied to detect binding of HlgA and LukE, respectively.

Collectively, these results show that in contrast to C5aR1, susceptibility to leukocidins in CXCR2 expressing cells is mostly driven by sialylation instead of sulfation. For LukED, sialylation drives cytotoxicity by facilitating the binding of the S-component, combined with a modulation of receptor expression levels. For HlgAB, the impact of sialylation on cellular susceptibility is likely mediated by receptor surface expression levels only. Sulfation has a limited impact on HlgAB and LukED cytotoxicity of CXCR2-expressing cells, by affecting the binding of the leucocidin S-components. When taken together, our genome-wide screenings highlight a conserved yet divergent role of PTMs in refining susceptibility of human phagocytic cells to the staphylococcal leukocidins.

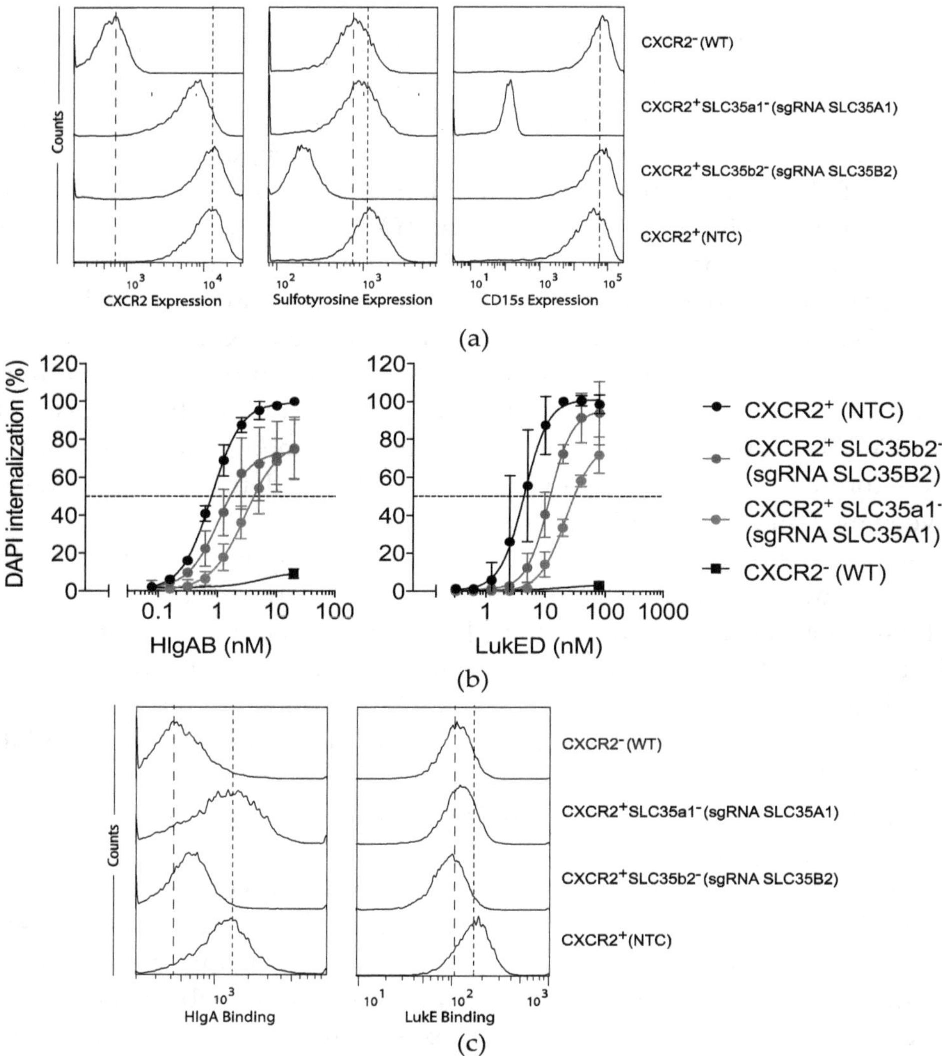


**Figure 5.** Sialylation and sulfation of CXCR2 refine susceptibility to LukED and HlgAB. (**a**) Anti-CXCR2, anti-sulfotyrosine and anti-CD15s antibodies were used to assess the expression of CXCR2, total sulfotyrosine, and CD15s on U937-CXCR2-SpCas9 cell lines transduced with sgRNA for *SLC35A1* (CXCR2$^+$ SLC35a1$^-$), *SLC35B2* (CXCR2$^+$ SLC35b2$^-$), non-targeting control sgRNA (NTC, CXCR2$^+$), and U937-SpCas9 (WT, CXCR2$^-$) cells. Antibody binding was determined by a fluorescent secondary antibody and the fluorescence measured and analyzed by flow cytometry. Dashed line: expression in U937-SpCas9 (WT, CXCR2$^-$) cells; dotted line: expression in NTC (CXCR2$^+$) U937 cells. Histograms depict representative examples of two independently repeated experiments. (**b**) Susceptibility of U937-CXCR2-SpCas9 cell lines transduced with sgRNA for *SLC35B2* (CXCR2$^+$ SLC35b2$^-$), *SLC35A1* (CXCR2$^+$ SLC35a1$^-$), non-targeting control sgRNA (NTC, CXCR2$^+$), and U937-SpCas9 (WT, CXCR2$^-$) cells to HlgAB and LukED. As a readout for cell permeability, internalization of DAPI was measured during 30 min post-toxin treatment on a monochromator-based microplate reader and expressed in relation to the area under the curve for NTC sgRNA transduced U937-CXCR2-SpCas9 cells at 20 nM HlgAB or LukED. (**c**) Binding of Alexa Fluor 647 maleimide-based labeled HlgA and polyhistidine-tagged LukE to U937-CXCR2-SpCas9 cell lines transduced with sgRNA for *SLC35B2* (CXCR2$^+$ SLC35b2$^-$), *SLC35A1* (CXCR2$^+$ SLC35a1$^-$), non-targeting control sgRNA (NTC, CXCR2$^+$), and U937-SpCas9 (WT, CXCR2$^-$) cells. Cells were subsequently incubated with anti-histidine-FITC secondary antibodies and/or the fluorescence was directly measured and analyzed by flow cytometry. Dashed line: binding to U937-SpCas9 (WT, CXCR2$^-$) cells; dotted line: binding to NTC (CXCR2$^+$) U937 cells. Histograms depict representative examples of two independently repeated experiments.

## 3. Discussion

We used a genome-wide CRISPR/Cas9 based approach to screen for cellular factors involved in susceptibility towards GPCR-targeting leukocidins, and identify two PTM pathways (sulfation and sialylation) driving GPCR-mediated susceptibility of phagocytes to leukocidins. Post-translational enzymatic reactions modify specific protein backbones and sidechains, thereby modulating protein function [17]. PTMs of GPCRs are important for regulating structure, function and association of the receptors with their natural ligands [18,19]. In addition, PTM moieties on GPCRs have also been implied in the interaction with different pathogens [20–22]. Hemolytic activity of the *Streptococcus pneumoniae* cytolysin Ply was recently shown to be mediated by its interaction with sialylated Lewis X histo-blood group antigen [22]. Neutralization of HlgCB cytotoxicity by pre-incubation with the ganglioside GM1, a sialylated oligosaccharide [23] further suggests a role for sialylation in leukocidin toxicity. Using a synthetic N-terminal C5aR1 peptide, it was previously proposed that tyrosine-sulfation facilitates binding of LukS-PV [7]. However, it has remained undetermined to what extent GPCR-sulfation contributes to pore formation. In addition, it is unknown if PTMs define a conserved signature for the interaction between the leukocidins and their respective GPCRs.

Disruption of the sulfation pathway results in a reduced S-component binding to GPCR-expressing U937 cells, supporting the notion that GPCR-sulfation enhances target cell susceptibility to leukocidins by strictly facilitating binding of the S-components [7]. Disruption of the sialylation pathway however likely results in a combination of a reduced binding as well as a compromised overall receptor expression. The observed reduction in binding of the S-components to the sialylation-mutant cell lines is limited (if reduced at all), and no monoclonal antibodies are currently available for specific detection of C5aR1 or CXCR2 in a sialylation-dependent manner. Therefore, it cannot be excluded that the interaction of the toxins with CD45 (the receptor for LukF-PV [12]) or elusive receptors of the other leukocidin F-components is mediated by sialylation as well.

C5aR1-expressing sulfation-deficient cells display a reduced susceptibility towards both PVL and HlgCB. In addition, our data indicate that posttranslational decoration of C5aR1-expressing cells with sialylation-moieties facilitates PVL and HlgCB cytotoxicity. Compared to sulfation, however, the contribution of sialylation to C5aR1-interacting leukocidin susceptibility is limited. Overall, this screening identifies sulfation-mediated receptor-employment as a major and conserved feature for C5aR1-interacting leukocidins. In contrast to C5aR1, sulfation of CXCR2 expressing cells does not have a major impact on LukED and HlgAB cytotoxicity. Although the CXCR2 N-terminus contains two tyrosines, it lacks acidic sulfation modification motifs that mediate TPST2 activity. DARC, the erythrocyte receptor for HlgAB and LukED and a close homologue of CXCR2, has been suggested to interact with both leukocidins in an N-terminal sulfated tyrosine dependent manner [11]. In contrast to CXCR2, the DARC N-terminus does contain tyrosine sulfation determinants. Likely, CXCR2 is sulfated with less efficiency [24,25]. This indicates a divergent role for GPCR-sulfation in leukocidin susceptibility, in which LukED and HlgAB interact sulfation-dependent in a receptor-specific manner. Opposed to our findings in C5aR1, our data identify sialylation rather than sulfation as the major PTM motif facilitating cytotoxicity of CXCR2-targeting leukocidins.

Multiple extracellular domains of the leukocidin receptors are involved during pore formation, and conformational changes of the GPCRs are likely to occur in the process of binding, hetero-oligomerization, and pore formation [16]. In addition to directly enhancing the binding of leukocidins and regulating overall receptor expression levels, PTMs possibly also affect the GPCR structures and their conformational plasticity [19]. These observations identify PTMs of GPCRs as potential targets for pharmacological interference during infection. Editing of individual genes encoding proteins involved in the sialylation (*CMAS*, *SLC35A1*) or sulfation pathway (*SLC35B2*, *PAPSS1*, *TPST2*) all resulted in disruption of overall sialylation or sulfation in GPCR-expressing cells, implying that the respective genes are part of a non-redundant and sequential pathway. To the best of our knowledge, no pharmacological compounds targeting the tyrosine sulfation pathway are currently being studied. As *TPST1* and *TPST2* double knockout mice die in the early postnatal phase [26], pharmaceutical interference with the sulfation pathway will likely be challenging due to the essentiality of this pathway in eukaryotic cell homeostasis. Although competitive small-molecule inhibitors of sialyltranferases are being developed, toxic side effects of currently available compounds will limit their applicability in humans on short term [27,28].

The necessity of the seemingly redundant range of phagocyte-targeting toxins produced by *S. aureus* is incompletely understood [5]. Recent studies have revealed similarities and differences in the interactions between the leukocidins and their respective GPCRs [8,11,16]. The differences in the impact of PTM moieties on leukocidin toxicity highlight an additional layer of refinement and divergence of the staphylococcal host-pathogen interaction. These differences further challenge the apparent functional redundancy of the leukocidins, and enhance our understanding of their cellular tropism and species specificity.

Expression of the leukocidin S-component receptors is essential for pore formation [5]. The PTM pathways identified in this study serve as refining mechanisms that further enhance susceptibility of leukocidin-receptor expressing cells to pore formation. PTMs display a wide heterogeneity in terms of cell type and tissue specificity [19,29]. While *S. aureus* is capable of causing a plethora of infections, its tissue tropism is poorly understood. Cell type and organ specific variations in sulfation and sialylation, contributing to local cellular susceptibility towards the leukocidins, may possibly contribute to organ specific *S. aureus* infections.

The mechanisms for the predisposition of otherwise healthy individuals to severe infections with *S. aureus* are poorly understood [1,30–32]. Variations in the genes encoding PTM pathways could provide insight into observed differences in susceptibility of humans to severe infections with *S. aureus*. It remains to be established if specific variants of the sialylation and sulfation pathway genes alter susceptibility to *S. aureus* infections in humans. The combination of a preceding Influenza-virus infection and a subsequent *S. aureus* pneumonia is well known for its poor outcome of disease [30,33–35]. The interaction of Influenza-virus neuraminidase with sialic acids in the lungs vs. the sialylation-mediated employment of GPCRs by the staphylococcal leukocidins deserves further study.

## 4. Materials and Methods

### *4.1. Cell Lines and Constructs*

U937 human monocytic cells were obtained from ATCC (American Type Culture Collection) and cultured in RPMI supplemented with penicillin/streptomycin and 10% fetal calf serum. Cell lines were constructed as previously described [12]. Briefly, to sensitize the cells to PVL and HlgCB or LukED and HlgCB, human C5aR1 (CD88; NM_001736) and CXCR2 (CD182; NM_001168298.1) were stably expressed in U937 cells using a lentiviral expression system (U937-C5aR1, U937-CXCR2 cells). We cloned the human *C5AR1* and human *CXCR2* cDNA in a dual promoter lentiviral vector (BIC-PGKZeo- T2a-mAmetrine; RP172), derived from no.2025.pCCLsin.PPT.pA.CTE.4x-scrT. eGFP.mCMV.hPGK.NG-FR.pre as described elsewhere [36]. The transfection of 293T cells with the C5aR1 and CXCR2 lentiviral expression systems, and subsequent transduction of U937-cells were performed as previously described [12]. To allow screening, a codon-optimized nuclear-localized *S. pyogenes* cas9 gene was subsequently transduced in U937-C5aR1 as described elsewhere [12]. The genome-wide sgRNA CRISPR/Cas9 library was designed as previously described [12]. The following sgRNA sequences were cloned in sgLenti to generate SLC35B2 (fw- 5′ TTGGACAGGCTGGCAAAG-GAGTAC 3′, rv 3′ AAACGTACTCCTTTGCCAGCCTGT 5′), PAPSS1 (fw- 5′ TTGGGCAAGTTGTG-GAACTTCTAC 3′, rv- 3′ AAACGTAGAAGTTCCACAACTTGC 5′), TPST2 (fw- 5′ TTGGGGCCCG-CGTGCTCTGCAACA 3′, rv- 3′ AAACTGTTGCAGAGCACGCGGG-CC 5′), SLC35A1 (fw- 5′ TTG-GACATACAAGAAGAGTACCCA 3′, rv- 3′ AAACTGGGTACTCTT-CTTGTATGT 5′) and CMAS (fw- 5′ TTGGGAGAATGTGGCCAAACAATT 3′, rv- 3′ AAACAATT-GTTTGGCCACATTCTC 5′) knockout cell lines. U937-SpCas9, U937-C5aR1-SpCas9 or

U937-CXCR2-SpCas9 cells were subsequently transduced with the sgRNA-expression viruses and selected to purity by puromycin treatment (2 μg/mL) to enrich for knocked-out cells.

### 4.2. Genome-Wide CRISPR/Cas9 Library Screen in U937-C5aR1 Cells

The genome-wide CRISPR/Cas9 screen for leukocidin resistance was performed as described elsewhere [12]. Briefly, cells transduced with the CRISPR sgRNA library were selected to purity with puromycin (2 μg/mL) initiated at two days post transduction. Twelve days post transduction, $2 \times 10^8$ cells were incubated with 15 nM PVL or 15 nM HlgCB for 30 min at 37 °C, which resulted in depletion of >99.5% of the cells. Cells were washed to remove the toxin and allowed to recover in complete RPMI for 15 days to enrich for viable cells. Genomic DNA was isolated and sgRNA inserts were subsequently PCR amplified for 16 cycles with primers 5′ GGCTTGGATTTCTATAACTTCGT-ATAGCA 3′ and 5′ CGGGGACTGTGGGCGATGTG 3′ using the Titanium Taq PCR kit (Clontech, Göteborg, Sweden). The PCR products were pooled and amplified using primers containing Illumina adapter sequences and a unique index. PCR products were subsequently pooled in equimolar ratios and subjected to deep-sequencing using the Illumina NextSeq500 platform. Sequences were aligned to the sgRNA library by using Bowtie2 (Johns Hopkins University, Baltimore, MA, USA; PMID: 22388286) and the counts per sgRNA were calculated. We used the MaGeCk package (Dana-Farber Cancer Institute, Boston, MA, USA; PMID: 25476604; available from https://sourceforge.net/projects/ mageck/) as a computational tool to identify genes significantly enriched in the screens by comparing sgRNA read counts of control cells versus PVL- and HlgCB incubated cells. Raw data obtained from the PVL resistance screen were previously reported [12].

### 4.3. Recombinant Protein Production and Cell Permeability Assays

Polyhistidine-tagged LukS-PV, LukF-PV, HlgC, HlgA, HlgB, LukE, LukD and Alexa Fluor 647 maleimide-based labeled HlgA used during this study were cloned and expressed as described elsewhere [7,8,37,38]. For permeability assays, each U937 cell line ($1 \times 10^7$ cells per ml in RPMI/HSA) was exposed to recombinant toxins and measured for 30 min at 37 °C in a monochromator-based microplate reader (FLUOstar Omega, BMG Labtech, Ortenberg, Germany) in the presence of 2,5 μg/mL 4′,6-diamidino-2-phenylindole (DAPI, Molecular Probes/Thermo Fisher, Landsmeer, The Netherlands) to determine pore-formation [12]. As PVL, HlgCB, HlgAB and LukED are two-component toxins, equimolar concentrations of polyhistidine-tagged S- and F-components were used. Pore formation was defined as the concentration dependent collective DAPI-internalization during the course of 30 min. The area under the curve (AUC), corrected for toxin-independent DAPI-internalization, was calculated and subsequently related to the maximum AUC obtained in U937-C5aR1-SpCas9 or U937-CXCR2-SpCas9 cells transduced with a non-targeting control sgRNA that were treated with the highest respective toxin concentration (Figure S2) [12]. Cell lines lacking expression of C5aR1 or CXCR2 (U937-SpCas9 cell lines) are resistant to DAPI-internalization upon toxin-exposure.

### 4.4. Determination of Receptor Expression Levels and Binding Assays

Receptor expression levels were determined as described elsewhere [8]. Briefly, single U937 cell suspensions were stained with mouse anti-human C5aR1 (Sulfation independent clone S5/1, AbD Serotec, Kidlington, UK), mouse anti-human C5aR1 (Sulfation dependent clone 347214, R&D Systems, Abingdon, UK) mouse anti-human CXCR2 (clone 6D499, Abnova, Heidelberg, Germany), mouse-anti-sulfotyrosine (Clone Sulfo-1C-A2, EMD Millipore, Burlington, MA, USA), mouse-anti-CD15s (Clone CLSLEX1, BD Pharmingen, Erembodegem, Belgium), followed by APC-conjugated goat-anti-mouse antibody (Jackson Immunoresearch, Westgrove, PA, USA). Samples were subsequently measured using flow cytometry. For the S-component binding studies on U937 cells, cells were incubated with 10 μg/mL polyhistidine-tagged LukS-PV or HlgC for 30 min on ice followed by anti-his-FITC (LifeSpan BioSciences, Seattle, WA, USA), or 10 μg/mL polyhistidine-tagged LukE for 20 min at 20 °C followed by anti-his-FITC (LifeSpan BioSciences, Seattle, WA, USA), or with 10 μg/mL Alexa Fluor 647 maleimide-based labeled HlgA [38] for 20 min at 20 °C. Cells were subsequently measured using flow cytometry.

### 4.5. Statistical Analyses

Calculations of the area under the curves, calculations of half-maximal effective lytic concentrations using linear regression analyses, and all statistical analyses were performed using Prism 7.0 (GraphPad Software, San Diego, CA, USA). Statistical significance was calculated using ANOVA analysis of variance with Bonferroni posttest correction for multiple comparison where appropriate. Four-parametric non-linear regression analyses were performed to obtain half-maximum effective concentrations (EC50s). Flow cytometric data were analyzed with FlowJo (Tree Star Software, Erembodegem, Belgium).

**Supplementary Materials:**
Table S1 (Related to Figure 1): Screening results for resistance to PVL toxicity - CRISPR/Cas9 library screen for PVL resistance set up in U937-C5aR1-SpCas9 cells, selected after exposure to PVL. Table S2 (Related to Figure 1): Screening results for resistance to HlgCB toxicity - CRISPR/Cas9 library screen for HlgCB resistance set up in U937-C5aR1-SpCas9 cells, selected after exposure to HlgCB. Figure S1 (Related to Figures 2 and 3): Sulfation and sialylation in U937-SpCas9 cell lines. Figure S2 (Related to Figure 4 and Materials and methods): Time and concentration dependent DAPI-internalization.

**Author Contributions:** A.T.T., M.V.G., and A.N.S. conceptualized the study. A.T.T., M.V.G., B.W.B., R.J.L., P.-J.A.H., and A.N.S. designed the methodology. A.T.T., M.V.G., J.P.J., L.M.S., A.V., C.J.C.D.H., and K.P.M.V.K. conducted the investigations. M.B. and M.T.M., provided resources. J.A.G.V.S. provided funding. K.P.M.V.K., B.W.B., J.A.G.V.S, P.-J.A.H., and A.N.S. provided supervision. A.T.T. and A.N.S. wrote the manuscript. All authors have read and agreed to the published version of the manuscript.

**Acknowledgments:** The authors wish to thank A.M.J. Wensing, C.H.E. Boel, and M.J.M. Bonten (University Medical Center Utrecht, Utrecht, The Netherlands) for support.

## References

1. Lowy, F.D. *Staphylococcus aureus* infections. *N. Engl. J. Med.* **1998**, *339*, 520–532. [CrossRef] [PubMed]
2. Thwaites, G.E.; Edgeworth, J.D.; Gkrania-Klotsas, E.; Kirby, A.; Tilley, R.; Torok, M.E.; Walker, S.; Wertheim, H.F.; Wilson, P.; Llewelyn, M.J.; et al. Clinical management of *Staphylococcus aureus* bacteraemia. *Lancet Infect. Dis.* **2011**, *11*, 208–222. [CrossRef]
3. Beutler, B. Innate immunity: An overview. *Mol. Immunol.* **2004**, *40*, 845–859. [CrossRef] [PubMed]
4. Spaan, A.N.; Surewaard, B.G.; Nijland, R.; van Strijp, J.A. Neutrophils Versus *Staphylococcus aureus*: A Biological Tug of War. *Annu. Rev. Microbiol.* **2013**, *67*, 629–650. [CrossRef]
5. Spaan, A.N.; van Strijp, J.A.G.; Torres, V.J. Leukocidins: Staphylococcal bi-component pore-forming toxins find their receptors. *Nat. Rev. Microbiol.* **2017**, *15*, 435. [CrossRef]
6. Alonzo, F., 3rd; Torres, V.J. The Bicomponent Pore-Forming Leucocidins of *Staphylococcus aureus*. *Microbiol. Mol. Biol. Rev.* **2014**, *78*, 199–230. [CrossRef]
7. Spaan, A.N.; Henry, T.; van Rooijen, W.J.; Perret, M.; Badiou, C.; Aerts, P.C.; Kemmink, J.; de Haas, C.J.; van Kessel, K.P.; Vandenesch, F.; et al. The staphylococcal toxin Panton-Valentine Leukocidin targets human C5a receptors. *Cell Host Microbe* **2013**, *13*, 584–594. [CrossRef]
8. Spaan, A.N.; Vrieling, M.; Wallet, P.; Badiou, C.; Reyes-Robles, T.; Ohneck, E.A.; Benito, Y.; de Haas, C.J.; Day, C.J.; Jennings, M.P.; et al. The staphylococcal toxins gamma-haemolysin AB and CB differentially target phagocytes by employing specific chemokine receptors. *Nat. Commun.* **2014**, *5*, 5438. [CrossRef]
9. Reyes-Robles, T.; Alonzo, F., 3rd; Kozhaya, L.; Lacy, D.B.; Unutmaz, D.; Torres, V.J. *Staphylococcus aureus* Leukotoxin ED Targets the Chemokine Receptors CXCR1 and CXCR2 to Kill Leukocytes and Promote Infection. *Cell Host Microbe* **2013**, *14*, 453–459. [CrossRef]
10. Alonzo, F., 3rd; Kozhaya, L.; Rawlings, S.A.; Reyes-Robles, T.; DuMont, A.L.; Myszka, D.G.; Landau, N.R.; Unutmaz, D.; Torres, V.J. CCR5 is a receptor for *Staphylococcus aureus* leukotoxin ED. *Nature* **2013**, *493*, 51–55. [CrossRef]

11. Spaan, A.N.; Reyes-Robles, T.; Badiou, C.; Cochet, S.; Boguslawski, K.M.; Yoong, P.; Day, C.J.; de Haas, C.J.; van Kessel, K.P.; Vandenesch, F.; et al. *Staphylococcus aureus* Targets the Duffy Antigen Receptor for Chemokines (DARC) to Lyse Erythrocytes. *Cell Host Microbe* **2015**, *18*, 363–370. [CrossRef] [PubMed]
12. Tromp, A.T.; Van Gent, M.; Abrial, P.; Martin, A.; Jansen, J.P.; De Haas, C.J.C.; Van Kessel, K.P.M.; Bardoel, B.W.; Kruse, E.; Bourdonnay, E.; et al. Human CD45 is an F-component-specific receptor for the staphylococcal toxin Panton-Valentine leukocidin. *Nat. Microbiol.* **2018**, *3*, 708–717. [CrossRef] [PubMed]
13. Virreira Winter, S.; Zychlinsky, A.; Bardoel, B.W. Genome-wide CRISPR screen reveals novel host factors required for *Staphylococcus aureus* alpha-hemolysin-mediated toxicity. *Sci. Rep.* **2016**, *6*, 24242. [CrossRef]
14. Yang, Y.S.; Wang, C.C.; Chen, B.H.; Hou, Y.H.; Hung, K.S.; Mao, Y.C. Tyrosine sulfation as a protein post-translational modification. *Molecules* **2015**, *20*, 2138–2164. [CrossRef] [PubMed]
15. Bhide, G.P.; Colley, K.J. Sialylation of N-glycans: Mechanism, cellular compartmentalization and function. *Histochem. Cell Biol.* **2017**, *147*, 149–174. [CrossRef] [PubMed]
16. Spaan, A.N.; Schiepers, A.; de Haas, C.J.; van Hooijdonk, D.D.; Badiou, C.; Contamin, H.; Vandenesch, F.; Lina, G.; Gerard, N.P.; Gerard, C.; et al. Differential Interaction of the Staphylococcal Toxins Panton-Valentine Leukocidin and gamma-Hemolysin CB with Human C5a Receptors. *J. Immunol.* **2015**, *195*, 1034–1043. [CrossRef] [PubMed]
17. Walsh, C.T.; Garneau-Tsodikova, S.; Gatto, G.J., Jr. Protein posttranslational modifications: The chemistry of proteome diversifications. *Angew. Chem. Int. Ed. Engl.* **2005**, *44*, 7342–7372. [CrossRef]
18. Farzan, M.; Schnitzler, C.E.; Vasilieva, N.; Leung, D.; Kuhn, J.; Gerard, C.; Gerard, N.P.; Choe, H. Sulfated tyrosines contribute to the formation of the C5a docking site of the human C5a anaphylatoxin receptor. *J. Exp. Med.* **2001**, *193*, 1059–1066. [CrossRef]
19. Ludeman, J.P.; Stone, M.J. The structural role of receptor tyrosine sulfation in chemokine recognition. *Br. J. Pharmacol.* **2014**, *171*, 1167–1179. [CrossRef]
20. Park, R.J.; Wang, T.; Koundakjian, D.; Hultquist, J.F.; Lamothe-Molina, P.; Monel, B.; Schumann, K.; Yu, H.; Krupzcak, K.M.; Garcia-Beltran, W.; et al. A genome-wide CRISPR screen identifies a restricted set of HIV host dependency factors. *Nat. Genet.* **2017**, *49*, 193–203. [CrossRef]
21. Choe, H.; Moore, M.J.; Owens, C.M.; Wright, P.L.; Vasilieva, N.; Li, W.; Singh, A.P.; Shakri, R.; Chitnis, C.E.; Farzan, M. Sulphated tyrosines mediate association of chemokines and *Plasmodium vivax* Duffy binding protein with the Duffy antigen/receptor for chemokines (DARC). *Mol. Microbiol.* **2005**, *55*, 1413–1422. [CrossRef] [PubMed]
22. Poole, J.; Day, C.J.; von Itzstein, M.; Paton, J.C.; Jennings, M.P. Glycointeractions in bacterial pathogenesis. *Nat. Rev. Microbiol.* **2018**, *16*, 440–452. [CrossRef] [PubMed]
23. Noda, M.; Kato, I.; Hirayama, T.; Matsuda, F. Fixation and inactivation of staphylococcal leukocidin by phosphatidylcholine and ganglioside GM1 in rabbit polymorphonuclear leukocytes. *Infect. Immun.* **1980**, *29*, 678–684. [PubMed]
24. Moussouras, N.A.; Getschman, A.E.; Lackner, E.R.; Veldkamp, C.T.; Dwinell, M.B.; Volkman, B.F. Differences in Sulfotyrosine Binding amongst CXCR1 and CXCR2 Chemokine Ligands. *Int. J. Mol. Sci.* **2017**, *18*, 1894. [CrossRef] [PubMed]
25. Hartmann-Fatu, C.; Bayer, P. Determinants of tyrosylprotein sulfation coding and substrate specificity of tyrosylprotein sulfotransferases in metazoans. *Chem. Biol. Interact.* **2016**, *259*, 17–22. [CrossRef] [PubMed]
26. Westmuckett, A.D.; Hoffhines, A.J.; Borghei, A.; Moore, K.L. Early postnatal pulmonary failure and primary hypothyroidism in mice with combined TPST-1 and TPST-2 deficiency. *Gen. Comp. Endocrinol.* **2008**, *156*, 145–153. [CrossRef]
27. Nicholls, J.M.; Moss, R.B.; Haslam, S.M. The use of sialidase therapy for respiratory viral infections. *Antivir. Res.* **2013**, *98*, 401–409. [CrossRef]
28. Macauley, M.S.; Arlian, B.M.; Rillahan, C.D.; Pang, P.C.; Bortell, N.; Marcondes, M.C.; Haslam, S.M.; Dell, A.; Paulson, J.C. Systemic blockade of sialylation in mice with a global inhibitor of sialyltransferases. *J. Biol. Chem.* **2014**, *289*, 35149–35158. [CrossRef]
29. Mishiro, E.; Sakakibara, Y.; Liu, M.C.; Suiko, M. Differential enzymatic characteristics and tissue-specific expression of human TPST-1 and TPST-2. *J. Biochem.* **2006**, *140*, 731–737. [CrossRef]

30. Gillet, Y.; Issartel, B.; Vanhems, P.; Fournet, J.C.; Lina, G.; Bes, M.; Vandenesch, F.; Piemont, Y.; Brousse, N.; Floret, D.; et al. Association between *Staphylococcus aureus* strains carrying gene for Panton-Valentine leukocidin and highly lethal necrotising pneumonia in young immunocompetent patients. *Lancet* **2002**, *359*, 753–759. [CrossRef]
31. Alcais, A.; Abel, L.; Casanova, J.L. Human genetics of infectious diseases: Between proof of principle and paradigm. *J. Clin. Investig.* **2009**, *119*, 2506–2514. [CrossRef] [PubMed]
32. Casanova, J.L. Human genetic basis of interindividual variability in the course of infection. *Proc. Natl. Acad. Sci. USA* **2015**, *112*, E7118–E7127. [CrossRef]
33. Deleo, F.R.; Otto, M.; Kreiswirth, B.N.; Chambers, H.F. Community-associated meticillin-resistant *Staphylococcus aureus*. *Lancet* **2010**, *375*, 1557–1568. [CrossRef]
34. Niemann, S.; Ehrhardt, C.; Medina, E.; Warnking, K.; Tuchscherr, L.; Heitmann, V.; Ludwig, S.; Peters, G.; Loffler, B. Combined Action of Influenza Virus and *Staphylococcus aureus* Panton-Valentine Leukocidin Provokes Severe Lung Epithelium Damage. *J. Infect. Dis.* **2012**, *206*, 1138–1148. [CrossRef] [PubMed]
35. DeLeo, F.R.; Otto, M. An antidote for *Staphylococcus aureus* pneumonia? *J. Exp. Med.* **2008**, *205*, 271–274. [CrossRef]
36. Van de Weijer, M.L.; Bassik, M.C.; Luteijn, R.D.; Voorburg, C.M.; Lohuis, M.A.; Kremmer, E.; Hoeben, R.C.; LeProust, E.M.; Chen, S.; Hoelen, H.; et al. A high-coverage shRNA screen identifies TMEM129 as an E3 ligase involved in ER-associated protein degradation. *Nat. Commun.* **2014**, *5*, 3832. [CrossRef]
37. Perret, M.; Badiou, C.; Lina, G.; Burbaud, S.; Benito, Y.; Bes, M.; Cottin, V.; Couzon, F.; Juruj, C.; Dauwalder, O.; et al. Cross-talk between *S. aureus* leukocidins-intoxicated macrophages and lung epithelial cells triggers chemokine secretion in an inflammasome-dependent manner. *Cell. Microbiol.* **2012**, *14*, 1019–1036. [CrossRef]
38. Haapasalo, K.; Wollman, A.J.M.; de Haas, C.J.C.; van Kessel, K.P.M.; van Strijp, J.A.G.; Leake, M.C. *Staphylococcus aureus* toxin LukSF dissociates from its membrane receptor target to enable renewed ligand sequestration. *FASEB J.* **2019**, *33*, 3807–3824. [CrossRef]

# 2

# Intracellular Growth and Cell Cycle Progression are Dependent on (p)ppGpp Synthetase/ Hydrolase in *Brucella abortus*

**Mathilde Van der Henst, Elodie Carlier and Xavier De Bolle ***

Unité de Recherche en Biologie des Microorganismes (URBM), University of Namur, 61 rue de Bruxelles, 5000 Namur, Belgium; mathilde.vanderhenst@unamur.be (M.V.d.H.); elodie.carlier@unamur.be (E.C.)

* Correspondence: xavier.debolle@unamur.be

**Abstract:** *Brucella abortus* is a pathogenic bacterium able to proliferate inside host cells. During the first steps of its trafficking, it is able to block the progression of its cell cycle, remaining at the G1 stage for several hours, before it reaches its replication niche. We hypothesized that starvation mediated by guanosine tetra- or penta-phosphate, (p)ppGpp, could be involved in the cell cycle arrest. Rsh is the (p)ppGpp synthetase/hydrolase. A *B. abortus* Δ*rsh* mutant is unable to grow in minimal medium, it is unable to survive in stationary phase in rich medium and it is unable to proliferate inside RAW 264.7 macrophages. A strain producing the heterologous constitutive (p)ppGpp hydrolase Mesh1b is also unable to proliferate inside these macrophages. Altogether, these data suggest that (p)ppGpp is necessary to allow *B. abortus* to adapt to its intracellular growth conditions. The deletion of *dksA*, proposed to mediate a part of the effect of (p)ppGpp on transcription, does not affect *B. abortus* growth in culture or inside macrophages. Expression of a gene coding for a constitutively active (p)ppGpp synthetase slows down growth in rich medium and inside macrophages. Using an mCherry–ParB fusion able to bind to the replication origin of the main chromosome of *B. abortus*, we observed that expression of the constitutive (p)ppGpp synthetase gene generates an accumulation of bacteria at the G1 phase. We thus propose that (p)ppGpp accumulation could be one of the factors contributing to the G1 arrest observed for *B. abortus* in RAW 264.7 macrophages.

**Keywords:** Brucella; cell cycle; (p)ppGpp; *rsh*

## 1. Introduction

Bacteria from the *Brucella* genus are the causative agents of brucellosis, a neglected disease which constitutes a worldwide anthropozoonosis. *Brucella* spp. are Gram negative alphaproteobacteria belonging to the Rhizobiales order [1]. *Brucella abortus* causes severe symptoms in mammals such as abortion in pregnant females and sterility in males. In humans, the disease is characterized by an undulant fever, also named Malta fever, and in the long term the infection leads to chronicity and symptoms such as arthritis, endocarditis and can have a fatal outcome without treatment [1]. In their hosts, *Brucellae* invade, survive and replicate inside professional and non-professional phagocytic cells such as macrophages and trophoblasts. Inside host cells, *Brucellae* are found in vacuoles named *Brucella* containing vacuoles (BCVs). In the first part of their trafficking, they successively harbor markers of early and late endosomes, a phase of the trafficking in which the bacterium does not proliferate [2,3]. This compartment presents a pH of about 4.0 to 4.5 and this acidification is essential for the successful establishment of *B. suis* infection [4]. Afterwards, the bacteria are found in BCVs having markers of the endoplasmic reticulum, where they replicate [5]. Later in the cellular infection, bacteria are found in vacuoles characterized by autophagy-related proteins [6].

Different cellular models for in vitro study of *B. abortus* infection of have been developed, such as the use of RAW 264.7 macrophages and HeLa epithelial cells. Some years ago, the investigations of the *B. abortus* infection process in these models revealed that the cell cycle regulation of *B. abortus* is linked to its virulence [2]. Indeed, for bacteria that did not segregate duplicated replication origins, the so-called G1 cells are more infectious than the S or G2 phase bacteria, i.e., bacteria currently replicating their genome or at the stage between the completion of genome replication genome and cell vision, respectively. More importantly, after internalization, bacteria remain in the G1 stage for up to 8 h, depending the host cell type, in Lamp1-positive compartments before reaching the endoplasmic reticulum where *B. abortus* can restart its cell cycle, its DNA replication and actively proliferate [2]. During the first hours of the infection, in BCVs with endocytic markers, *B. abortus* encounters harsh conditions such as acidic stress [7] and alkylating stress [8]. In addition, it was already proposed that *B. abortus* has to face a starvation conditions inside host cells [9]. Starvation is the most obvious condition that could explain why *B. abortus* is blocked at the G1 stage of the cell cycle during the first phase of its intracellular trafficking in HeLa cells and RAW 264.7 macrophages. Starvation sensing is classically involving the synthesis of (p)ppGpp (guanosine penta- or tetra-phosphate), also called alarmone. The synthesis and degradation of (p)ppGpp are catalyzed by enzymes of the RelA/SpoT family, also called Rsh enzymes. It was found that *rsh* mutants, which should be not able to produce (p)ppGpp anymore, are strongly impaired during in vitro infection as well as during murine infection [10,11].

The alarmone (p)ppGpp is widely used by bacteria to quickly adapt to stress conditions such as nutrient starvation. The production and accumulation of this alarmone induces pleiotropic effects, modulating transcription and translation, that commonly result in cell cycle and DNA replication delay [12–15]. The ability to produce (p)ppGpp has been associated with virulence in bacterial pathogens belonging to relatively distant phylogenetic groups, such as *Legionella pneumophila* [16], *Vibrio cholerae* [17], and *Mycobacterium tuberculosis* [18]. In *Escherichia coli*, during the stringent response induced by starvation, (p)ppGpp binds directly to a site located at the interface between the $\beta'$ and $\omega$ subunits of the RNA polymerase [19]. A second distinct site between the $\beta'$ subunit and the DksA transcription factor has been shown to be bound by (p)ppGpp as well [20]. This interaction has been shown to enhance the transcriptional effects of DksA on the RNA polymerase, suggesting synergistic effects of DksA and (p)ppGpp together [20].

The RelA/SpoT homolog proteins are responsible for (p)ppGpp homeostasis. In *E. coli*, there are two enzymes of the Rsh family, RelA and SpoT [21]. SpoT contains a synthetase domain, a hydrolase domain, and two C-terminal regulatory domains; thus, this enzyme can both catalyze the production and the degradation of (p)ppGpp, respectively. RelA contains similar domains, however the functionality of the hydrolase domain of RelA has been lost during evolution, leading to a monofunctional enzyme that can only synthesize the alarmone [22]. In most alphaproteobacteria, including *B. abortus*, the production and the degradation of (p)ppGpp depends on one enzyme named Rsh (for RelA SpoT homolog) [10,21].

In the present study, we analyzed the impact of alterations in (p)ppGpp synthesis or degradation on the growth, the cell cycle and the infection process of *B. abortus*. We show that mutants either unable to produce (p)ppGpp or producing a (p)ppGpp hydrolase are impaired for the infection process. In addition, our results show that expression of a constitutive (p)ppGpp synthetase negatively impacts growth and DNA replication of *B. abortus*, and also leads to a strong proliferation defect during infection of RAW 264.7 macrophages. We also observed that a *B. abortus dksA* null mutant was able to proliferate inside host cells as the wild type (WT) strain, suggesting that DksA is not crucially involved in the (p)ppGpp-dependent phenotypes observed during infection. These results suggest that adjustment of (p)ppGpp levels are crucial for the infection process in *B. abortus*.

## 2. Results

### 2.1. *rsh* Deletion Drastically Impacts Growth in Minimal Medium and the Infection Process

We generated a Δ*rsh* strain by allelic replacement in *B. abortus* 544 and we assayed the growth of this strain in rich culture medium (2YT) as well as in Plommet minimal medium [23] supplemented with erythritol as a carbon source. The growth of Δ*rsh* in 2YT was similar to the WT strain during the exponential phase, but the shift into the stationary phase occurred later and at a higher optical density (OD) compared to the wild type strain (Figure 1A). To evaluate bacterial viability, we counted the colony forming units (CFUs) throughout the culture in liquid rich medium (Figure 2). The Δ*rsh* strain showed a clear survival defect during the stationary phase, marked by a decrease in CFUs between 24 h and 48 h compared to the WT and the complemented strain.

Because it has been shown in other bacteria that the stringent response is linked to nutrient availability, we tested the growth of the Δ*rsh* strain in Plommet minimal medium, supplemented with erythritol as a carbon source, to mimic starvation conditions. The Δ*rsh* showed a clear growth defect compared to the WT as the OD rapidly decreased during the mid-exponential phase, indicating that Δ*rsh* cannot grow and survive in this medium, as expected for a mutant unable to produce (p)ppGpp (Figure 1B).

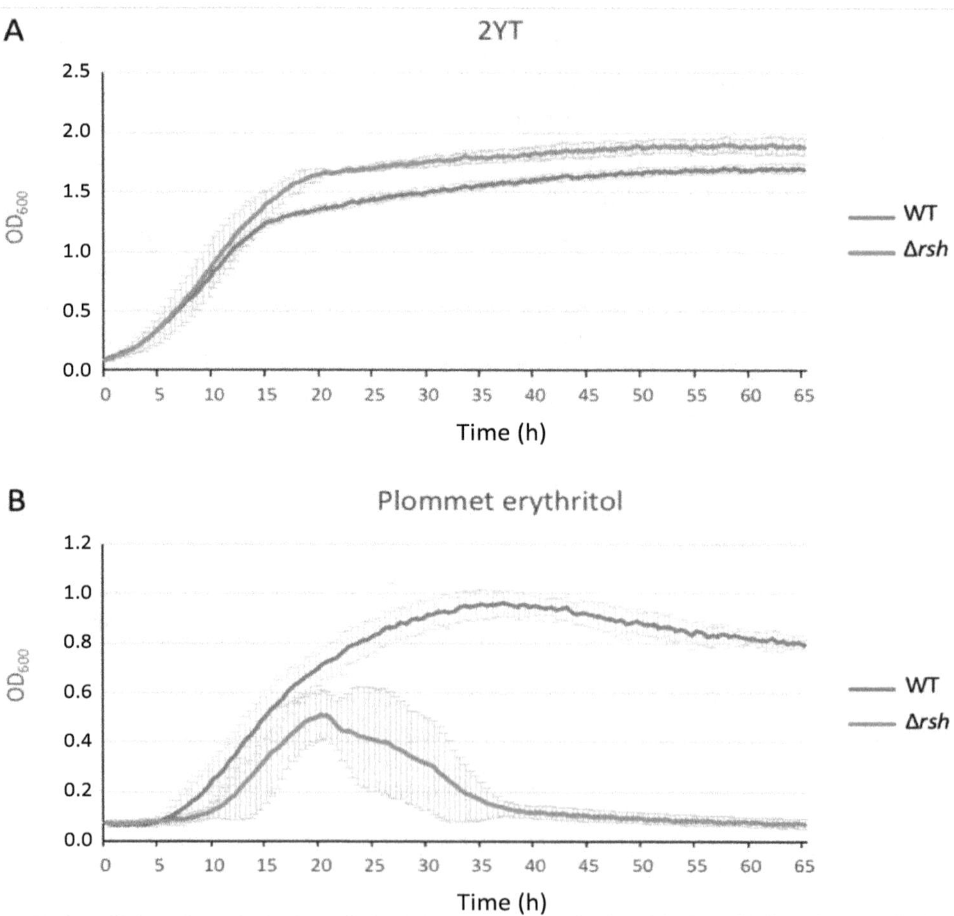


**Figure 1.** Growth of the Δ*rsh* mutant in 2YT rich medium (**A**) and in Plommet erythritol minimal medium (**B**). Strains were grown in liquid culture overnight in order to reach exponential phase. Cultures were then diluted at an optical density (OD) of 0.1 in 2YT medium. The OD of each strain was measured every 30 min. The graph represents the means of a biological quadruplicate. The error bars represent the standard deviation for each time point. WT: wild type.

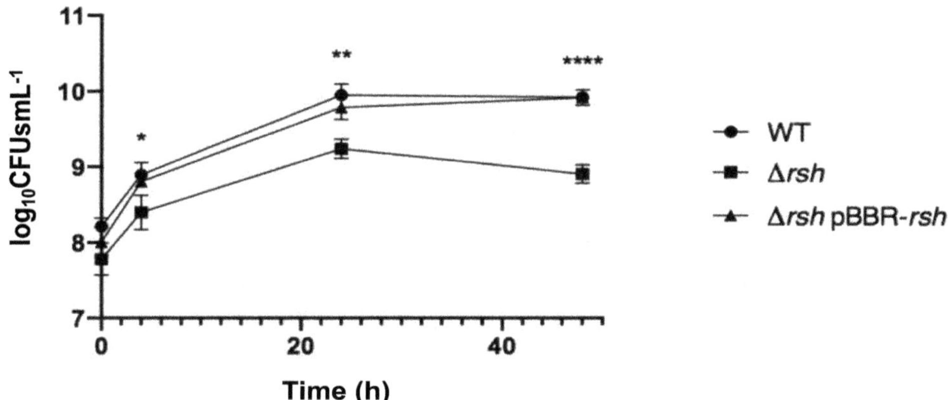


**Figure 2.** Survival and growth of *B. abortus* WT, Δ*rsh* and Δ*rsh* pBBR-*rsh* in 2YT rich medium. Strains were grown in liquid culture overnight in order to reach exponential phase. Cultures were then diluted at an OD of 0.1 ($3 \times 10^8$ bacteria/mL for the WT strain) in 2YT medium. The numbers of live bacteria ($log_{10}$CFUs $mL^{-1}$) were determined at 0 h, 4 h, 24 h and 48 h by plating serial dilutions. Values represent the means of three independent experiments and the error bars represent the standard deviation. The asterisks mean significant for $p < 0.05$ (*) $p < 0.01$ (**); $p < 0.0001$ (****), and the $p$ values were calculated by one-way ANOVA.

We tested the ability of Δ*rsh* to infect and multiply inside RAW 264.7 macrophages compared to the WT strain and the complemented strain by performing CFU counting throughout the cellular infection (Figure 3). The Δ*rsh* mutant showed a significant decrease in CFUs at 24 h post-infection compared to the WT strain, suggesting that the *rsh* gene is required for intracellular proliferation. A slight but significant difference was also observed at 2 h post-infection between the WT and complemented strain (Figure 3), probably highlighting a low toxicity of the vector or a *rsh* overexpression effect.

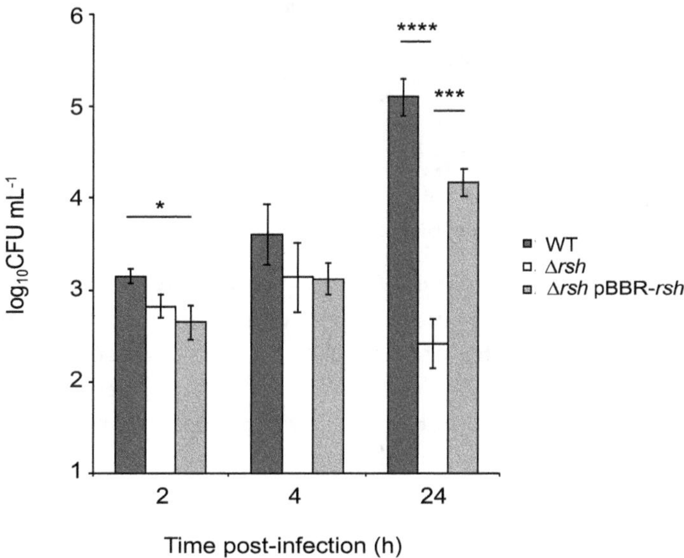


**Figure 3.** Survival and growth of *B. abortus* WT, Δ*rsh* and Δ*rsh* pBBR-*rsh* during infection of RAW 264.7 macrophages. Strains were grown in liquid culture overnight in order to reach exponential phase. Cultures were then diluted in Dulbecco's Modified Eagle's Medium (DMEM) to obtain a multiplicity of infection (MOI) of 50. The numbers of live bacteria ($log_{10}$CFUs $mL^{-1}$ of cellular lysate, 0.5 mL per well) were determined at 2 h, 4 h, and 24 h by plating serial dilutions. Values represent the means of three independent experiments and the error bars represent the standard deviations. A one-way ANOVA test was performed as statistical analysis. The asterisks mean significant for $p < 0.05$ (*); $p < 0.001$ (***); $p < 0.0001$ (****).

### 2.2. *The Artificial Hydrolysis of (p)ppGpp Leads to a Δrsh Phenotype during Infection*

Since Rsh is responsible for (p)ppGpp homeostasis and an Δ*rsh* mutant failed to proliferate inside RAW 264.7 cells, we tested the involvement of (p)ppGpp in the infection process. However, it is known that Rsh is involved in regulation networks through protein–protein contacts [14,24] in other bacteria. Therefore, we cannot rule out that the absence of the Rsh protein, rather than the absence of (p)ppGpp, would be responsible for the defect observed in infection. This is reinforced by the observation that mutants for homologs of the glutamine-dependent control pathway of Rsh are also attenuated in RAW 264.7 macrophages [25]. We thus generated a strain in which (p)ppGpp is hydrolyzed by a strong (p)ppGpp hydrolase, a product of the *mesh1* gene from *Drosophila melanogaster* [26]. Indeed, it was shown that Mesh1 was active in vitro and in vivo [26]. We thus expect this heterologous enzyme to be constitutive in *B. abortus*. We adapted the *mesh1* coding sequence for the codon bias of *B. abortus* and expressed the resulting coding sequence on a medium copy replicative plasmid, leading to the *B. abortus* pBBRi-*mesh1b* strain. Interestingly, this strain showed a clear decrease in CFUs at 24 h post-infection of RAW 264.7 macrophages (Figure 4), which is consistent with a crucial role played by (p)ppGpp to allow growth inside host cells, as suggested above.

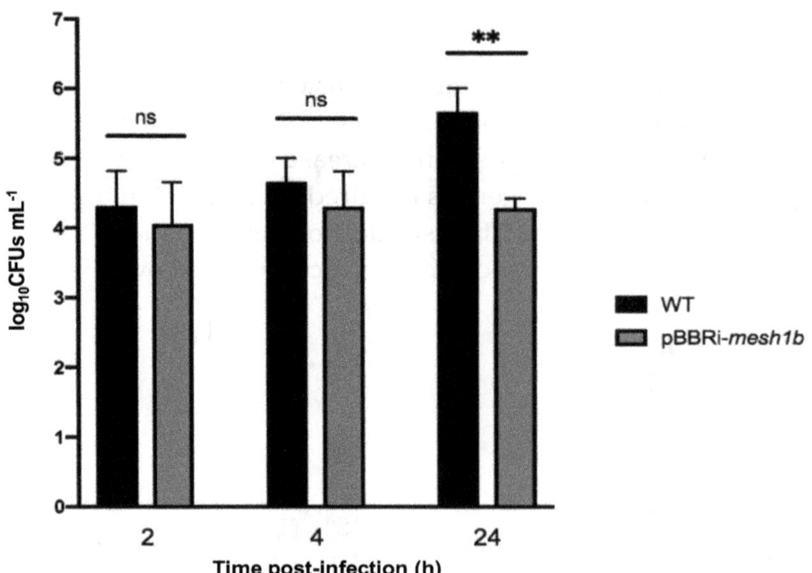


**Figure 4.** Survival and growth of *B. abortus* WT and pBBRi-*mesh1b* during infection of RAW 264.7 macrophages. Strains were grown in liquid culture overnight in order to reach exponential phase. Cultures were then diluted in DMEM to obtain a MOI of 50. The numbers of live bacteria ($\log_{10}$CFUs $mL^{-1}$) were determined at 2 h, 4 h, and 24 h post-infection by plating serial dilutions. Values represent the means of three independent experiments and the error bars represent the standard deviation. A Student *t* test was performed for the comparison of the two strains. The asterisks mean significant for $p < 0.01$ (**) and "ns" means "not significant".

### 2.3. *Expression of a Constitutive Allele for a (p)ppGpp Synthetase Impacts Bacterial Growth and Chromosome Replication*

In order to get more insight about the role of (p)ppGpp in *B. abortus*, we constructed a strain that artificially produces this alarmone. We used a truncated version of the *relA* gene from *E. coli*, *relA'* [12] that removes the C-terminal regulatory domains of the encoded protein. The *relA'* coding sequence was inserted downstream of an isopropyl β-D-1-thiogalactoside (IPTG)-inducible promoter on the pSRK replicative plasmid [27]. The resulting strain, named *pSRK-relA'*, is supposed to produce (p)ppGpp synthetase when IPTG is added to the medium. As a negative control, we used the *pSRK-relA′** strain containing the point mutation E335Q, which leads to a catalytically dead protein. Since the detection of (p)ppGpp levels using $^{32}P$ is not compatible with our biosafety level 3 set up, we tried to gain indirect evidence that (p)ppGpp is indeed produced when the expression of *relA′* is induced. We assayed the growth of the *pSRK-relA′* and *pSRK-relA′** strains in rich culture medium with or without IPTG induction. The *pSRK-relA′*, *pSRK-relA′** and WT strains grew equally in 2YT; however, when IPTG was added to the medium, a growth delay was only observed for the *pSRK-relA′* strain (Figure 5). This observation is consistent with the production of (p)ppGpp levels that are sufficient to limit growth when *relA′* expression is induced.

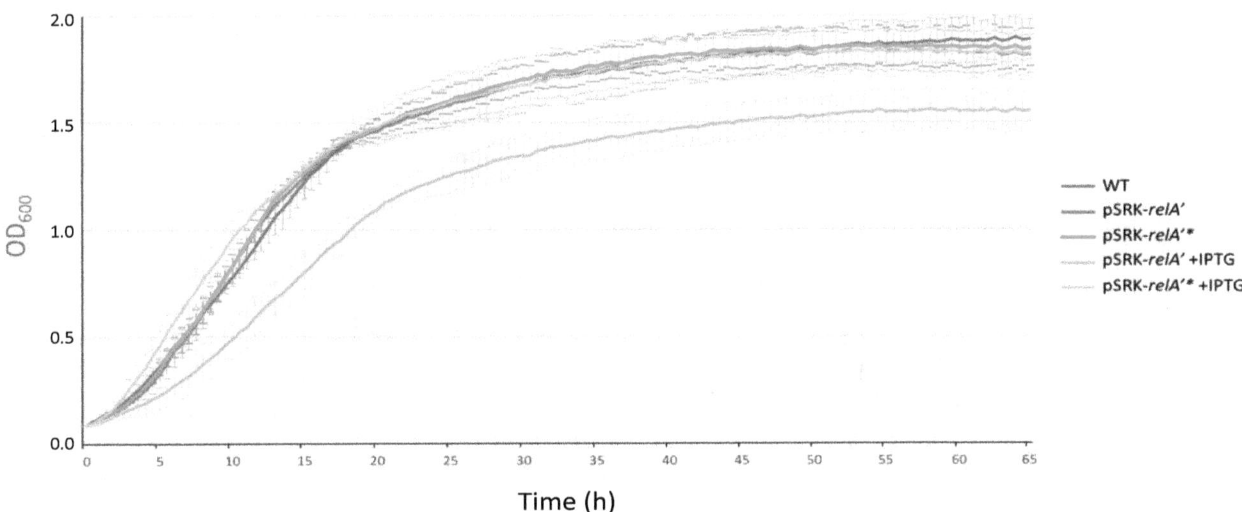


**Figure 5.** Growth curve in rich culture medium for the WT, *pSRK-relA′* and *pSRK-relA′** with or without IPTG. Strains were grown in liquid culture (2YT medium) overnight in order to reach exponential phase. Cultures were then diluted at an OD of 0.1 in 2YT medium supplemented or not with IPTG. The OD of each strain was measured every 30 min. The graph represents the means of a biological triplicate. The error bars represent the standard deviation for 3 biological replicates for each time point.

Since it was already reported that (p)ppGpp has an impact on DNA replication in *C. crescentus* and *E. coli* [12–15], we took advantage of a *B. abortus* strain allowing us to monitor the chromosomal replication status at the single cell level in order to study the impact of (p)ppGpp overproduction on DNA replication. This strain expresses an *mCherry-parB* allele that allows us to highlight the segregation of replication origin(s) of chromosome I. In this strain, one mCherry focus means that segregation has not yet started and the bacterium is probably in the G1 phase of the cell cycle, and two mCherry foci correspond to two segregated replication origins, meaning that the bacterium has already started replication and is thus in the S or G2 phase of the cell cycle [2]. The *pSRK-relA′* and *pSRK-relA′** plasmids were inserted in a *B. abortus mCherry-parB* strain and we counted the number of G1 bacteria every two hours for 6 h after the inoculation of bacteria in rich medium, with or without IPTG. Interestingly, we observed an increase in the proportion of G1 bacteria over the time of induction with IPTG for the *pSRK-relA′* strain (Supplementary Figure S3). The proportion of G1 bacteria of the non-induced *pSRK-relA′* and both the induced or non-induced *pSRK-relA′** remained stable after the addition of IPTG (Figure 6). These results strongly suggested that artificial induction of (p)ppGpp synthesis could delay the transition between the G1 phase to the S phase and subsequently have an impact on the initiation of chromosomal replication in *B. abortus*.

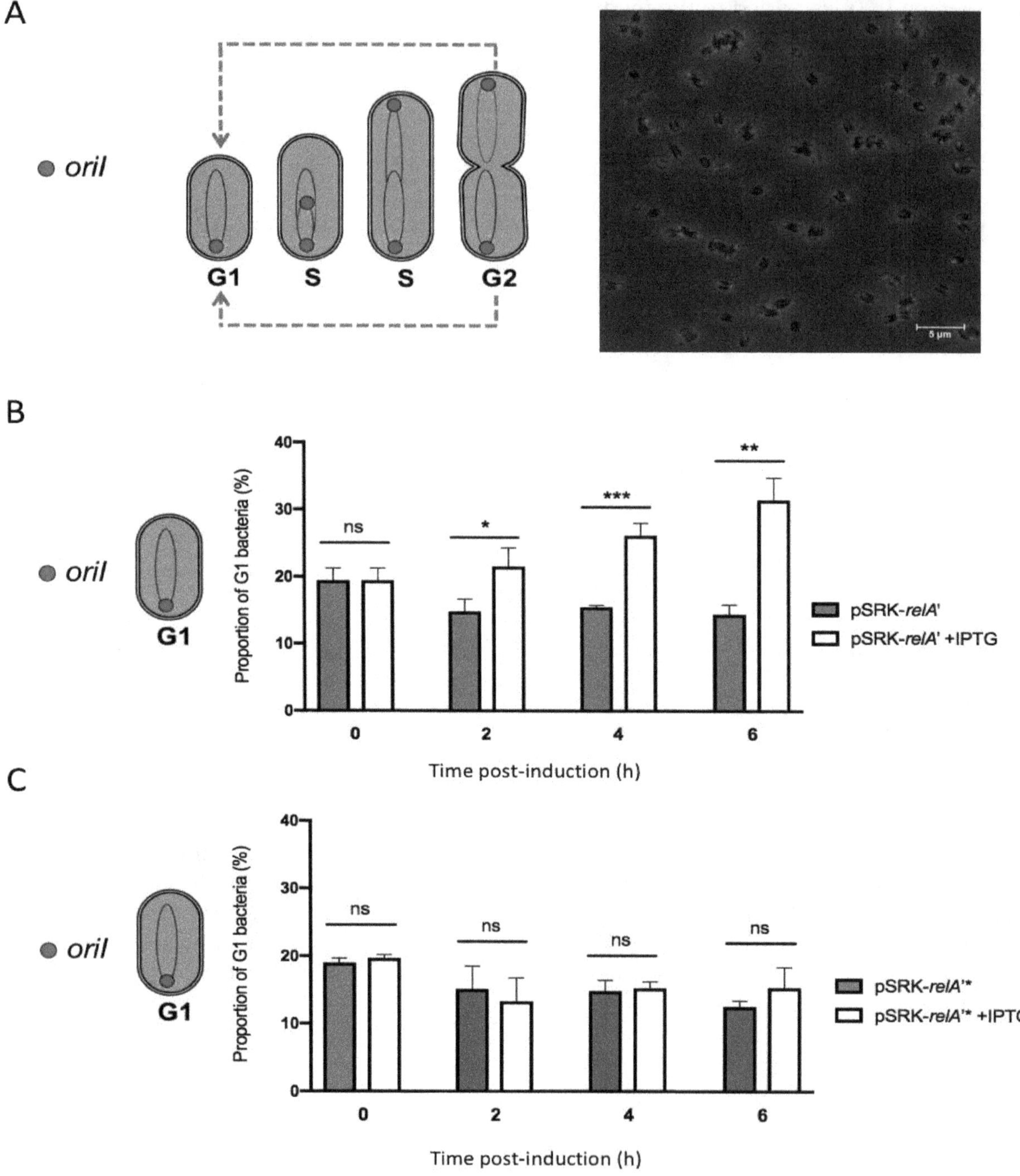


**Figure 6.** Proportion of G1 bacteria in rich culture medium with or without IPTG for the *pSRK-relA′* and *pSRK-relA′** strains. (**A**) Schematic drawing of the mCherry-ParB localization throughout the cell cycle [2] and fluorescence microscopy of the *pSRK-relA′ mCherry-parB* strain. Scale bar represents 5 µm. (**B**) Strains were grown in liquid culture (2YT medium) overnight in order to reach exponential phase. Cultures were then diluted to an OD of 0.1 in 2YT medium supplemented or not with IPTG. Samples were taken every 2 h, placed on a phosphate-buffered saline (PBS) agarose pad and observed with a fluorescence microscope. Bacteria in G1 phase (presenting only one focus of mCherry-ParB) were counted for each time post-induction. Error bars represent the standard deviation from the means of three independent experiments (biological triplicates). The significant differences are indicated by $p < 0.05$ (*), $p < 0.01$ (**) and $p < 0.001$ (***); "ns" means not significant. The number of bacteria considered in these triplicate experiments are detailed in Supplementary Table S1.

### 2.4. Induced Production of a Constitutive (p)ppGpp Synthetase Leads to a Proliferation Defect during Infection

Since (p)ppGpp overproduction seemed to have an impact on replication, i.e., an increase of the proportion of G1 cells in the bacterial population, and that the G1 bacteria are more infectious, we decided to investigate the effect of overproduction of (p)ppGpp on the infection process. We infected RAW 264.7 macrophages with the *pSRK-relA′* strain induced or not with IPTG (Figure 7). The IPTG was kept in the cell culture medium during the infection for the induced condition. We first observed that bacterial internalization is not enhanced by the increase in the proportion of bacteria in the G1 phase of the cell cycle. We also observed that induction of *pSRK-relA′* induced a strong defect in intracellular proliferation compared to the WT and uninduced *pSRK-relA′* conditions. This result suggests that overproduction of (p)ppGpp during infection prevents growth in the intracellular niche.

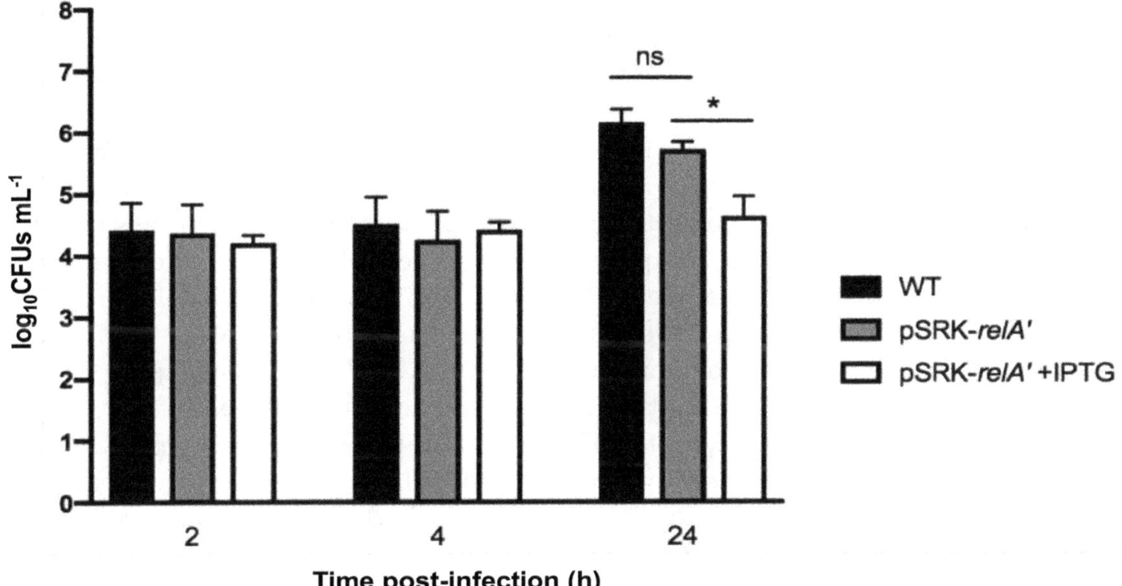


**Figure 7.** Survival of *B. abortus* WT and *pSRK-relA′* strains with and without IPTG during infection of RAW 264.7 macrophages. Strains were grown in liquid culture overnight in order to reach exponential phase. Cultures were then diluted at an OD of 0.1 with or without IPTG (1 mM) and were incubated for 3 h at 37 °C. Cultures were then diluted in DMEM with or without IPTG (10 mM) to obtain a MOI of 50. Concentrations of live bacteria ($\log_{10}$CFUs mL$^{-1}$) were determined at 0 h, 4 h, and 24 h post-infection by plating serial dilutions. Values represent the means of three independent experiments and the error bars represent the standard deviation. A Student's t test was performed as statistical analysis. The asterisks mean significant for $p < 0.05$ (*) and "ns" means "not significant".

### 2.5. DksA Is Not Required during the Infection Process

Because (p)ppGpp seemed important during host infection and DksA is involved in a part of the (p)ppGpp transcriptional response in other species, we tested the ability of Δ*dksA* to infect and proliferate inside RAW 264.7 macrophages. No difference in CFUs was observed between WT and Δ*dksA* strains (Supplementary Figure S1), meaning that DksA is not crucially involved in the infection process and that the phenotype observed for (p)ppGpp-deprived mutants (Δ*rsh* and *pBBRi-mesh1b*) is probably not mediated by DksA.

## 3. Discussion

*Brucella abortus* is able to control its cell cycle progression when it is inside host cells, particularly the replication and segregation of its replication origins [2]. However, the molecular mechanisms involved in this control are unknown. Since convincing data show that the ability to adapt to starvation is a key factor for the success of cellular infections by *Brucella melitensis* and *Brucella suis* [10,11], we further investigate the role of the (p)ppGpp, the alarmone produced in the presence of starvation conditions, and the Rsh enzyme that is proposed to synthesize this alarmone. We first confirmed that in *B. abortus* 544, like in other *Brucella* strains, Rsh is crucial for the success of a cellular infection (Figure 3), for the survival in stationary phase (Figure 2) and the growth in minimal medium (Figure 1). In agreement with the absence of proliferation of the *rsh* mutant in macrophages, a strain constitutively producing a (p)ppGpp hydrolase (Mesh1b) from *Drosophila melanogaster* is also unable to grow in RAW 264.7 macrophages. We observed that the survival of *pBBR-mesh1b* strain is less severely impacted during infection than the Δ*rsh* strain. One could imagine that this intermediate phenotype is due to the presence of residual alarmone in the *pBBR-mesh1b* strain, while it is not the case in the Δ*rsh* strain since the (p)ppGpp synthetase domain is not present. Another explanation for these different phenotypes could be that Rsh plays additional role(s) for survival in infection than the regulation of (p)ppGpp homeostasis. Interestingly, a mutant overproducing (p)ppGpp is also unable to proliferate in these cells, suggesting that the (p)ppGpp level should be in a specific range of concentration to allow cellular infection; having too much or not enough (p)ppGpp would be detrimental for the success of the cellular infection.

What does (p)ppGpp control and how is Rsh regulated? It was shown that the absence of *rsh* in *B. suis* affects the transcription of genes known to be involved in virulence [28], such as *pyrB*, which was shown to be essential for *B. abortus* proliferation in RAW 264.7 macrophages [25]. It is thus likely that a minimum level of (p)ppGpp would be required for the success of a cellular infection by *B. abortus*. It was shown that the glutamine pool modulates Rsh (called SpoT) in the model alpha-proteobacterium *Caulobacter crescentus* through the phosphotransferase system (PTS) [14]. Interestingly, mutants for components of this system were found to be attenuated in RAW 264.7 macrophages [25]. Moreover, PTS and the two-component regulator BvrR control the expression of the *virB* operon [29,30], coding for a type IV secretion system that is crucial for intracellular proliferation in most cell types [31]. These data indicate that a quite complex regulation network is probably linking Rsh control and virulence. However, the molecular mechanisms controlling Rsh activity in *B. abortus* are unknown and deserve further investigation.

One striking conclusion of our data is the moderate effect that (p)ppGpp overproduction has on the proportion of bacteria at the G1 stage of the cell cycle. Indeed, while overproduction seems to be sufficient to impair growth inside host cells (Figure 7), the proportion of G1 after 6 h of induction is about a third of the culture while it is approximately 15% in the absence of induction (Figure 6). Since the cell cycle takes about 3 h in the conditions tested, it is likely that only a fraction of the bacteria arrested their cell cycle at the G1 stage. Inside host cells, the proportion of G1 cells is about 75% and remains stable for 2 to 4 h at least [2], suggesting that other mechanisms are probably involved in the control of the cell cycle in host cells, early in the trafficking. These mechanisms could involve the acidic nature of the BCV, or the diffusion sensing proposed to occur through a regulation system homologous to quorum sensing [32]. More investigations are thus needed to discover the multiple factors involved in the cell cycle control of *B. abortus* inside host cells.

## 4. Materials and Methods

### 4.1. Strains and Growth Conditions

The reference strain *B. abortus* 544 was used for all experiments and was grown on solid or in liquid 2YT medium (LB 32 g/L Invitrogen, Yeast Extract 5g/L, BD and Peptone 6 g/L, BD) at 37 °C. *E. coli* strain DH10B was used for plasmid constructs and the conjugative strain *E. coli* S17-1 was used for mating with *B. abortus*. Both strains were cultivated in LB medium (Luria Bertani, Casein Hydrolysate 10g/L, NaCl 5g/L, Yeast Extract 5g/L) at 37 °C. Depending on the plasmid used, different selection antibiotics were added to the culture medium: ampicillin (100 μg/mL); carbenicillin (100 μg/mL); kanamycin (50 μg/mL for the replicative plasmid, and 10 μg/mL for the integrated plasmid); nalidixic acid (25 μL/mL); chloramphenicol (20 μg/mL for the replicative plasmid and 4 μg/mL for the integrated plasmid). Isopropyl β-d-1-thiogalactopyranoside (IPTG) was used at a concentration of 1 mM in bacterial liquid culture and at 10 mM in the mammalian cell culture medium during cellular infections. When the Δ*rsh* mutant was constructed, we added casamino acids 0.5% (BactoTM Casamino Acids from Thermo Fisher, Waltham, MA, USA) in the conjugation medium.

### 4.2. Strains Construction

Deletion strains were constructed by allelic exchange using the pNPTS138 vectors (M. R. K. Alley, Imperial College of Science, London, UK) carrying the upstream and the downstream regions of the targeted gene. The primer sequences used for amplification of the upstream region of the *dksA* gene were 5′-ttGGATCCcaagcgccagatcttca-3′ and 5′-ttGAATTCttcactcattctgaatcacccc-3′. The primer sequences used for amplification of the downstream region of the *dksA* gene were 5′-ttGAATTCtgatatcgaataatggtttggaaa-3′ and 5′-ttAAGCTTcgcccagcttcaaattac-3′.

We used the *rsh* deletion plasmid pMQ203 (provided by M. Quebatte, Biozentrum, Basel), containing the upstream and downstream regions of *rsh* amplified with the following hybridization sequences: 5′-ccggatgatctgaaggaa-3′, 5′-gcgcatcatctgccgaaa-3′ and 5′-gtctgggacctcaagcat-3′, 5′-cccgtggtgacgatatct-3′.

The Δ*rsh* pBBR-*rsh* strain was generated by inserting the pBBR-*rsh* in the Δ*rsh* strain. The pBBR-*rsh* was constructed by cloning the endogenous promoter of *rsh* and the *rsh* coding sequence, amplified with the primers 5′-aaaCTCGAGgcgagattgccgatgaga-3′ and 5′-aaaCTGCAGctatccgttcacacgctttg-3′.

The pBBRi-*mesh1b* strain was constructed by inserting the coding sequence *mesh1b* in the pBBRi plasmid. The sequence of *mesh1b* was adapted to the codon usage of *Brucella*, and is available in Supplementary Figure S2.

The *mCherry-parB* strains containing pSRK-*relA′* and pSRK-*relA′** were created using the Tn7 system [33] which consists in transposition of mini-Tn7 expressing *mCherry-parB* under the control of the *PgidA* promoter as previously reported [2] and the resistance cassette to ampicillin/carbenicillin under the control of *Pbla* promoter at the *glmS* locus of *B. abortus*. The primer sequences used for the amplification of *PgidA*-mCherry-*parB* and *Pbla-amp* were 5′-cgcggatcctctgtggaatcctgtttgttg-3′, 5′-AGCGGATACATATTTGAActagctttgaagacggcg-3′ and 5′-TTCAAATATGTATCCGCTCATGA-3′, 5′-cgggatccTTACCAATGCTTAATCAGTGAGG-3′.

### 4.3. Growth Assays

The bacterial growth curves were performed using a bioscreen (Epoch2 Microplate *Photospectrometer* from BioTek). Bacterial cultures in the exponential phase of growth were washed two times with PBS and were normalized at an OD of 0.1 in a given medium. A 200 μL aliquot of the normalized culture was transferred to a plate and each condition was performed in technical triplicate (3× 200 μL). The plate was incubated at 37 °C with shaking and the $OD_{600}$ of each well was measured every 30 min. One biological replicate constitutes the mean of three technical replicates and experiments were repeated at least three times to obtain biological triplicates.

### 4.4. Survival Assays

Bacterial cultures in the exponential phase of growth were normalized at an OD of 0.1 in 2YT liquid medium. Serial dilutions were plated on 2YT solid medium at different time points and plates were incubated at 37 °C.

### 4.5. Infections of RAW 264.7 Macrophages

RAW macrophages were put in wells in DMEM medium (with decomplemented bovine serum, glucose, glutamine, and no pyruvate, Gibco®) to have $1 \times 10^5$ cells/mL. *B. abortus* 544 was grown in 2YT at 37 °C until exponential phase. The OD of the bacterial culture was measured, and dilutions were performed to have a MOI equal to 50 (50 times more bacteria than macrophages). An input control was performed for each condition by plating bacteria on a 2YT agar plate before infecting cells.

Cell medium was removed to add the appropriate bacterial dilution. The mix was centrifuged for 10 min at 1200 rpm (4 °C) and then incubated at 37 °C with 5% $CO_2$ (this time point is set as time zero). After one hour of incubation, medium was removed and replaced by medium containing gentamycin (50 µg/mL) for 1 h in order to kill extracellular bacteria, and then by medium containing gentamycin (10 µg/mL). Note that for the experiments using IPTG, the IPTG (10 mM) was kept during all the steps of the infection. At either 2 h, 4 h or 24 h post infection, cells were first washed with sterile PBS and were then incubated in PBS + Triton 0.1% at 37 °C for 10 min in order to lyse the cells while keeping bacteria alive. After that, cells were flushed and lysates were harvested. Serial dilutions were performed and each dilution was spotted on 2YT agar plates and incubated at 37 °C.

### 4.6. Infections of HeLa Cells

HeLa cells were plated in wells in DMEM medium (with sodium pyruvate, non-essential amino acid, glucose, glutamine, and no pyruvate, Gibco®) at $4 \times 10^4$ cells/mL. *B. abortus* 544 was grown in 2YT at 37 °C until exponential phase, the OD of the bacterial culture was measured, and dilutions were performed to have a MOI equal to 300. An input control was performed for each condition by plating bacteria on a 2YT agar plate before infecting cells. Prior to infections in the presence of IPTG (see below), *relA'* expression was induced 3 h before infection with IPTG (1 mM in YT medium). Cell medium was removed to add the appropriate bacterial dilution. The mix was centrifuged for 10 min at 1200 rpm (4 °C) and incubated at 37 °C with 5% $CO_2$ (this time point is set as time zero). After one hour of incubation, medium was removed and replaced by medium containing gentamycin (50 µg/mL) in order to kill extracellular bacteria, and then gentamycin (10 µg/mL). Note that for the experiments using IPTG, the IPTG (10 mM) was kept during all the steps of the infection. At either 2 h, 4 h or 24 h post infection, cells were first washed with sterile PBS and were then incubated in PBS + Triton 0.1% at 37 °C for 10 min in order to lyse the cells while keeping bacteria alive. After that, cells were flushed and lysates were harvested. Serial dilutions were performed and each dilution was spotted on 2YT agar plates and incubated at 37 °C.

### 4.7. G1 Counting

Bacteria in exponential phase of growth were diluted to an OD of 0.1 in 2YT liquid medium with or without IPTG. At each time point, 200 µL of the culture was washed two times in PBS and bacteria were loaded onto a PBS agarose pad to be observed and counted by fluorescence microscopy.

**Author Contributions:** M.V.d.H. and X.D.B. designed the work and wrote the manuscript; M.V.d.H. and E.C. performed the experiments. All authors have read and agreed to the published version of the manuscript.

**Acknowledgments:** We thank Maxime Quebatte (Biozentrum, Basel) for the generous gift of the pMQ203 plasmid. We thank Severin Ronneau for the generous gift of the pSRK-*relA'* plasmid and for the numerous fruitful discussions on (p)ppGpp. M.V.d.H. was supported by a FRIA Ph.D. fellowship from FRS-FNRS. We thank the University of Namur for logistic support.

## References

1. Moreno, E.; Moriyon, I. The Genus *Brucella*. *Prokaryotes* **2006**, *5*, 315–456.
2. Deghelt, M.; Mullier, C.; Sternon, J.F.; Francis, N.; Laloux, G.; Dotreppe, D.; Van der Henst, C.; Jacobs-Wagner, C.; Letesson, J.J.; De Bolle, X. The newborn *Brucella abortus* blocked at the G1 stage of its cell cycle is the major infectious bacterial subpopulation. *Nat. Commun.* **2014**, *5*, 4366. [CrossRef] [PubMed]
3. Starr, T.; Ng, T.W.; Wehrly, T.D.; Knodler, L.A.; Celli, J. *Brucella* intracellular replication requires trafficking through the late endosomal/lysosomal compartment. *Traffic* **2008**, *9*, 678–694. [CrossRef] [PubMed]
4. Porte, F.; Liautard, J.P.; Kohler, S. Early acidification of phagosomes containing *Brucella suis* is essential for intracellular survival in murine macrophages. *Infect. Immun.* **1999**, *67*, 4041–4047. [CrossRef] [PubMed]

5. Pizarro-Cerda, J.; Meresse, S.; Parton, R.G.; van der Goot, G.; Sola-Landa, A.; Lopez-Goni, I.; Moreno, E.; Gorvel, J.P. *Brucella abortus* transits through the autophagic pathway and replicates in the endoplasmic reticulum of nonprofessional phagocytes. *Infect. Immun.* **1998**, *66*, 5711–5724. [CrossRef]
6. Starr, T.; Child, R.; Wehrly, T.D.; Hansen, B.; Hwang, S.; Lopez-Otin, C.; Virgin, H.W.; Celli, J. Selective subversion of autophagy complexes facilitates completion of the Brucella intracellular cycle. *Cell Host Microbe* **2012**, *11*, 33–45. [CrossRef] [PubMed]
7. Roop, R.M., 2nd; Gaines, J.M.; Anderson, E.S.; Caswell, C.C.; Martin, D.W. Survival of the fittest: How Brucella strains adapt to their intracellular niche in the host. *Med. Microbiol. Immunol.* **2009**, *198*, 221–238. [CrossRef]
8. Poncin, K.; Roba, A.; Jimmidi, R.; Potemberg, G.; Fioravanti, A.; Francis, N.; Willemart, K.; Zeippen, N.; Machelart, A.; Biondi, E.G.; et al. Occurrence and repair of alkylating stress in the intracellular pathogen *Brucella abortus*. *Nat. Commun.* **2019**, *10*, 4847. [CrossRef]
9. Kohler, S.; Foulongne, V.; Ouahrani-Bettache, S.; Bourg, G.; Teyssier, J.; Ramuz, M.; Liautard, J.P. The analysis of the intramacrophagic virulome of *Brucella suis* deciphers the environment encountered by the pathogen inside the macrophage host cell. *Proc. Natl. Acad. Sci. USA* **2002**, *99*, 15711–15716. [CrossRef]
10. Dozot, M.; Boigegrain, R.A.; Delrue, R.M.; Hallez, R.; Ouahrani-Bettache, S.; Danese, I.; Letesson, J.J.; De Bolle, X.; Kohler, S. The stringent response mediator Rsh is required for *Brucella melitensis* and *Brucella suis* virulence, and for expression of the type IV secretion system virB. *Cell Microbiol.* **2006**, *8*, 1791–1802. [CrossRef]
11. Kim, S.; Watanabe, K.; Suzuki, H.; Watarai, M. Roles of *Brucella abortus* SpoT in morphological differentiation and intramacrophagic replication. *Microbiology* **2005**, *151*, 1607–1617. [CrossRef] [PubMed]
12. Gonzalez, D.; Collier, J. Effects of (p)ppGpp on the progression of the cell cycle of *Caulobacter crescentus*. *J. Bacteriol.* **2014**, *196*, 2514–2525. [CrossRef] [PubMed]
13. Lesley, J.A.; Shapiro, L. SpoT regulates DnaA stability and initiation of DNA replication in carbon-starved *Caulobacter crescentus*. *J. Bacteriol.* **2008**, *190*, 6867–6880. [CrossRef] [PubMed]
14. Ronneau, S.; Petit, K.; De Bolle, X.; Hallez, R. Phosphotransferase-dependent accumulation of (p)ppGpp in response to glutamine deprivation in *Caulobacter crescentus*. *Nat. Commun.* **2016**, *7*, 11423. [CrossRef] [PubMed]
15. Schreiber, G.; Ron, E.Z.; Glaser, G. ppGpp-mediated regulation of DNA replication and cell division in *Escherichia coli*. *Curr. Microbiol.* **1995**, *30*, 27–32. [CrossRef]
16. Dalebroux, Z.D.; Edwards, R.L.; Swanson, M.S. SpoT governs *Legionella pneumophila* differentiation in host macrophages. *Mol. Microbiol.* **2009**, *71*, 640–658. [CrossRef]
17. Das, B.; Pal, R.R.; Bag, S.; Bhadra, R.K. Stringent response in *Vibrio cholerae*: Genetic analysis of *spoT* gene function and identification of a novel (p)ppGpp synthetase gene. *Mol. Microbiol.* **2009**, *72*, 380–398. [CrossRef]
18. Dahl, J.L.; Kraus, C.N.; Boshoff, H.I.; Doan, B.; Foley, K.; Avarbock, D.; Kaplan, G.; Mizrahi, V.; Rubin, H.; Barry, C.E., 3rd. The role of RelMtb-mediated adaptation to stationary phase in long-term persistence of *Mycobacterium tuberculosis* in mice. *Proc. Natl. Acad. Sci. USA* **2003**, *100*, 10026–10031. [CrossRef]
19. Paul, B.J.; Barker, M.M.; Ross, W.; Schneider, D.A.; Webb, C.; Foster, J.W.; Gourse, R.L. DksA: A critical component of the transcription initiation machinery that potentiates the regulation of rRNA promoters by ppGpp and the initiating NTP. *Cell* **2004**, *118*, 311–322. [CrossRef]
20. Ross, W.; Sanchez-Vazquez, P.; Chen, A.Y.; Lee, J.H.; Burgos, H.L.; Gourse, R.L. ppGpp Binding to a Site at the RNAP-DksA Interface Accounts for Its Dramatic Effects on Transcription Initiation during the Stringent Response. *Mol. Cell* **2016**, *62*, 811–823. [CrossRef]
21. Ronneau, S.; Hallez, R. Make and break the alarmone: Regulation of (p)ppGpp synthetase/hydrolase enzymes in bacteria. *FEMS Microbiol. Rev.* **2019**, *43*, 389–400. [CrossRef]
22. Mittenhuber, G. Comparative genomics and evolution of genes encoding bacterial (p)ppGpp synthetases/hydrolases (the Rel, RelA and SpoT proteins). *J. Mol. Microbiol. Biotechnol.* **2001**, *3*, 585–600. [PubMed]
23. Plommet, M. Minimal requirements for growth of *Brucella suis* and other *Brucella* species. *Zentralblatt Bakteriologie* **1991**, *275*, 436–450. [CrossRef]
24. Ronneau, S.; Caballero-Montes, J.; Coppine, J.; Mayard, A.; Garcia-Pino, A.; Hallez, R. Regulation of (p)ppGpp hydrolysis by a conserved archetypal regulatory domain. *Nucleic Acids Res.* **2019**, *47*, 843–854. [CrossRef]

25. Sternon, J.F.; Godessart, P.; Goncalves de Freitas, R.; Van der Henst, M.; Poncin, K.; Francis, N.; Willemart, K.; Christen, M.; Christen, B.; Letesson, J.J.; et al. Transposon Sequencing of *Brucella abortus* Uncovers Essential Genes for Growth In Vitro and Inside Macrophages. *Infect. Immun.* **2018**, *86*. [CrossRef]
26. Sun, D.; Lee, G.; Lee, J.H.; Kim, H.Y.; Rhee, H.W.; Park, S.Y.; Kim, K.J.; Kim, Y.; Kim, B.Y.; Hong, J.I.; et al. A metazoan ortholog of SpoT hydrolyzes ppGpp and functions in starvation responses. *Nat. Struct. Mol. Biol.* **2010**, *17*, 1188–1194. [CrossRef]
27. Khan, S.R.; Gaines, J.; Roop, R.M., 2nd; Farrand, S.K. Broad-host-range expression vectors with tightly regulated promoters and their use to examine the influence of TraR and TraM expression on Ti plasmid quorum sensing. *Appl. Environ. Microbiol.* **2008**, *74*, 5053–5062. [CrossRef]
28. Hanna, N.; Ouahrani-Bettache, S.; Drake, K.L.; Adams, L.G.; Kohler, S.; Occhialini, A. Global Rsh-dependent transcription profile of *Brucella suis* during stringent response unravels adaptation to nutrient starvation and cross-talk with other stress responses. *BMC Genom.* **2013**, *14*, 459. [CrossRef]
29. Dozot, M.; Poncet, S.; Nicolas, C.; Copin, R.; Bouraoui, H.; Maze, A.; Deutscher, J.; De Bolle, X.; Letesson, J.J. Functional characterization of the incomplete phosphotransferase system (PTS) of the intracellular pathogen *Brucella melitensis*. *PLoS ONE* **2010**, *5*. [CrossRef]
30. Martinez-Nunez, C.; Altamirano-Silva, P.; Alvarado-Guillen, F.; Moreno, E.; Guzman-Verri, C.; Chaves-Olarte, E. The two-component system BvrR/BvrS regulates the expression of the type IV secretion system VirB in *Brucella abortus*. *J. Bacteriol.* **2010**, *192*, 5603–5608. [CrossRef]
31. Lacerda, T.L.; Salcedo, S.P.; Gorvel, J.P. Brucella T4SS: The VIP pass inside host cells. *Curr. Opin. Microbiol.* **2013**, *16*, 45–51. [CrossRef] [PubMed]
32. Terwagne, M.; Mirabella, A.; Lemaire, J.; Deschamps, C.; De Bolle, X.; Letesson, J.J. Quorum sensing and self-quorum quenching in the intracellular pathogen *Brucella melitensis*. *PLoS ONE* **2013**, *8*, e82514. [CrossRef] [PubMed]
33. Choi, K.H.; Gaynor, J.B.; White, K.G.; Lopez, C.; Bosio, C.M.; Karkhoff-Schweizer, R.R.; Schweizer, H.P. A Tn7-based broad-range bacterial cloning and expression system. *Nat. Methods* **2005**, *2*, 443–448. [CrossRef] [PubMed]

# 3

# Molecular Characterization of *Staphylococcus aureus* Isolated from Chronic Infected Wounds in Rural Ghana

**Manuel Wolters [1], Hagen Frickmann [2,3], Martin Christner [1], Anna Both [1], Holger Rohde [1], Kwabena Oppong [4], Charity Wiafe Akenten [4], Jürgen May [5,6] and Denise Dekker [6,7,*]**

1. Institute of Medical Microbiology, Virology and Hygiene, Universitiy Medical Center Hamburg-Eppendorf (UKE), 20251 Hamburg, Germany; m.wolters@uke.de (M.W.); mchristner@uke.de (M.C.); a.both@uke.de (A.B.); Rohde@uke.de (H.R.)
2. Department of Microbiology and Hospital Hygiene, Bundeswehr Hospital Hamburg, 20359 Hamburg, Germany; frickmann@bnitm.de
3. Institute for Medical Microbiology, Virology and Hygiene, University Medicine Rostock, 18057 Rostock, Germany
4. Kumasi Centre for Collaborative Research in Tropical Medicine (KCCR), Kumasi, Ghana; oppong@kccr.de (K.O.); danquah01@yahoo.co.uk (C.W.A.)
5. Tropical Medicine II, Universitiy Medical Center Hamburg-Eppendorf (UKE), 20251 Hamburg, Germany; j.may@uke.de
6. Department of Infectious Disease Epidemiology, Bernhard Nocht Institute for Tropical Medicine (BNITM), 20359 Hamburg, Germany; may@bnitm.de
7. German Centre for Infection Research (DZIF), Hamburg-Lübeck-Borstel-Riems, 38124 Braunschweig, Germany

* Correspondence: dekker@bnitm.de

**Abstract:** Background: Globally, *Staphylococcus aureus* is an important bacterial pathogen causing a wide range of community and hospital acquired infections. In Ghana, resistance of *S. aureus* to locally available antibiotics is increasing but the molecular basis of resistance and the population structure of *S. aureus* in particular in chronic wounds are poorly described. However, this information is essential to understand the underlying mechanisms of resistance and spread of resistant clones. We therefore subjected 28 *S. aureus* isolates from chronic infected wounds in a rural area of Ghana to whole genome sequencing. Results: Overall, resistance of *S. aureus* to locally available antibiotics was high and 29% were Methicillin resistant *Staphylococcus aureus* (MRSA). The most abundant sequence type was ST88 (29%, 8/28) followed by ST152 (18%, 5/28). All ST88 carried the *mecA* gene, which was associated with this sequence type only. Chloramphenicol resistance gene *fexB* was exclusively associated with the methicillin-resistant ST88 strains. Panton-Valentine leukocidin (PVL) carriage was associated with ST121 and ST152. Other detected mechanisms of resistance included *dfrG*, conferring resistance to trimethoprim. Conclusions: This study provides valuable information for understanding the population structure and resistance mechanisms of *S. aureus* isolated from chronic wound infections in rural Ghana.

**Keywords:** rural Ghana; molecular epidemiology; chronic wounds; *Staphylococcus aureus*

## 1. Introduction

*Staphylococcus aureus* is an important bacterial pathogen in all parts of the world, causing both community and hospital acquired infections. In particular methicillin-resistant *S. aureus* (MRSA) has evolved as a global health threat due to its resistance to beta lactam and other classes of antibiotics [1]. In the last 20 years the prevalence of MRSA appears to be increasing in many African countries as suggested by data from the first decade of the present century [2]. More recent reviews indicate ongoing epidemiological relevance of this resistance type in Africa [3], with increased reporting of outbreak-association in Western African Ghana [4]. In Ghana, the abundance of MRSA in carriage studies or clinical samples demonstrated large geographical differences [5–7]. Moreover, resistance of *S. aureus* to a variety of other locally available oral antibiotics such as tetracyclines, trimethoprim/sulfamethoxazole and penicillins is frequently observed in Ghana [8,9]. However, the underlying mechanisms of resistance in this region are not well understood and to our knowledge has not been described for *S. aureus* from chronic wounds.

Effective surveillance of antimicrobial resistance in bacteria including *S. aureus* is essential for estimating the burden of resistance and molecular strain typing provides important information for understanding the spread of resistant clones. However, in Africa both surveillance and strain typing information are scarce due to the limited diagnostic microbiology infrastructure generally available in large parts of the continent. Molecular typing including *spa*-typing, multi-locus sequencing and also whole genome sequence typing has been applied in only a few studies in *S. aureus* in humans and livestock in Ghana [3,10–14], identifying strains of multiple clonal clusters [14]. In particular, the MRSA clone sequence type (ST)88-IV (2B) is not only abundant in Ghana, but also in other African counties: Angola, Cameroon, Gabon, Madagascar, Nigeria, as well as São Tomé e Príncipe [15]. In Ghana, reported rates range between 24.2–83.3% of all MRSA isolates [12].

In a previous study *S. aureus* was isolated from 14.0% ($n$ = 28) of samples from patients with chronic wounds in Ghana [5]. In that study, a high frequency of methicillin-resistance (29%) was noted. Moreover, resistance to other commonly used antibiotics like penicillins, tetracyclines and trimethoprim/sulfamethoxazole was frequently observed. For Ghanaian *S. aureus* isolates from wounds, data on prevalent clones, resistance mechanisms and pathogenicity-associated genetic determinants is limited. To fill this gap in the global epidemiological picture, we have subjected the 28 *S. aureus* isolates from our previous study to whole genome sequencing (WGS), aiming at analyzing the underlying molecular basis of antimicrobial resistance and the population structure of this strain collection.

## 2. Materials and Methods

### 2.1. Sample Collection, Microbiology and Antibiotic Susceptibility Testing

*S. aureus* was isolated from female and male patients ≥15 years with an infected wound at the Outpatient Department of the Agogo Presbyterian Hospital, in the Asante Akim North District of Ghana from January to November 2016. Sample collection and microbiological investigations were reported previously [5]. Antibiotic susceptibility was tested by the disk diffusion method and interpreted following the European Committee on Antimicrobial Susceptibility Testing (EUCAST) guidelines v.10.0 (http://www.eucast.org).

### 2.2. Whole Genome Sequencing and Data Analysis

Whole genome sequencing of the isolates was performed using the Illumina NextSeq platform. WGS data were analyzed using the Nullarbor pipeline (vers. 2.0.20181010; Seemann T, available at: https://github.com/tseemann/nullarbor). Reads were assembled with spades [16] (vers. 3.13.1) and annotated with Prokka [17] (vers. 1.13.3). The resistance and virulence gene profiles were determined with ABRicate (https://github.com/tseemann/abricate) (vers 0.9.9) employing NCBI AMR (7th October 2020; 5283 sequences), Resfinder [18] (3077 sequences) and VFDB [19] (2597 sequences) databases. SCCmec types were determined with the SCCmecFinder web tool (https://cge.cbs.dtu.dk/services/SCCmedFinder/10.1128/mSphere.00612-17).

The MLST sequence types were extracted from the WGS data using the MLST tool (vers. 2.16.1). Sequencing reads have been deposited in NCBI's small reads archive (BioProject: PRJNA670821).

### 2.3. Ethical Considerations

The Committee on Human Research, Publications and Ethics, School of Medical Science, Kwame Nkrumah University of Science and Technology in Kumasi, Ghana, approved this study (approval number CHRPE/AP/078/16) on 14th December, 2015.

## 3. Results

The identified MLST sequence types and selected associated resistance and virulence genes are summarized in Table 1; the complete MLST, virulence factor and resistance gene datasets (VFDB, Resfinder and NCBI AMR), as well as AST results are available in the supplemental material (Table S1) The most abundant sequence type was ST88 (8/28). All ST88 isolates were *mecA* positive, SCC*mec* type IV(2B), *agr* type 3 and negative for the Pantone–Valentine Leukocidin toxin (PVL) genes *lukS-PV/lukF-PV*. Seven of eight ST88 isolates carried the *fexB* gene, which confers resistance to chloramphenicol [20]. The five isolates of the second most abundant clone ST152 were all *mecA* negative and carried the PVL genes *lukS-PV/lukF-PV*.

**Table 1.** Sequence types and associated virulence and resistance genes.

| Sequence Type | n | *lukS-PV/lukF-PV* | *mecA* | *fexB* |
|---|---|---|---|---|
| ST88 | 8 | 0 | 8 | 7 |
| ST152 | 5 | 5 | 0 | 0 |
| ST15 | 3 | 0 | 0 | 0 |
| ST1 | 2 | 0 | 0 | 0 |
| ST5 | 2 | 0 | 0 | 0 |
| ST45 | 2 | 0 | 0 | 0 |
| ST2434 | 2 | 0 | 0 | 0 |
| ST72 | 1 | 0 | 0 | 0 |
| ST121 | 1 | 1 | 0 | 0 |
| ST3248 | 1 | 0 | 0 | 0 |
| ST3249 | 1 | 0 | 0 | 0 |
| Totals | 28 | 6 | 8 | 7 |

Highest rates of phenotypic antimicrobial resistance were detected for penicillin (100%, 28/28), tetracycline (57%, 16/28) and trimethoprim/sulfamethoxazole (39%, 11/28) (Table 2). All isolates were susceptible to linezolid, rifampicin, fosfomycin, tigecycline, ciprofloxacin, levofloxacin and daptomycin. WGS resistance gene profiling identified corresponding acquired resistance genes in 97% (65/67) of all phenotypically detected antimicrobial resistances (Table 2). In all but one penicillin-resistant isolate penicillinase-encoding *blaZ* was detected. The single penicillin resistant, *blaZ* negative strain carried *mecA*, reasonably explaining the observed betalactam-resistant phenotype. All oxacillin-resistant isolates were *mecA* positive and belonged to ST88. *MecC* was not detected in the strain collection. Two main mechanisms of resistance to tetracyclines have been described in *S. aureus*: active efflux, resulting from plasmid-located *tetK* and *tetL* genes and ribosomal protection mediated by *tetM* or *tetO* genes. In our collection tetracycline resistant isolates carried either tetK (7/16) alone or *tetL* (8/16) in combination with *tetM* (8/16). The *tetL* and *tetM* genes were exclusively detected in the ST88 MRSA isolates, while tet*K* was found in various clonal backgrounds. Beside mutation of the chromosomal dihydrofolate reductase (DHFR) gene, three acquired dihydrofolate reductase gene variants are known to confer resistance to trimethoprim in *S. aureus* of human origin: *dfrA*, *dfrG* and *dfrK*. In our collection all trimethoprim/sulfamethoxazole resistant isolates carried *dfrG*. Of the two erythromycin resistant isolates one carried *msrA* and one *ermC*. As expected, the latter isolate was also resistant to clindamycin. Gentamicin resistance is most commonly conferred by aminoglycoside-modifying enzymes. However, in the single case of a gentamicin-resistant isolate no corresponding resistance gene was found.

**Table 2.** Phenotypic antibiotic resistance and associated genetic resistance markers.

| Antibiotic | Phenotypic AST Resistant n (%) | WGS Resistance Gene or Mutation | Positive n (%) |
|---|---|---|---|
| Penicillin | 28 (100) | *blaZ* | 27 (96) |
| | | *blaZ* neg, *mecA* pos | 1 (4) |
| Tetracycline | 16 (57) | *tetK* | 7 (44) |
| | | *tetL* | 8 (50) |
| | | *tetM* | 8 (50) |
| Trimethoprim/ Sulfamethoxazole | 11 (39) | *dfrG* | 11 (100) |
| Oxacillin | 8 (29) | *mecA* | 8 (100) |
| Erythromycin | 2 (7) | *msrA* | 1 (50) |
| | | *ermC* | 1 (50) |
| Clindamycin | 1 (4) | *ermC* | 1 (100) |
| Gentamicin | 1 (4) | not detected | none |

## 4. Discussion

In this study we describe the molecular epidemiology of *S. aureus* isolated from chronic infected wounds in outpatients in a rural area of Ghana.

Overall, the frequency of MRSA (29%) was high, comparable to other clinical studies conducted in Ghana [6,21,22]. Nevertheless, the frequencies of MRSA seen in patients in Ghana seem subject to geographical variations [5–7]. Moreover, the previously reported high rates of resistance to orally available antibiotics including penicillin, tetracycline and cotrimoxazole in *S. aureus* [9] were confirmed by our study results. The particularly high frequencies of penicillin resistance might be attributed to the fact that penicillin-based antibiotics are amongst the most frequently prescribed drugs in Ghana and available over the counter without prescription [23]. High rates of resistance inevitably reduce effective antibiotic treatment in areas where resources are scarce. This favors the use of cleaning and disinfecting procedures for the management of wound infections whenever clinically possible.

All isolated MRSA belonged to ST88, described as the dominant clone in various African countries, including Ghana [15]. In addition, all but one MRSA strain also carried the *fexB* gene, conferring resistance to chloramphenicol, which was not found in any of the other sequence types. Previously, *fexB* has only been described in *S. aureus* strains from Ghanaian patients with Buruli ulcer [20], also a chronic wound. As previously described in 2007 [24], chloramphenicol has been extensively prescribed in Africa, although a low risk of 0.002% for chloramphenicol-induced aplastic anemia had been described in Nigeria as early as in 1993 with an associated recommendation for strict risk-benefit-assessments prior to its prescription [25]. To the authors best knowledge, little has changed in the meantime and the substance is still in broad use in Sub-Saharan Africa, as it is readily available and shows excellent penetration even in difficult to reach compartments including bradytroph tissue like bone [26]. It is therefore possible that that frequent application of chloramphenicol in patients that did not respond to beta-lactam antibiotics, due to *mecA*-carriage of their *S. aureus* strains, might have facilitated the selection of *fexB* carrying bacteria. Due to the lack of reliable clinical data on previous antibiotic treatment this hypothesis could not be confirmed. Other mechanisms of resistance detected included *dfrG*, conferring resistance to trimethoprim, frequently found in strains isolated from Ghana [27,28]. Earlier this was regarded as an infrequent cause of trimethoprim resistance in *S. aureus* isolated from patients but is now widespread in Africa and common in *S. aureus* from ill travelers returning to Europe [27].

Panton-Valentine Leukocidin (PVL), which has been proposed as an epidemiological marker for severe skin infections [29], was encoded in strains of the ST121 and ST152 clonal lineages. The ST152 clonal lineage, in particular, is both known to be associated with PVL expression [30] and wide distribution in Ghana [10,31].

## 5. Conclusions

This study provides insight into the molecular epidemiology of *S. aureus* sequence types found in chronic infected wounds in a rural area of Ghana.

However, the number of samples used was quite small, and they were taken from outpatients in one hospital, so they may not be representative of the community or the wider area. Moreover, we did not have reliable information about prior use of antibiotics in these patients.

Nevertheless, this study stipulates valuable information for understanding the spread of resistant clones found in patients visiting the study hospital, which is important for effective surveillance of antibiotic resistant *S. aureus* and vital for estimating the burden of resistance.

**Supplementary Materials:**

Table S1: Complete MLST, *agr* and *SCCmec* typing, virulence factor and resistance gene datasets (VFDB, Resfinder and NCBI AMR) and phenotypic AST results. VFDB, Resfinder and NCBI AMR: The numbers represent the percent identity in the alignment between the best matching resistance or virulence gene in ResFinder or VFDB and the corresponding sequence in the input genome. A perfect alignment is 100% and must also cover the entire length of the resistance gene in the database.

**Author Contributions:** M.W., D.D. and J.M. designed and coordinated this study. M.W., M.C., A.B. and H.R. performed data analyses analysis. H.F., D.D. and M.W. wrote the first draft of this manuscript. K.O. conducted and supervised fieldwork. C.W.A. conducted and supervised lab work. M.C. and H.R. supported writing and editing the manuscript. All authors have read and agreed to the published version of the manuscript.

**Acknowledgments:** We are grateful to all patients, who participated in this study and to the personnel at the Agogo Presbyterian Hospital. Without their efforts, this research study would not have been possible.

## References

1. Andersson, H.; Lindholm, C.; Fossum, B. MRSA—Global threat and personal disaster: Patients' experiences. *Int. Nurs. Rev.* **2011**, *58*, 47–53. [CrossRef] [PubMed]
2. Falagas, M.E.; Karageorgopoulos, D.E.; Leptidis, J.; Korbila, I.P. MRSA in Africa: Filling the global map of antimicrobial resistance. *PLoS ONE* **2013**, *8*, e68024. [CrossRef] [PubMed]
3. Osei Sekyere, J.; Mensah, E. Molecular epidemiology and mechanisms of antibiotic resistance in *Enterococcus* spp., *Staphylococcus* spp., and *Streptococcus* spp. in Africa: A systematic review from a One Health perspective. *Ann. N. Y. Acad. Sci.* **2020**, *1465*, 29–58. [CrossRef] [PubMed]
4. Donkor, E.S.; Dayie, N.T.K.D.; Tette, E.M.A. Methicillin-Resistant *Staphylococcus aureus* in Ghana: Past, Present, and Future. *Microb. Drug. Resist.* **2019**, *25*, 717–724. [CrossRef] [PubMed]
5. Krumkamp, R.; Oppong, K.; Hogan, B.; Strauss, R.; Frickmann, H.; Wiafe-Akenten, C.; Boahen, K.G.; Rickerts, V.; McCormick Smith, I.; Groß, U.; et al. Spectrum of antibiotic resistant bacteria and fungi isolated from chronically infected wounds in a rural district hospital in Ghana. *PLoS ONE* **2020**, *15*, e0237263. [CrossRef]
6. Amissah, N.A.; van Dam, L.; Ablordey, A.; Ampomah, O.W.; Prah, I.; Tetteh, C.S.; van der Werf, T.S.; Friedrich, A.W.; Rossen, J.W.; van Dijl, J.M.; et al. Epidemiology of *Staphylococcus aureus* in a burn unit of a tertiary care center in Ghana. *PLoS ONE* **2017**, *12*, e0181072. [CrossRef]
7. Janssen, H.; Janssen, I.; Cooper, P.; Kainyah, C.; Pellio, T.; Quintel, M.; Monnheimer, M.; Groß, U.; Schulze, M.H. Antimicrobial-Resistant Bacteria in Infected Wounds, Ghana, 2014. *Emerg Infect. Dis.* **2018**, *24*, 916–919. [CrossRef]
8. Dekker, D.; Wolters, M.; Mertens, E.; Boahen, K.G.; Krumkamp, R.; Eibach, D.; Schwarz, N.G.; Adu-Sarkodie, Y.; Rohde, H.; Christner, M.; et al. Antibiotic resistance and clonal diversity of invasive *Staphylococcus aureus* in the rural Ashanti Region, Ghana. *BMC Infect. Dis.* **2016**, *16*, 720. [CrossRef]
9. Newman, M.J.; Frimpong, E.; Donkor, E.S.; Opintan, J.A.; Asamoah-Adu, A. Resistance to antimicrobial drugs in Ghana. *Infect. Drug Resist.* **2011**, *4*, 215–220.

10. Egyir, B.; Hadjirin, N.F.; Gupta, S.; Owusu, F.; Agbodzi, B.; Adogla-Bessa, T.; Addo, K.K.; Stegger, M.; Larsen, A.R.; Holmes, M.A. Whole-genome sequence profiling of antibiotic-resistant *Staphylococcus aureus* isolates from livestock and farm attendants in Ghana. *J. Glob. Antimicrob. Resist.* **2020**, *22*, 527–532. [CrossRef]
11. Donkor, E.S.; Jamrozy, D.; Mills, R.O.; Dankwah, T.; Amoo, P.K.; Egyir, B.; Badoe, E.V.; Twasam, J.; Bentley, S.D. A genomic infection control study for Staphylococcus aureus in two Ghanaian hospitals. *Infect. Drug Resist.* **2018**, *11*, 1757–1765. [CrossRef] [PubMed]
12. Kpeli, G.; Buultjens, A.H.; Giulieri, S.; Owusu-Mireku, E.; Aboagye, S.Y.; Baines, S.L.; Seemann, T.; Bulach, D.; Gonçalves da Silva, A.; Monk, I.R.; et al. Genomic analysis of ST88 community-acquired methicillin resistant *Staphylococcus aureus* in Ghana. *Peer J.* **2017**, *5*, e3047. [CrossRef] [PubMed]
13. Egyir, B.; Guardabassi, L.; Monecke, S.; Addo, K.K.; Newman, M.J.; Larsen, A.R. Methicillin-resistant *Staphylococcus aureus* strains from Ghana include USA300. *J. Glob. Antimicrob. Resist.* **2015**, *3*, 26–30. [CrossRef] [PubMed]
14. Egyir, B.; Guardabassi, L.; Sørum, M.; Nielsen, S.S.; Kolekang, A.; Frimpong, E.; Addo, K.K.; Newman, M.J.; Larsen, A.R. Molecular epidemiology and antimicrobial susceptibility of clinical *Staphylococcus aureus* from healthcare institutions in Ghana. *PLoS ONE* **2014**, *9*, e89716. [CrossRef] [PubMed]
15. Abdulgader, S.M.; Shittu, A.O.; Nicol, M.P.; Kaba, M. Molecular epidemiology of Methicillin-resistant *Staphylococcus aureus* in Africa: A systematic review. *Front. Microbiol.* **2015**, *6*, 34. [CrossRef]
16. Bankevich, A.; Nurk, S.; Antipov, D.; Gurevich, A.; Dviorkin, M.; Kulikov, A.S.; Lesin, V.M.; Nikolenko, S.I.; Pham, S.; Prjibelski, A.D.; et al. SPAdes: A new genome assembly algorithm and its applications to single-cell sequencing. *J. Comput Biol.* **2012**, *19*, 455–477. [CrossRef]
17. Seeman, T. Prokka: Rapid prokaryotic genome annotation. *Bioinformatics* **2014**, *30*, 2068–2069. [CrossRef]
18. Zankari, E.; Hasman, H.; Cosentino, S.; Vestergaard, M.; Rasmussen, S.; Lund, O.; Aarestrup, F.M.; Larsen, M.V. Identification of acquired antimicrobial resistance genes. *J. Antimicrob. Chemother.* **2012**, *67*, 2640–2644. [CrossRef]
19. Chen, L.; Zheng, D.; Liu, B.; Yang, J.; Jin, Q. VFDB 2016: Hierarchical and refined dataset for big data analysis—10 years on. *Nucleic Acids Res.* **2016**, *44*, 694–697. [CrossRef]
20. Amisssah, N.A.; Chlebowicz, A.A.; Ablordey, A.; Sabat, A.J.; Tetteh, C.S.; Prah, I.; van der Wrf, T.S.; Fruedrich, A.W.; van Djil, J.M.; Rossen, J.W.; et al. Molecular characterization of *Staphylococcus aureus* isolates transmitted between patients with Buruli ulcer. *PLoS ONE Trop Dis.* **2015**, *9*, e0004049.
21. Opintan, J.A.; Newman, M.J.; Arhin, R.E.; Donkor, E.S.; Gyansa-Lutterodt, M.; Mills-Pappoe, W. Laboratory-based nationwide surveillance of antimicrobial resistance in Ghana. *Infect. Drug Resist.* **2015**, *8*, 379–389. [CrossRef] [PubMed]
22. Amissah, N.A.; Buultjens, A.H.; Ablordey, A.; van Dam, L.; Opoku-Ware, A.; Baines, S.L.; Bulach, D.; Tetteh, C.S.; Prah, I.; van der Werf, T.S.; et al. Methicillin Resistant *Staphylococcus aureus* Transmission in a Ghanaian Burn Unit: The Importance of Active Surveillance in Resource-Limited Settings. *Front. Microbiol.* **2017**, *8*, 1906. [CrossRef] [PubMed]
23. Tagoe, D.N.A.; Attah, C.O. A study of antibiotic use and abuse in Ghana: A case study of the Cape Coast Metropolis. *Internet J. Health* **2010**, *11*. [CrossRef]
24. Falagas, M.E.; Kopterides, P. Old antibiotics for infections in critically ill patients. *Curr. Opin. Crit. Care* **2007**, *13*, 592–597. [CrossRef] [PubMed]
25. Durosinmi, M.A.; Ajayi, A.A. A prospective study of chloramphenicol induced aplastic anaemia in Nigerians. *Trop. Geogr. Med.* **1993**, *45*, 159–161.
26. Summersgill, J.T.; Schupp, L.G.; Raff, M.J. Comparative penetration of metronidazole, clindamycin, chloramphenicol, cefoxitin, ticarcillin, and moxalactam into bone. *Antimicrob. Agents Chemother.* **1982**, *21*, 601–603. [CrossRef]
27. Nurjadi, D.; Olalekan, A.O.; Layer, F.; Shittu, A.O.; Alabi, A.; Ghebremedhin, B.; Schaumburg, F.; Hofmann-Eifler, J.; Van Genderen, P.J.; Caumes, E.; et al. Emergence of trimethoprim resistance gene *dfrG* in *Staphylococcus aureus* causing human infection and colonization in sub-Saharan Africa and its import to Europe. *J. Antimicrob. Chemother.* **2014**, *69*, 2361–2368. [CrossRef]
28. Nurjadi, D.; Schäfer, J.; Friedrich-Jänicke, B.; Mueller, A.; Neumayr, A.; Calvo-Cano, A.; Goorhuis, A.; Molhoek, N.; Lagler, H.; Kantele, A.; et al. Predominance of *dfrG* as determinant of trimethoprim resistance in imported *Staphylococcus aureus*. *Clin. Microbiol Infect.* **2015**, *21*, 1095.e5–1095.e9. [CrossRef]

29. Tong, A.; Tong, S.Y.; Zhang, Y.; Lamlertthon, S.; Sharma-Kuinkel, B.K.; Rude, T.; Ahn, S.H.; Ruffin, F.; Llorens, L.; Tamarana, G.; et al. Panton-Valentine leukocidin is not the primary determinant of outcome for *Staphylococcus aureus* skin infections: Evaluation from the CANVAS studies. *PLoS ONE* **2012**, *7*, e37212. [CrossRef]
30. Ruimy, R.; Maiga, A.; Armand-Lefevre, L.; Maiga, I.; Diallo, A.; Koumaré, A.K.; Ouattara, K.; Soumaré, S.; Gaillard, K.; Lucet, J.C.; et al. The carriage population of *Staphylococcus aureus* from Mali is composed of a combination of pandemic clones and the divergent Panton-Valentine leukocidin-positive genotype ST152. *J. Bacteriol.* **2008**, *190*, 3962–3968. [CrossRef]
31. Eibach, D.; Nagel, M.; Hogan, B.; Azuure, C.; Krumkamp, R.; Dekker, D.; Gajdiss, M.; Brunke, M.; Sarpong, N.; Owusu-Dabo, E.; et al. Nasal Carriage of *Staphylococcus aureus* among Children in the Ashanti Region of Ghana. *PLoS ONE* **2017**, *12*, e0170320. [CrossRef] [PubMed]

4

# *Brucella abortus*-Stimulated Platelets Activate Brain Microvascular Endothelial Cells Increasing Cell Transmigration through the Erk1/2 Pathway

**Ana María Rodríguez [1,†], Aldana Trotta [2,†], Agustina P. Melnyczajko [1], M. Cruz Miraglia [1,§], Kwang Sik Kim [3], M. Victoria Delpino [1], Paula Barrionuevo [2,‡] and Guillermo Hernán Giambartolomei [1,*,‡]**

1 Instituto de Inmunología, Genética y Metabolismo (INIGEM), CONICET, Facultad de Farmacia y Bioquímica, Universidad de Buenos Aires, Buenos Aires C1120AAD, Argentina; amrodriguez@fmed.uba.ar (A.M.R.); amelnyczajko@uade.edu.ar (A.P.M.); miraglia.maria@inta.gob.ar (M.C.M.); mdelpino@ffyb.uba.ar (M.V.D.)
2 Instituto de Medicina Experimental (IMEX) (CONICET-Academia Nacional de Medicina), Buenos Aires C1425ASU, Argentina; aldana.trotta@unahur.edu.ar (A.T.); pbarrionuevo@fbmc.fcen.uba.ar (P.B.)
3 Division of Pediatric Infectious Diseases, Department of Pediatrics, Johns Hopkins University School of Medicine, Baltimore, MD 21287, USA; kwangkim@jhmi.edu
* Correspondence: ggiambart@ffyb.uba.ar
† These authors have contributed equally to this work.
‡ These authors share senior authorship.
§ Present address: Instituto de Virología, Centro de Investigación en Ciencias Veterinarias y Agronómicas, Instituto Nacional de Tecnología Agropecuaria (INTA), Hurlingham B1686, Argentina.

**Abstract:** Central nervous system invasion by bacteria of the genus *Brucella* results in an inflammatory disorder called neurobrucellosis. A common feature associated with this pathology is blood–brain barrier (BBB) activation. However, the underlying mechanisms involved with such BBB activation remain unknown. The aim of this work was to investigate the role of *Brucella abortus*-stimulated platelets on human brain microvascular endothelial cell (HBMEC) activation. Platelets enhanced HBMEC activation in response to *B. abortus* infection. Furthermore, supernatants from *B. abortus*-stimulated platelets also activated brain endothelial cells, inducing increased secretion of IL-6, IL-8, CCL-2 as well as ICAM-1 and CD40 upregulation on HBMEC compared with supernatants from unstimulated platelets. Outer membrane protein 19, a *B. abortus* lipoprotein, recapitulated *B. abortus*-mediated activation of HBMECs by platelets. In addition, supernatants from *B. abortus*-activated platelets promoted transendothelial migration of neutrophils and monocytes. Finally, using a pharmacological inhibitor, we demonstrated that the Erk1/2 pathway is involved in the endothelial activation induced by *B. abortus*-stimulated platelets and also in transendothelial migration of neutrophils. These results describe a mechanism whereby *B. abortus*-stimulated platelets induce endothelial cell activation, promoting neutrophils and monocytes to traverse the BBB probably contributing to the inflammatory pathology of neurobrucellosis.

**Keywords:** *Brucella abortus*; neurobrucellosis; platelets; brain microvascular endothelial cells; endothelial cells

## 1. Introduction

Blood–brain barrier (BBB) integrity is necessary to protect the brain from injuries such as toxins and germs, as well as to help in maintaining central nervous system (CNS) homeostasis [1]. BBB activation and dysfunction contributes to several brain pathologies. Many factors are able to induce BBB dysfunction such as inflammatory mediators, matrix metalloproteinases, free radicals, and vascular endothelial growth factor, among others [2].

Bacteria of the genus *Brucella* produce several types of inflammatory disorders [3]. Neurobrucellosis is a neurodegenerative inflammatory disorder caused by invasion of the CNS by *Brucella* spp. and constitutes the most morbid pathology associated with this infection [4]. One of the most characteristic clinical signs of this disease is pleocytosis; i.e., the presence of leukocytes in the cerebrospinal fluid [4].

We have recently described a putative mechanism employed by *Brucella abortus* to gain access to the CNS. We have demonstrated, using an in vitro model, that *B. abortus* traverses the BBB into the cerebral parenchyma inside infected monocytes, by a mechanism known as "Trojan horse". Moreover, infected monocytes act as bacterial source for de novo infection of glial cells [5]. In addition, we have described that activation of glial cells by *B. abortus* is crucial in neurobrucellosis pathology [6]. *B. abortus*-activated astrocytes and microglia secrete pro-inflammatory mediators such as tumor necrosis factor (TNF)-$\alpha$, interleukin (IL)-6, IL-1$\beta$, C-C motif chemokine ligand 2 CCL-2, C-X-C motif chemokine ligand 1 (CXCL1), metalloproteinase (MMP)-9, and nitric oxide (NO), among others [6–8]. These inflammatory mediators, and astrocytes/microglia-secreted IL-1$\beta$ in particular, activate the BBB, allowing monocyte and neutrophil transmigration [9]. The effect of glial cell activation on BBB cells is well known [9–11]. However, whether peripheral inflammation induced by *Brucella*-activated cells can modify the BBB remains unknown.

Platelets are well characterized as responsible for maintaining vascular integrity in addition to being hemostatic mediators [12]. In the last few years, the immune function of platelets has been described in both homeostasis and pathology [12–14]. Platelets express several immune receptors such as toll-like receptors (TLR) and Fc receptors, which allow the recognition of different pathogens [12]. Upon pathogen recognition, platelets can be activated and secrete a wide variety of immunomodulatory mediators present in their granules [15–17]. The immunoregulatory functions of pathogen-activated platelets have been described recently, as well as their ability to activate endothelial cells, including the microvascular endothelium of the BBB [14,18–20]. We have recently described the interactions between platelets and *B. abortus* [14]. *B. abortus* is able to invade, infect platelets, and activate them. As a consequence of this interaction, platelets establish complexes with *B. abortus*-infected monocytes, increasing the efficiency of the infection and modulating monocyte and neutrophil functions [14,21]. Moreover, we demonstrated that complexes between platelets and both monocytes and neutrophils are more abundant in patients with brucellosis than in healthy donors [21].

We have already described the effect of glial cell activation on BBB cells during neurobrucellosis; however, whether platelets activated by *Brucella* can modify the BBB remains unexplored. Therefore, the aim of this work was to elucidate whether *B. abortus*-activated platelets can activate the BBB and affect transmigration of monocytes and neutrophils. Here, we demonstrated that *B. abortus*-activated platelets activate brain microvascular endothelial cells, and other endothelial cells, through Erk1/2 signaling pathway, leading monocytes and neutrophils to traverse polarized brain microvascular endothelial monolayers.

## 2. Results

### 2.1. Interaction with Platelets Enhances the Activation of Endothelial Cells in the Context of B. abortus Infection

We decided to evaluate the capacity of *B. abortus* to activate human brain microvascular endothelial cells (HBMECs) in presence of platelets. For this, HBMECs were co-cultured with platelets (cell:platelet ratio, 1:100) and infected with *B. abortus* (MOI of 100) for 24 h. As control, HBMECs were cultured with platelets alone or they were infected in the absence of platelets. We measured ICAM-1 (intercellular adhesion molecule 1, also known as CD54) expression to determine the level of activation of endothelial cells. Endothelial ICAM-1 plays a critical role at different steps of neutrophil migration into inflamed tissues [22]. As we have previously reported, infection with *B. abortus* in the absence of platelets induced a slight activation of HBMECs, measured as the upregulation of surface ICAM-1 [9]. The presence of platelets in the absence of infection also induced a slight activation of HBMECs. However, these effects were not significant. On the other hand, the presence of platelets during the infection of HBMECs induced a significant ($p < 0.0005$) upregulation of ICAM-1 (Figure 1A). These results demonstrate that the presence of platelets enhances *B. abortus*-induced activation of microvascular brain endothelial cells.

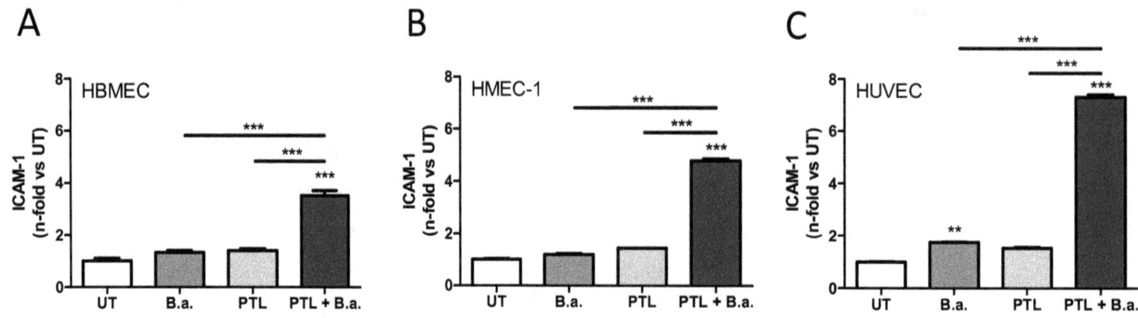


**Figure 1.** Interaction with platelets enhances the activation of endothelial cells in the context of *Brucella abortus* infection. Endothelial cells lines human brain microvascular endothelial cell (HBMEC) (**A**), human microvascular endothelial cells (HMEC-1) (**B**), and human umbilical vein endothelial cell (HUVEC) primary culture (**C**) were incubated with *B. abortus* (B.a.), in the presence or the absence of platelets (PTL), or with PTL alone for 24 h. ICAM-1 (intercellular adhesion molecule 1) expression on the cell surface was assessed by flow cytometry. Bars represent the mean ± SEM of duplicates. Data shown are from a representative experiment out of at least three performed. ** $p < 0.005$, *** $p < 0.0005$ vs. untreated cells (UT).

To investigate whether the effect of platelets during *B. abortus* infection also occurs in other endothelia, human microvascular endothelial cells (HMEC-1) and human umbilical vein endothelial cell (HUVEC) were infected with *B. abortus* in the presence or absence of platelets. The presence of platelets during the infection of both types of endothelial cells induced a significant upregulation of ICAM-1 expression compared to infected cells or cells incubated with platelets alone ($p < 0.0005$) (Figure 1B,C). These data demonstrate that the presence of platelets enhances *B. abortus*-activation of different endothelial cell types.

### 2.2. Supernatants from B. abortus-Stimulated Platelets Activate Brain Microvascular Endothelial Cells

To investigate whether this effect was due to direct interaction between platelets and endothelial cells or factors released by *B. abortus*-activated platelets, we performed experiments using conditioned media. First, platelets were stimulated with or without *B. abortus* (platelets: *B. abortus* ratio, 1:1) for 24 h. Then, culture supernatants were collected and filtered to eliminate platelets and bacteria. Finally, cell-free supernatants were used to stimulate HBMECs for an additional 24 h. Stimulation of HBMECs with supernatants from *B. abortus*-stimulated platelets induced a significant ($p < 0.005$) upregulation of ICAM-1 surface expression (Figure 2A). These results demonstrate that supernatants from *B. abortus*-stimulated platelets are able to activate microvascular brain endothelial cells. Furthermore, in order to expand our results, we investigated whether these supernatants could also activate other endothelial cell types. For this, HMEC-1 and HUVEC were stimulated with supernatants collected from *B. abortus*-stimulated platelets. We observed an upregulation of ICAM-1 surface expression on both cell types (Figure 2B,C). Collectively, these data demonstrated that supernatants from *B. abortus*-stimulated platelet are able to activate several types of endothelial cells.

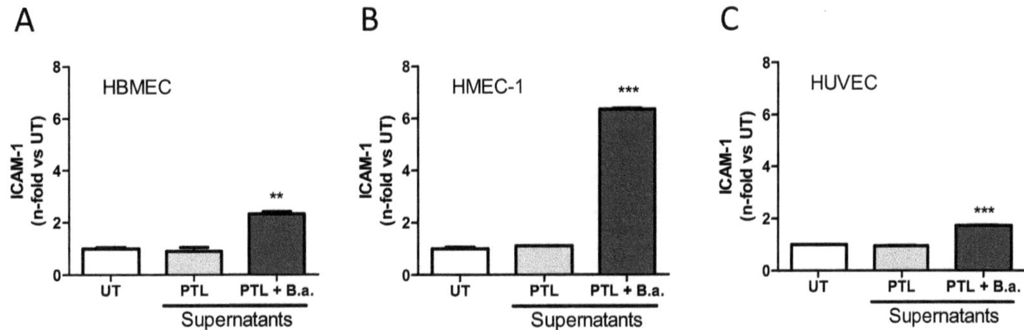


**Figure 2.** Supernatants from *B. abortus*-stimulated platelets activate endothelial cells. Supernatants collected from *B. abortus*-stimulated platelets (PTL + B.a.) or platelets alone (PTL) were used to stimulate the endothelial cell lines HBMEC (**A**), HMEC-1 (**B**), and HUVEC primary culture (**C**) for 24 h (dilution 1:2). ICAM-1 expression was measured on the cell surface by flow cytometry. ** $p < 0.005$, *** $p < 0.0001$ vs. untreated cells (UT).

Next, we studied in depth the HBMEC activation induced by supernatants from *B. abortus*-stimulated platelets. Activation of HBMECs induced by supernatants from *B. abortus*- stimulated platelets was dose dependent (Figure 3A) and, more importantly, it was achieved by using supernatants from different platelet donors (Figure 3B). CD40 expression in endothelial cells has been implicated in several pathologic conditions of the CNS including Alzheimer's disease and human immunodeficiency virus 1 (HIV-1) encephalitis, where an important role of CD40 has been demonstrated in BBB disruption [23]. Besides the upregulation of ICAM-1, the activated phenotype induced by *B. abortus*-infected platelet supernatants also included the significant upregulation of CD40 surface expression ($p < 0.05$) (Figure 3C) and the secretion of significant ($p < 0.0005$) amounts of IL-6, IL-8, and CCL-2 (Figure 3D–F) when compared with untreated HBMECs or HBMECs treated with supernatants from unstimulated platelets. Levels of activation were comparable to those obtained when HBMECs were stimulated with culture supernatants from *Brucella*-infected astrocytes [9] or IL-1β used as positive controls. Importantly, the concentrations of IL-6, IL-8, and CCL-2 measured in supernatants from *B. abortus*-stimulated platelets used for stimulation were negligible (<200 pg/mL in all cases, data not shown). These results demonstrate that supernatants from *B. abortus*-stimulated platelets induce an activated phenotype in microvascular brain endothelial cells, characterized by the upregulation of surface molecules such as ICAM-1 and CD40, and the secretion of both cytokines and chemokines.

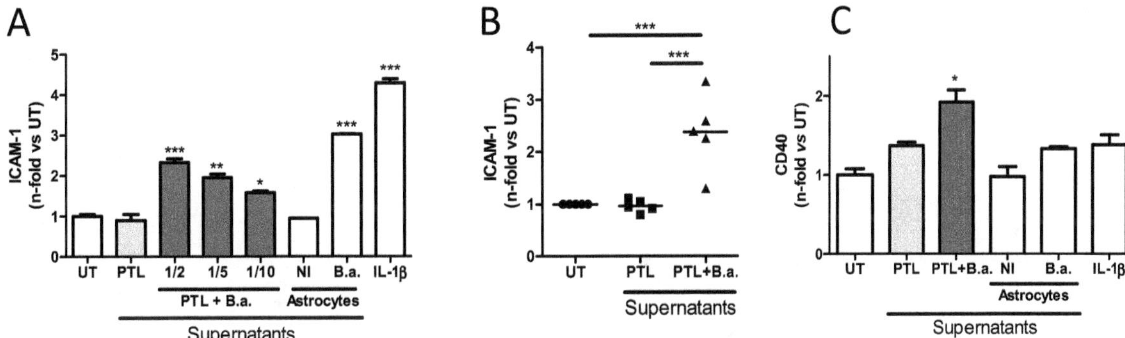


**Figure 3.** *Cont.*

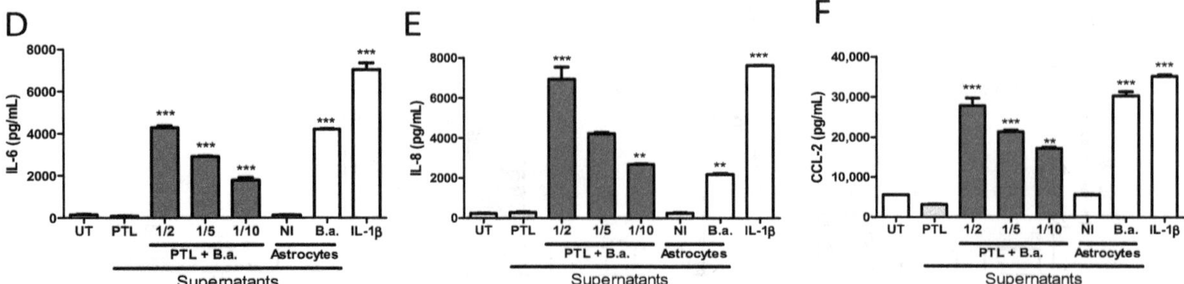


**Figure 3.** Supernatants from *B. abortus*-infected platelets induce an activated phenotype in HBMECs. Platelets were incubated with *B. abortus* (PTL + B.a.) or alone (PTL) for 24 h. Culture supernatants from *B. abortus*-infected astrocytes and interleukin (IL)-1β were used as control. Platelets supernatants were collected, filtered, and used to stimulate HBMECs for 24 h at the indicated dilutions. (**A**) ICAM-1 expression was measured by flow cytometry on HBMEC surface. (**B**) ICAM-1 expression obtained by stimulating HBMECs with supernatants obtained from five independent PTL donors. (**C**) CD40 expression was measured on HBMEC surface by flow cytometry. The secretion of IL-6 (**D**), IL-8 (**E**), and C-C motif chemokine ligand 2 (CCL-2) (**F**) was determined from HBMEC treated culture supernatants by ELISA. Bars represent the mean ± SEM of duplicates. Data shown are from a representative experiment out of at least three performed, except B. * $p < 0.05$, ** $p < 0.005$, *** $p < 0.0005$ vs. untreated cells (UT). NI: non infected.

### *2.3. Secreted Factors from B. abortus-Activated Platelets Activate HBMECs*

Previous studies have shown that *Brucella* spp. release outer-membrane vesicles (OMVs, also known as blebs) containing lipopolysaccharide (LPS), outer membrane proteins, and other bacterial components [24]. To rule out the possibility that OMVs were implicated in HBMEC activation, they were removed from the supernatants by ultracentrifugation, as previously described [24]. HBMECs were then incubated with OMVs-free supernatants for 24 h and the activation of HBMECs was evaluated. There were no significant differences ($p > 0.05$) between non-depleted and OMVs-free supernatants regarding HBMEC activation, measured as ICAM-1 expression (Figure 4A) and IL-6, IL-8, and CCL-2 secretion (Figure 4B–D, respectively). To discard any putative participation of *Brucella*-secreted factors on HBMEC activation, *B. abortus* was incubated alone in the same culture conditions for 24 h. Then, culture supernatants were filtered and ultracentrifuged as described above. These platelet-free *B. abortus* culture supernatants were unable to activate HBMECs (Figure 4). Altogether, these results indicate that secreted factors from *B. abortus*-activated platelets are responsible for HBMEC activation.

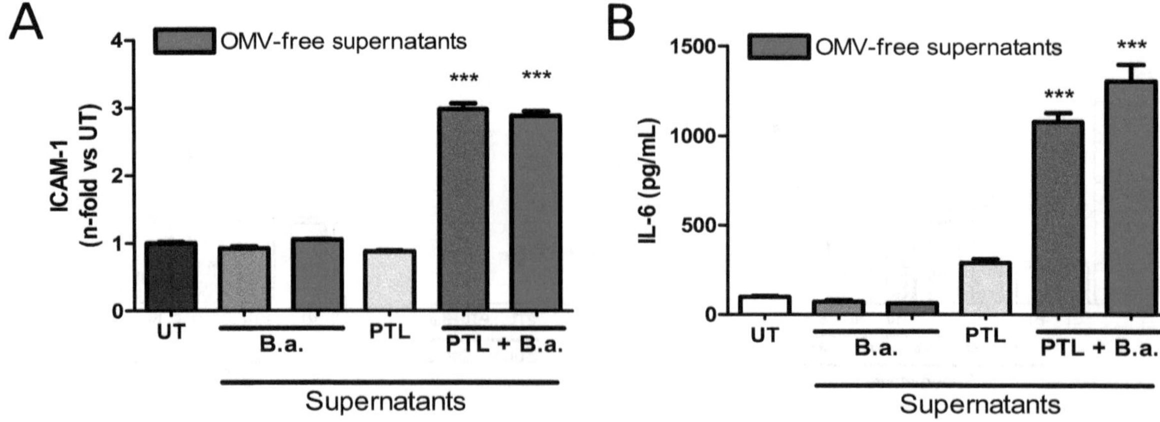


**Figure 4.** *Cont.*

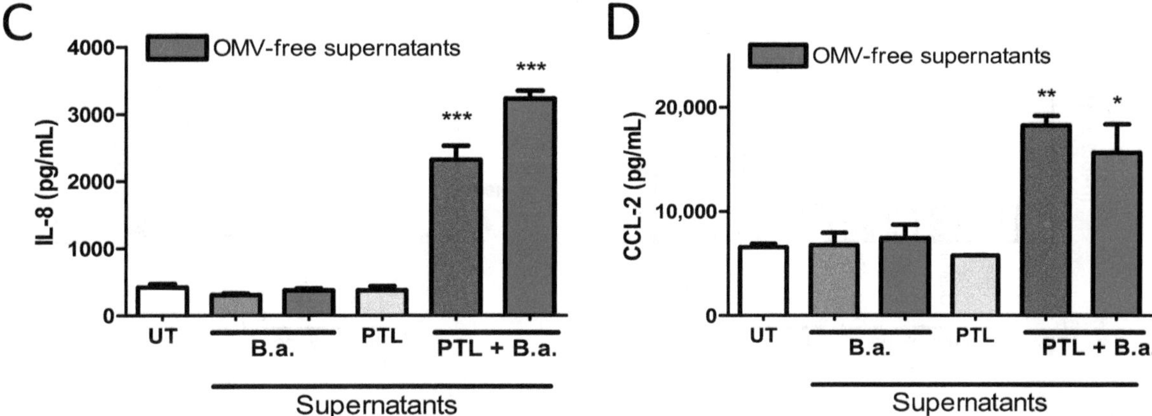


**Figure 4.** Secreted factors from *B. abortus*-stimulated platelets activate HBMECs. Supernatants from *B. abortus*-activated platelets (PLT + B.a.) or from *B. abortus* alone were ultracentrifuged (outer-membrane vesicle (OMV)-free supernatants) or not and used to stimulate HBMECs for 24 h. Supernatants from non-ultracentrifuged PTL alone were used as control. (**A**) ICAM-1 surface expression on HBMECs was assessed by flow cytometry. HBMEC secretion of IL-6 (**B**), IL-8 (**C**), and CCL-2 (**D**) was quantified by ELISA. Bars represent the mean ± SEM of duplicates from a representative experiment out of at least three performed. * $p < 0.05$, ** $p < 0.005$, *** $p < 0.0001$ vs. untreated cells (UT).

*2.4. B. abortus Lipoprotein-Stimulated Platelets Activate Brain Microvascular Endothelial Cells*

To test whether bacterial viability was necessary to induce the activation of platelets and consequently HBMEC activation, platelets were incubated for 24 h with heat-killed *B. abortus* (HKBA). Supernatants were then filtered and used as stimuli on HBMECs. HKBA supernatants were used to treat HBMECs as a negative control. As positive control, platelets were activated with thrombin. Supernatants from HKBA-stimulated platelets were able to activate HBMECs, inducing the upregulation of ICAM-1 (Figure 5A), and increasing the secretion of IL-6, IL-8, and CCL-2 (Figure 5B–D, respectively). We have previously demonstrated that different cell types can be activated by *B. abortus* lipoproteins [8,25,26]. Therefore, we further evaluated the contribution of lipoproteins in the induction of HBMEC activation by platelets. For this, platelets were incubated with *B. abortus* lipidated- or unlipidated-outer-membrane protein 19 (L-Omp19 or U-Omp19, respectively), used as a *Brucella* lipoprotein model [25]. Then, HBMECs were stimulated with the filtered supernatants for an additional 24 h, and the expression of ICAM-1 and secretion levels of IL-6, IL-8, and CCL-2 were evaluated. L-Omp19-activated platelets recapitulated HBMEC activation induced by supernatants from *B. abortus*-stimulated platelets. Furthermore, this activation was dependent on the lipidation of Omp19, as U-Omp19-stimulated platelets failed to induce HBMEC activation (Figure 5A–D). Culture supernatants from thrombin-activated platelets also induced partial activation of HBMECs. Neither HKBA-, L-Omp19-, nor U-Omp19-stimulated supernatants were able to activate HBMECs, demonstrating the presence of a platelet-secreted factor involved in the activation of HBMECs. Altogether, these results demonstrated that the presence of supernatants from platelets stimulated by structural components of *Brucella* (such as L-Omp19), independently of bacterial viability, are involved in the activation of HBMECs.

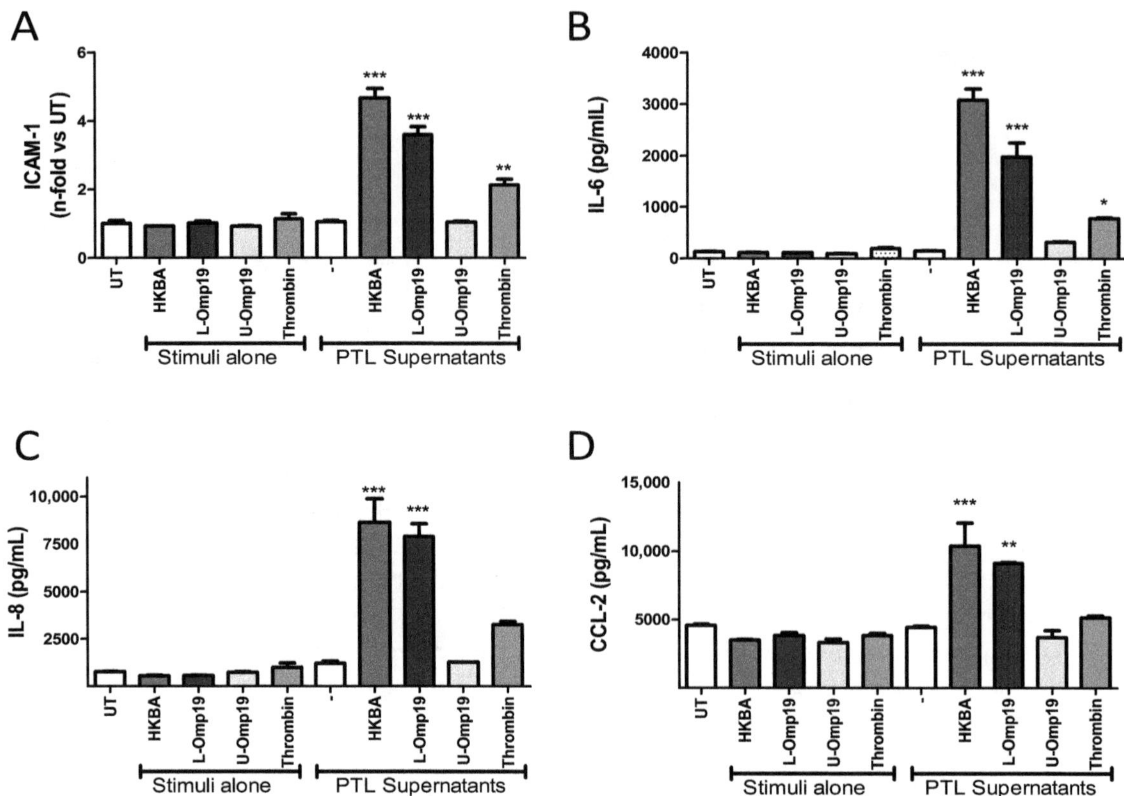


**Figure 5.** *B. abortus* lipoprotein-stimulated platelets activate HBMECs. Platelets were stimulated for 24 h with heat killed *B. abortus* (HKBA, $10^8$ bacteria/mL), lipidated-outer-membrane protein 19 (L-Omp19), unlipidated-outer-membrane protein 19 (U-Omp19) (500 ng/mL), or thrombin (0.1 U/mL). Then, supernatants were collected, filtered. and used to stimulate HBMECs for an additional 24 h. Supernatants from PTL and stimuli incubated alone at the same conditions were used as control. ICAM-1 (**A**) was determined on HBMEC surface by flow cytometry. The secretion of IL-6 (**B**), IL-8 (**C**), and CCL-2 (**D**) was determined by ELISA. Bars represent the mean ± SEM of duplicates from a representative experiment out of at least three performed. * $p < 0.05$, ** $p < 0.005$, *** $p < 0.0001$ vs. untreated cells (UT).

## *2.5. Platelet-Stimulated HBMECs Induce Transendothelial Migration of Neutrophils and Monocytes*

The presence of leukocytes in the cerebrospinal fluid and cerebral parenchyma has been described during neurobrucellosis [4]. This phenomenon, named pleocytosis, could be a consequence of the activation induced by *Brucella*-activated platelets on the blood–brain barrier [9]. To test this possibility, we used a previously established assay of transendothelial migration [9]. Briefly, HBMECs were seeded in the upper chamber of a Transwell plate, and they were cultured for 5 days to establish a monolayer. Then, HBMEC monolayers were treated for 24 h with supernatants from *B. abortus*-stimulated platelets. Culture supernatants from *Brucella*-infected astrocytes and human IL-1β were used as control. Finally, neutrophils or monocytes were seeded in the upper chamber and incubated for 3 h and the number of transmigrated cells to the lower chamber was quantified.

Monocyte as well as neutrophil migration increased when the HBMEC monolayer was treated with supernatants from *B. abortus*-stimulated platelets (Figure 6A,B), but not when HBMECs were stimulated with supernatants from platelets alone. Cellular transmigration was comparable to that obtained when HBMECs where stimulated with culture supernatants from *Brucella*-infected astrocytes or IL-1β used as positive controls. These results indicate that activation of brain endothelial cells by supernatants from *B. abortus*-stimulated platelets could induce transmigration of immune cells through a polarized brain endothelial cell monolayer. Taken together, these results suggest that activated platelets could be responsible for the induction of pleocytosis in the context of *B. abortus* CNS infection.

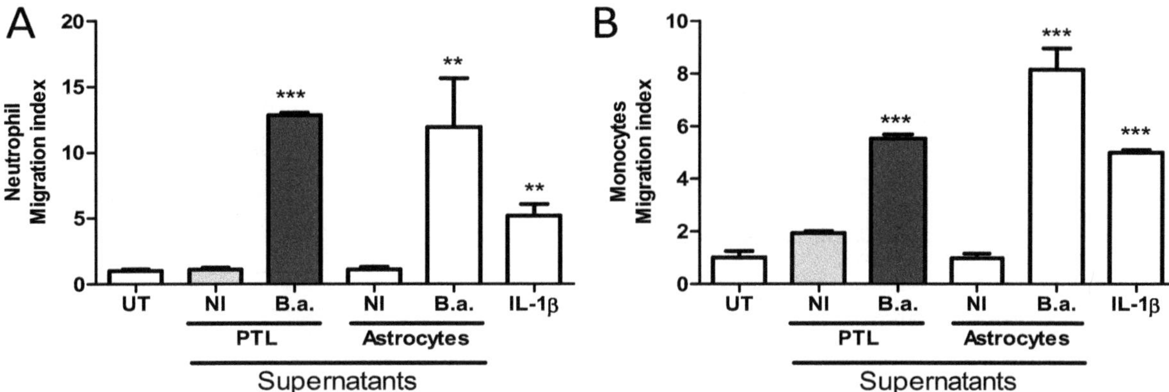


**Figure 6.** *B. abortus*-infected platelets activate HBMEC, promoting transendothelial migration of neutrophils and monocytes. HBMEC monolayers were established on the membrane of Transwell plates and then treated with supernatants from *B. abortus*-stimulated platelets for additional 24 h. Culture supernatants from *Brucella*-infected astrocytes and human IL-1β were used as control. Next, neutrophils (**A**) or monocytes (**B**) were seeded in the upper chamber and incubated for 3 h. Finally, media from the lower chamber were harvested and the number of migrated cells was quantified. Bars represent the mean ± SEM of duplicates from a representative experiment of at least three performed. ** $p < 0.005$, *** $p < 0.0005$ vs. untreated cells (UT) or indicated treatment. NI: non infected.

### *2.6. The Erk1/2 Pathway Is Involved in HMBEC Activation Induced by B. abortus-Stimulated Platelets and It Is Implicated in Transendothelial Migration of Neutrophils*

We decided to further investigate the molecular mechanisms involved in endothelial activation and cellular transcytosis. It was previously reported that the extracellular signal-regulated kinase (Erk)1/2 pathway is involved in the activation of brain endothelial cells [27]. Taking this into account, we investigated the participation of the Erk1/2 signaling pathway in the activation of HBMECs by supernatants from *B. abortus*-stimulated platelets. For this, we used the Erk1/2-specific inhibitor PD98059 to treat HBMEC cells. Inhibition of the Erk1/2 pathway partially reduced the upregulation of ICAM-1 induced by supernatants from *B. abortus*-stimulated platelets ($p < 0.005$) (Figure 7A). Moreover, our results showed that cytokine and chemokine secretion of activated HBMECs is also regulated by the Erk1/2 pathway, since the inhibition with PD98059 also partially diminished the secretion of IL-6, IL-8, and CCL-2, compared to non-treated cells (Figure 7B–D). These results indicate that the Erk1/2 pathway is involved in HBMEC activation by supernatants from *B. abortus*-activated platelets.

The involvement of the Erk1/2 pathway in ICAM-1 upregulation is particularly interesting since ICAM-1 is one of the immunoglobulin-like cell adhesion molecules implicated in the transendothelial migration of immune cells [22]. Thus, we investigated the involvement of the Erk1/2 pathway on the increasing transendothelial migration throughout HBMECs activated by supernatants from *B. abortus*-stimulated platelets. For this, HBMEC monolayers on Transwells were pre-treated with PD98059 2 h before and throughout treatment with *B. abortus*-stimulated platelet supernatants. 24 h later, neutrophils were seeded in the upper chamber for 3 h and the number of migrated cells to the lower chamber was quantified. Neutrophil migration was totally inhibited in HBMECs pre-treated with PD98059, demonstrating the implication of the Erk1/2 pathway on the increased transmigration of immune cells (Figure 7E).

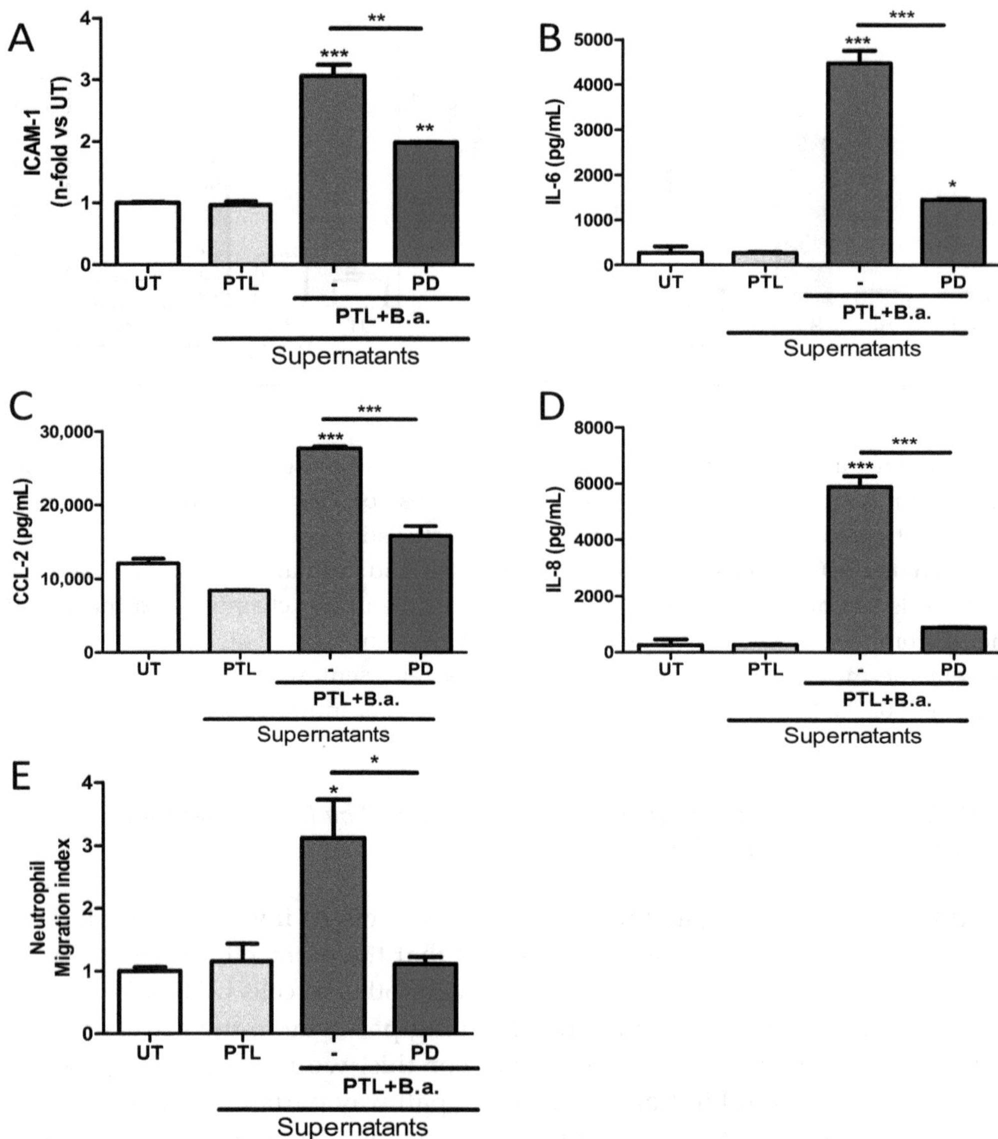


**Figure 7.** Extracellular signal-regulated kinase (Erk)1/2 pathway is involved in HBMEC activation induced by *B. abortus*-stimulated platelets and it is implicated in transendothelial migration of neutrophils. HBMEC were pre-incubated with the Erk1/2 inhibitor (PD98059) for 2 h before platelets–supernatants stimulation and kept throughout. ICAM-1 was determined on HBMECs surface by flow cytometry (**A**). The secretion of IL-6 (**B**), IL-8 (**C**), and CCL-2 (**D**) was determined by ELISA. HBMEC monolayers were established on the membrane of Transwell plates. HBMEC activation was inhibited by PD98059 and then treated with supernatants from *B. abortus*-stimulated platelets for additional 24 h. Next, neutrophils were seeded in the upper chamber and incubated for 3 h. Finally, media from the lower chamber was harvested and the number of migrated cells was quantified (**E**). Bars represent the mean ± SEM of duplicates from a representative experiment of at least three performed * $p < 0.05$, ** $p < 0.005$, *** $p < 0.0005$ vs. untreated cells (UT) or indicated treatment.

## 3. Discussion

Physiological and pathological immune responses are a continuum in which platelets are recognized as innate immune effector cells. Their activation stimulates interactions with endothelial cells and myeloid leukocytes in many pathologic inflammatory syndromes, as well as consequences in acute inflammation [28–31]. Platelets also have signaling functions in endothelial cells. These functions also contribute to critical inflammatory and immune responses [32,33].

Brain microvascular endothelium activation and BBB dysfunction is a significant contributor to the pathogenesis of a variety of brain pathologies [2], many of them of microbial origin [9,18,34]. We have previously described the ability of *B. abortus* to induce inflammation in the cerebral parenchyma, which leads to the activation of the endothelial cells that form the BBB [9]. In this study, we elucidated the role of platelets in brain microvascular endothelial cell activation mediated by *B. abortus*. Platelets enhance HBMEC activation in the context of *B. abortus* infection. These results correlate with the reported ability of other bacterial species to activate platelets and harm endothelial cells [35,36].

Interestingly, HBMEC activation does not require direct contact between platelets and brain endothelial cells, since supernatants of *B. abortus*-stimulated platelets recapitulated the HBMEC activation observed in the presence of platelets. Furthermore, HBMEC activation by secreted factors from *B. abortus*-stimulated platelets is sufficient to induce transmigration of both monocytes and neutrophils. Moreover, *B. abortus*-stimulated platelets also activate the HMEC-1 cell line and primary culture of HUVEC, underscoring the ability of *B. abortus*-stimulated platelets to activate any endothelium.

A long time ago, it was demonstrated that activated platelets increase CCL-2 secretion and ICAM-1 expression on HUVECs [37]. This indicates that activated platelets are able to change the chemotactic and adhesive properties of endothelial cells, increasing the ability to attract monocytes and neutrophils. Under physiological conditions, endothelial cells of the vasculature of non-inflamed tissues have as main functions the maintenance of blood fluidity and the control of vascular permeability [33]. Under these conditions, resting endothelial cells do not interact with circulating leukocytes since the proteins necessary for this interaction are mainly retained inside the cell [38]. Under acute inflammatory conditions, such as those induced by *B. abortus* infection, the vascular endothelium is rapidly activated, mobilizing these adhesion molecules to the extracellular membrane [33]. In accordance with this, we demonstrated that, although the infection with *B. abortus* induces a mild activation of HBMEC, HMEC-1, and HUVEC, the presence of platelets during the infection enhances its activation state upregulating the expression of ICAM-1 and CD40, thus stressing the amplifying role of platelets on endothelial inflammation [39]. In line with these results, other authors have shown that the presence of activated platelets significantly induces the expression of E-Selectin (CD62E), CD106 (VCAM-1), and ICAM-1 on the surface of HUVEC cells, even in the absence of others inflammatory agents [40]. In addition to the increase in adhesion molecules, we have demonstrated that the activation of HBMECs by supernatants from *B. abortus*-stimulated platelets increase IL-6, IL-8, and CCL-2 secretion. These results are in agreement with those previously published describing that HUVEC secrete IL-8 and CCL-2 after co-incubation with activated platelets [40]. In turn, in vivo experiments have shown that platelets are one of the first cellular components arrested in the inflamed endothelium, promoting their activation and allowing the subsequent arrest of leukocytes [41].

Platelet activation was also induced by exposure to heat-killed *B. abortus*, which indicated that it was not dependent on bacterial viability and suggests that it was elicited by a structural bacterial component. Our laboratory has been investigating for years the role of lipoproteins in inflammation generated by *Brucella*. We have described that *Brucella* LPS does not produce cellular activation, however, *Brucella* lipoproteins produce activation of several cell types [6,8,25,26]. Thus, we hypothesized that *B. abortus* lipoproteins might be the structural components involved in the observed phenomenon. L-Omp19, a prototypical *B. abortus* lipoprotein, recapitulated platelet stimulation and concomitant HBMEC activation. Acylation of Omp19 was required for its biological activity since U-Omp19 had no effect on platelet stimulation. The genome of *B. abortus* possesses no less than 80 genes encoding putative lipoproteins [42], and many of them are expressed in the outer membrane of the bacterium [43]. In this context, we posit that any surface-exposed *Brucella* lipoprotein may be significant beyond in vitro assays and not one lipoprotein but rather a combination of them may contribute to the platelet activation elicited by *B. abortus*.

Our recent work revealed a physiological mechanism employed by *B. abortus* to traverse the BBB. *Brucella* is incapable of traversing the BBB by itself, despite the ability to invade and replicate in endothelial cells of the brain microvasculature. Instead, it could cross a BBB model in vitro as a consequence of naturally migrating monocytes carrying viable bacteria, which serve as source of de novo infection to astrocytes and microglia [5]. Interestingly, we have also demonstrated that activated *B. abortus*-infected glial cells were able to increase the transmigration of monocytes through the secretion of inflammatory mediators [9]. These mediators would escalate the entering of infected cells from the peripheral circulation, increasing the infection and the subsequent BBB dysfunction through a pathological vicious circle. The capacity of secreted factors from *B. abortus*-stimulated platelets to increase neutrophil and monocyte transmigration through microvascular endothelial cells demonstrated in this paper would worsen this situation (Figure 8).

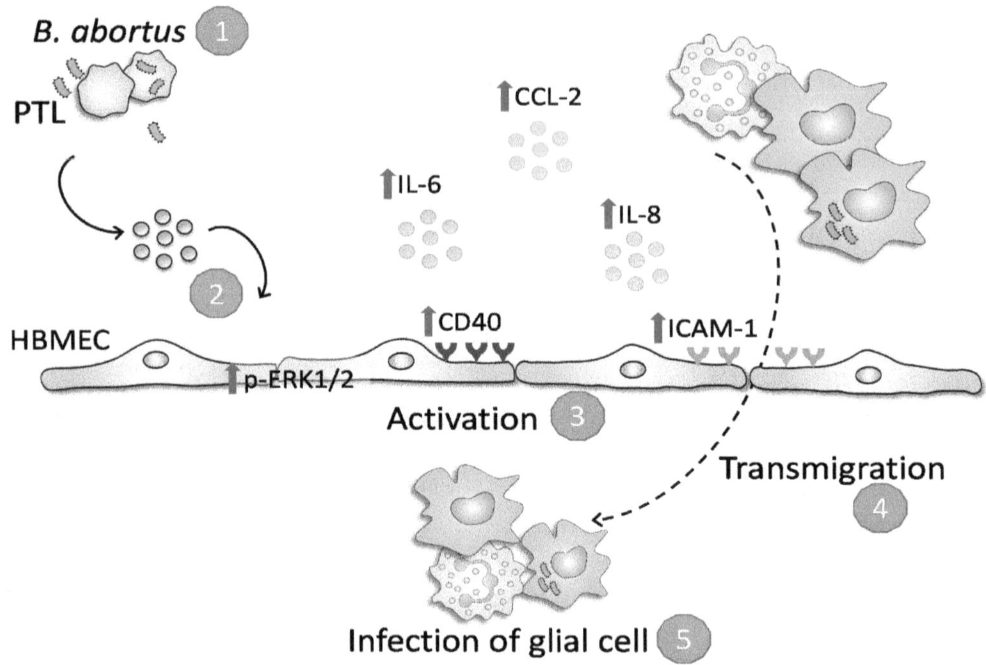


**Figure 8.** *B. abortus*-stimulated platelets (1) secret factors (2) that induce HBMEC activation, leading to ICAM-1 and CD40 upregulation, increasing the secretion of IL-6, IL-8, and CCL-2 (3), and promoting neutrophils and monocytes to traverse a polarized HMBEC monolayers (4). Platelet-induced activation would escalate the entering of infected cells from the peripheral circulation and the subsequent infection of glial cells (5), worsening the inflammatory signs of neurobrucellosis.

The mitogen-activated protein kinase (MAPK) pathway has been associated to several biological processes such as cell activation and proliferation, cell differentiation, and apoptosis [44]. In particular, the Erk1/2 pathway is involved in HBMEC activation [27] and endothelial permeability [45]. Experiments of pharmacological inhibition determined that Erk1/2 was involved in HBMEC activation induced by supernatants from *B. abortus*-activated platelets. In particular, it was involved in ICAM-1 upregulation and enhanced the transmigration of neutrophils. Since MAPK inhibitors, such as pyridinyl imidazole drugs, have been identified as putative drugs for anti-inflammatory therapies in the CNS [46], the data presented in this paper suggest that inhibiting such molecules (Erk1/2) may represent a pharmaceutical strategy to restrict BBB deterioration, thereby potentially reducing the morbidity associated with neurobrucellosis.

In summary, the results presented here describe a mechanism whereby *B. abortus*-stimulated platelets can induce HBMEC and other endothelial cell activation, promoting neutrophils and monocytes to traverse the BBB. Moreover, this could contribute to increase the infection of glial cells, generating and/or deteriorating neurobrucellosis and the inflammatory response motivated by glial activation (Figure 8).

## 4. Materials and Methods

### 4.1. Ethics Statement

Human platelets, monocytes, and neutrophils were isolated from the blood of healthy adult donors in agreement with the guidelines of the Ethical Committee of the Instituto de Medicina Experimental (protocol number: 20160518-M). All adult blood donors provided their informed consent prior to the study.

### 4.2. Bacteria and Lipoproteins

*B. abortus* S2308 was cultured in tryptic soy broth supplemented with yeast extract (Merck, Buenos Aires, Argentina). The number of bacteria on stationary-phase cultures was determined by comparing the optical density at 600 nm with a standard curve. All live *Brucella* manipulations were performed in biosafety level 3 facilities located at the Instituto de Investigaciones Biomédicas en Retrovirus y SIDA (INBIRS, Buenos Aires, Argentina). To obtain heat-killed *B. abortus* (HKBA), bacteria were washed five times for 10 min each in sterile phosphate-buffered saline (PBS), heat-killed at 70 °C for 20 min, aliquoted, and stored at −70 °C until used. The total absence of *B. abortus* viability after heat killing was verified by the absence of bacterial growth on tryptic soy agar.

*B. abortus* lipidated-outer membrane protein 19 (L-Omp19) and unlipidated-Omp19 (U-Omp19) were obtained as described [31]. Both recombinant proteins contained less than 0.25 endotoxin U/μg of protein as assessed by Limulus Amebocyte Lysates (Associates of Cape Cod Inc., Falmouth, MA, USA).

### 4.3. Cell Lines

HBMECs were isolated from a brain biopsy of an adult female with epilepsy as previously described [47]. These cells were positive for factor VIII–Rag, carbonic anhydrase IV, and Ulex europaeus agglutinin I. They took up fluorescently labeled low-density lipoprotein and expressed g-glutamyl transpeptidase, thus demonstrating their brain endothelial cell properties [47]. HBMECs were subsequently immortalized by transfection with SV40 large T Ag and maintained their morphological and functional characteristics for at least 30 passages [48]. The cells are polarized and exhibit a transendothelial electric resistance (TEER) of at least 100 ohms/cm$^2$ [49]. Cells (passage < 30) were cultured in tissue culture flasks in Roswell Park Memorial Institute (RPMI) medium 1640 (Life Technologies, Grand Island, NE, USA) supplemented 10% with heat-inactivated fetal bovine serum (FBS) (Life Technologies), 10% NuSerum IV (Becton Dickinson, Bedford, OH, USA), 1% modified Eagle's medium nonessential amino acids (Life Technologies), sodium pyruvate (1 mM), L-glutamine (2 mM), 1% MEM vitamin solution (Life Technologies), penicillin (100 U/mL) and streptomycin (100 μg/mL). Human microvascular endothelial cells (HMEC-1) were obtained from ATCC® (CRL-3243™, Manassas, VA, USA). Cells were grown in Dulbecco's Modified Eagle's (DMEM) medium (Life Technologies) containing 10% FBS (Natocor, Córdoba, Argentina), 10 μg/mL hydrocortisone, 1 ng/mL epidermal growth factor (BD Pharmingen, San Diego, CA, USA), L-glutamine (2 mM), penicillin (100 U/mL), and streptomycin (100 μg/mL). All cell cultures were incubated at 37 °C in a humidified atmosphere of 5% $CO_2$. Human umbilical vascular endothelial cells (HUVECs) were obtained as described previously [10,50]. Briefly, umbilical vascular tissue was treated with collagenase for digestion. Cells were seeded until confluence on 1% gelatin-coated 25 cm$^2$ tissue culture flasks and identified by their cobblestone morphology and von Willebrand factor (VWF) antibody (Immunotech, Ocala, FL, USA) binding. Cells were grown in RPMI 1640 medium (Gibco) supplemented with 10% FBS (Life Technologies), heparin (100 μg/mL), endothelial cell growth factor (50 μg/mL), sodium pyruvate (2 mM), L-glutamine (2 mM), penicillin (100 U/mL), and streptomycin (100 μg/mL) at 37 °C in a humidified 5% $CO_2$ incubator. HUVECs used for experiments were kept between the first and third culture passage.

### 4.4. Platelet Purification and Stimulation

Platelets were obtained from whole blood from healthy adult human donors as described previously [14]. Briefly, blood samples were collected into tubes containing sodium citrate (Merck) and centrifuged. The platelet-rich plasma was collected and centrifuged in presence of 75 nM prostaglandin I2 (Cayman Chemical, Ann Arbor, MI, USA). Platelets were then washed with RPMI 1640 medium. Finally, platelets were resuspended in RPMI 1640 medium. Platelets were incubated with *B. abortus* ($1 \times 10^7$/mL) (PLT:B.a. ratio of 1:1) for 24 h in RPMI 1640 medium with 10% FBS (Life Technologies) and L-glutamine (2 mM). In addition, platelets were incubated with HKBA ($1 \times 10^8$ bacteria/mL), L-Omp19, U-Omp19 (both 500 ng/mL), or thrombin (0.1 U/mL) (Sigma Aldrich, St. Louis, MO, USA). Then, supernatants were collected, sterilized by filtration, ultracentrifuged when mentioned (at 100,000× *g* for 5 h at 4 °C), and stored at −70 °C until they were used.

### 4.5. Endothelial Cell Treatment

HBMEC, HMEC-1, and HUVEC were cultured in 48 wells plate ($5 \times 10^4$/0.2 mL). To co-culture infection, platelets were added (cell:platelets ratio, 1:100) and endothelial cells–platelets cultures were infected by *B. abortus* (multiplicity of infection of 100). In all cases, the infection was performed for 2 h in medium containing no antibiotics. Then, cells were maintained for 24 h in the presence of antibiotics (100 μg/mL gentamicin and 50 μg/mL streptomycin) to kill the remaining extracellular bacteria. For experiments with platelet-conditioned media, HBMEC, HUVEC, and HMEC-1 cells were treated with 0.2 mL of diluted supernatants from *B. abortus*-stimulated platelets for 24 h. Culture supernatants from *Brucella*-infected astrocytes and recombinant human IL-1β were used as control. In all cases, cells were harvested to determine cell surface molecule expression by flow cytometry. Supernatants from stimulated endothelial cells were collected and stored at −70 °C until they were used.

### 4.6. Erk1/2 Signaling Pathway

HBMECs were treated with Erk1/2 MAPK pharmacological inhibitor PD98059 (50 μM) (Calbiochem, San Diego, CA, USA) or vehicle (dimethyl sulfoxide) 2 h before the stimulation with supernatants and the inhibitor were kept throughout the experiment, based on previous report [7].

### 4.7. Measurement of Cytokine and Chemokine Concentrations

Human IL-6, IL-8, and CCL-2 concentrations were quantified in supernatants harvested from HBMECs and HMEC-1 treated with supernatants from *B. abortus*-stimulated platelets by Sandwich ELISA using paired cytokine-specific mAbs according to the manufacturer's instructions (BD Pharmingen).

### 4.8. Determination of Cell Surface Molecules by Flow Cytometry

ICAM-1 and CD40 surface expression was determined by flow cytometry. For this, treated HBMECs or HMEC-1 were washed and stained with a PE-labeled antibody (Ab) against human ICAM-1 (CD54) (clone HA58, BD Pharmingen), PE-labeled Ab against human CD40 (clone 5C3; BioLegend, San Diego, CA, USA) or the PE-labeled isotype-matched control Ab (BD Pharmingen). Labeled cells were analyzed on a FACSCalibur flow cytometer (BD Biosciences, San Diego, CA, USA), and data were processed using FlowJo software.

### 4.9. Neutrophil and Monocytes Transendothelial Migration Assay

Peripheral blood mononuclear cells (PBMCs) and neutrophils were separated by Ficoll-Hypaque (GE Healthcare, Uppsala, Sweden) gradient centrifugation. Human neutrophils were isolated by sedimentation of erythrocytes in 6% dextran and hypotonic lysis as previously described [26]. Monocytes were then purified from PBMCs by Percoll (GE Healthcare) gradient. Both types of cells were resuspended in RPMI 1640 supplemented with 10% FBS. Cell purity was 90% as determined by flow cytometry for both populations. Viability of cells was more than 95% in all the experiments as measured by trypan blue exclusion test.

HBMEC monolayers were established from 20,000 cells per insert on 3-μm pore size membrane Transwell plates of 6.5-mm diameter insert (Corning-Costar, Acton, MA, USA) previously treated with rat tail collagen (50 mg/mL in 1% acetic acid) (BD Biosciences) and neutralized in a saturated atmosphere of ammonium hydroxide. After 5 days, when cellular confluence was reached TEER and passive diffusion of horseradish peroxidase was measured as an indication of monolayer integrity [5]. Then, monolayers were incubated for 24 h with supernatants from *B. abortus*-stimulated platelets. Supernatants from platelets alone as well as non-treated HBMECs were used as negative control. Culture supernatants from *Brucella*-infected astrocytes and recombinant human IL-1β were used as positive control. After that, monolayers were washed and neutrophils or monocytes ($1 \times 10^5$ cells) were added to the upper chamber in fresh medium. Plates were incubated for 3 h at 37 °C in 5% $CO_2$ and transmigrated cells to the lower chamber were counted on a hemocytometer.

### *4.10. Statistical Analysis*

Results were analyzed with one-way ANOVA followed by Tukey post-test using the GraphPad Prism 5.0 software.

**Author Contributions:** A.M.R., A.T., P.B., and G.H.G. conceived and designed the experiments. A.M.R., A.T., A.P.M., M.C.M., and M.V.D. performed the experiments. K.S.K. supported the work with key suggestions and helped with data interpretation. A.M.R. and G.H.G. analyzed the data and wrote the manuscript. G.H.G. supervised experiments, interpreted the data, and supervised the manuscript. All authors reviewed the manuscript. All authors have read and agreed to the published version of the manuscript.

**Acknowledgments:** We thank Horacio Salomón and the staff of the Instituto de Investigaciones Biomédicas en Retrovirus y SIDA (Universidad de Buenos Aires) for allowing us to use the biosafety level 3 laboratory facilities. A.M.R., M.V.D., P.B., and G.H.G. are members of the Research Career of CONICET (Argentina). A.T. and M.C.M. are recipients of fellowships from Consejo Nacional de Investigaciones Científicas y Técnicas (Argentina).

## References

1. Johansson, B.B. The Physiology of the Blood-Brain Barrier. *Neurotransm. Interact. Cogn. Funct.* **1990**, *274* 25–39. [CrossRef] ,
2. Almutairi, M.M.A.; Gong, C.; Xu, Y.G.; Chang, Y.; Shi, H. Factors controlling permeability of the blood–brain barrier. *Cell. Mol. Life Sci.* **2015**, *73*, 57–77. [CrossRef] [PubMed]
3. Young, E.J.; Corbel, M.J. *Brucellosis in Latin America. Brucellosis: Clinical and Laboratory Aspects*; CRC Press: Boca Raton, FL, USA, 1989; pp. 151–162.
4. Giambartolomei, G.H.; Wallach, J.C.; Baldi, P.C. Neurobrucellosis. In *Encephalitis: Diagnosis and Treatment* The Egerton Group: New York, NY, USA, 2008; Volume 14, pp. 255–272. ;
5. Miraglia, M.C.; Rodriguez, A.M.; Barrionuevo, P.; Rodríguez, J.; Kim, K.S.; Dennis, V.A.; Delpino, M.V.; Giambartolomei, G.H. Brucella abortus Traverses Brain Microvascular Endothelial Cells Using Infected Monocytes as a Trojan Horse. *Front. Microbiol.* **2018**, *8*, 200. [CrossRef] [PubMed]
6. Samartino, C.G.; Delpino, M.V.; Godoy, C.P.; Di Genaro, M.S.; Pasquevich, K.A.; Zwerdling, A.; Barrionuevo, P.; Mathieu, P.; Cassataro, J.; Pitossi, F.; et al. Brucella abortus Induces the Secretion of Proinflammatory Mediators from Glial Cells Leading to Astrocyte Apoptosis. *Am. J. Pathol.* **2010**, *176*, 1323–1338. [CrossRef]

7. Miraglia, M.C.; Scian, R.; Samartino, C.G.; Barrionuevo, P.; Rodriguez, A.M.; E Ibañez, A.; Coria, L.M.; Velásquez, L.N.; Baldi, P.C.; Cassataro, J.; et al. Brucella abortus induces TNF-$\alpha$-dependent astroglial MMP-9 secretion through mitogen-activated protein kinases. *J. Neuroinflamm.* **2013**, *10*, 47. [CrossRef]
8. Rodríguez, A.M.; Delpino, M.V.; Miraglia, M.C.; Franco, M.M.C.; Barrionuevo, P.; Dennis, V.A.; Oliveira, S.C.; Giambartolomei, G.H. Brucella abortus-activated microglia induce neuronal death through primary phagocytosis. *Glia* **2017**, *65*, 1137–1151. [CrossRef]
9. Miraglia, M.C.; Franco, M.M.C.; Rodriguez, A.M.; Bellozi, P.M.Q.; Ferrari, C.C.; Farias, M.I.; Dennis, V.A.; Barrionuevo, P.; de Oliveira, A.C.P.; Pitossi, F.; et al. Glial Cell-Elicited Activation of Brain Microvasculature in Response to Brucella abortus Infection Requires ASC Inflammasome-Dependent IL-1beta Production. *J. Immunol.* **2016**, *196*, 3794–3805. [CrossRef]
10. Landoni, V.I.; Schierloh, P.; Nebel, M.D.C.; Fernandez, G.C.; Calatayud, C.; Lapponi, M.J.; Isturiz, M.A. Shiga Toxin 1 Induces on Lipopolysaccharide-Treated Astrocytes the Release of Tumor Necrosis Factor-alpha that Alter Brain-Like Endothelium Integrity. *PLoS Pathog.* **2012**, *8*, e1002632. [CrossRef]
11. Abbott, N.J. Astrocyte-endothelial interactions and blood-brain barrier permeability. *J. Anat.* **2002**, *200*, 629–638. [CrossRef]
12. Holinstat, M. Normal platelet function. *Cancer Metastasis Rev.* **2017**, *36*, 195–198. [CrossRef]
13. Carestia, A.; Kaufman, T.; Rivadeneyra, L.; Landoni, V.I.; Pozner, R.G.; Negrotto, S.; D'Atri, L.P.; Gómez, R.M.; Schattner, M. Mediators and molecular pathways involved in the regulation of neutrophil extracellular trap formation mediated by activated platelets. *J. Leukoc. Biol.* **2015**, *99*, 153–162. [CrossRef] [PubMed]
14. Trotta, A.; Velásquez, L.N.; Milillo, M.A.; Delpino, M.V.; Rodriguez, A.M.; Landoni, V.I.; Giambartolomei, G.H.; Pozner, R.G.; Barrionuevo, P. Platelets Promote Brucella abortus Monocyte Invasion by Establishing Complexes with Monocytes. *Front. Immunol.* **2018**, *9*, 1000. [CrossRef] [PubMed]
15. Blair, P.; Flaumenhaft, R. Platelet $\alpha$-granules: Basic biology and clinical correlates. *Blood Rev.* **2009**, *23*, 177–189. [CrossRef] [PubMed]
16. May, A.E.; Seizer, P.; Gawaz, M. Platelets: Inflammatory Firebugs of Vascular Walls. *Arter. Thromb. Vasc. Biol.* **2008**, *28*, s5–s10. [CrossRef]
17. Fréchette, J.-P.; Martineau, I.; Gagnon, G. Platelet-rich Plasmas: Growth Factor Content and Roles in Wound Healing. *J. Dent. Res.* **2005**, *84*, 434–439. [CrossRef]
18. Cox, D.; McConkey, S.; McConkey, S. The role of platelets in the pathogenesis of cerebral malaria. *Cell. Mol. Life Sci.* **2009**, *67*, 557–568. [CrossRef]
19. Davidson, D.C.; Hirschman, M.P.; Sun, A.; Singh, M.V.; Kasischke, K.; Maggirwar, S.B. Excess Soluble CD40L Contributes to Blood Brain Barrier Permeability In Vivo: Implications for HIV-Associated Neurocognitive Disorders. *PLoS ONE* **2012**, *7*, e51793. [CrossRef]
20. Garciarena, C.D.; McHale, T.M.; Watkin, R.L.; Kerrigan, S.W. Coordinated Molecular Cross-Talk between Staphylococcus aureus, Endothelial Cells and Platelets in Bloodstream Infection. *Pathogens* **2015**, *4*, 869–882. [CrossRef]
21. Trotta, A.; Milillo, M.A.; Serafino, A.; Castillo, L.A.; Weiss, F.B.; Delpino, M.V.; Giambartolomei, G.H.; Fernández, G.C.; Barrionuevo, P. Brucella abortus–infected platelets modulate the activation of neutrophils. *Immunol. Cell Biol.* **2020**. [CrossRef]
22. Lyck, R.; Enzmann, G. The physiological roles of ICAM-1 and ICAM-2 in neutrophil migration into tissues. *Curr. Opin. Hematol.* **2015**, *22*, 53–59. [CrossRef]
23. Ramirez, S.H.; Fan, S.; Dykstra, H.; Reichenbach, N.; Del Valle, L.; Potula, R.; Phipps, R.P.; Maggirwar, S.B.; Persidsky, Y. Dyad of CD40/CD40 ligand fosters neuroinflammation at the blood-brain barrier and is regulated via JNK signaling: Implications for HIV-1 encephalitis. *J. Neurosci.* **2010**, *30*, 9454–9464. [CrossRef] [PubMed]
24. Pollak, C.N.; Delpino, M.V.; Fossati, C.A.; Baldi, P.C. Outer Membrane Vesicles from Brucella abortus Promote Bacterial Internalization by Human Monocytes and Modulate Their Innate Immune Response. *PLoS ONE* **2012**, *7*, e50214. [CrossRef]
25. Giambartolomei, G.H.; Zwerdling, A.; Cassataro, J.; Bruno, L.; Fossati, C.A.; Philipp, M.T. Lipoproteins, not lipopolysaccharide, are the key mediators of the proinflammatory response elicited by heat-killed Brucella abortus. *J. Immunol.* **2004**, *173*, 4635–4642. [CrossRef] [PubMed]
26. Zwerdling, A.; Delpino, M.V.; Pasquevich, K.A.; Barrionuevo, P.; Cassataro, J.; Samartino, C.G.; Giambartolomei, G.H. Brucella abortus activates human neutrophils. *Microbes Infect.* **2009**, *11*, 689–697. [CrossRef] [PubMed]

27. Miao, Z.; Dong, Y.; Fang, W.; Shang, D.; Liu, D.; Zhang, K.; Li, B.; Chen, Y.-H. VEGF Increases Paracellular Permeability in Brain Endothelial Cells via Upregulation of EphA2. *Anat. Rec. Adv. Integr. Anat. Evol. Biol.* **2014**, *297*, 964–972. [CrossRef] [PubMed]
28. Weyrich, A.S.; Zimmerman, G.A. Platelets: Signaling cells in the immune continuum. *Trends Immunol.* **2004**, *25*, 489–495. [CrossRef] [PubMed]
29. Iwasaki, A.; Medzhitov, R. Regulation of Adaptive Immunity by the Innate Immune System. *Science* **2010**, *327*, 291–295. [CrossRef]
30. Gawaz, M.; Langer, H.; May, A.E. Platelets in inflammation and atherogenesis. *J. Clin. Investig.* **2005**, *115*, 3378–3384. [CrossRef]
31. Semple, J.W.; Italiano, J.E.; Freedman, J. Platelets and the immune continuum. *Nat. Rev. Immunol.* **2011**, *11*, 264–274. [CrossRef]
32. Danese, S.; Dejana, E.; Fiocchi, C. Immune regulation by microvascular endothelial cells: Directing innate and adaptive immunity, coagulation, and inflammation. *J. Immunol.* **2007**, *178*, 6017–6022. [CrossRef]
33. Pober, J.S.; Sessa, W.C. Evolving functions of endothelial cells in inflammation. *Nat. Rev. Immunol.* **2007**, *7*, 803–815. [CrossRef] [PubMed]
34. Singh, M.V.; Davidson, D.C.; Jackson, J.W.; Singh, V.B.; Silva, J.; Ramirez, S.H.; Maggirwar, S.B. Characterization of platelet-monocyte complexes in HIV-1-infected individuals: Possible role in HIV-associated neuroinflammation. *J. Immunol.* **2014**, *192*, 4674–4684. [CrossRef] [PubMed]
35. Cox, D.; Kerrigan, S.W.; Watson, S. Platelets and the innate immune system: Mechanisms of bacterial-induced platelet activation. *J. Thromb. Haemost.* **2011**, *9*, 1097–1107. [CrossRef] [PubMed]
36. Fitzgerald, J.R.; Foster, T.J.; Cox, D. The interaction of bacterial pathogens with platelets. *Nat. Rev. Genet.* **2006**, *4*, 445–457. [CrossRef] [PubMed]
37. Gawaz, M.; Neumann, F.-J.; Dickfeld, T.; Koch, W.; Laugwitz, K.-L.; Adelsberger, H.; Langenbrink, K.; Page, S.; Neumeier, D.; Schoömig, A.; et al. Activated Platelets Induce Monocyte Chemotactic Protein-1 Secretion and Surface Expression of Intercellular Adhesion Molecule-1 on Endothelial Cells. *Circulation* **1998**, *98*, 1164–1171. [CrossRef]
38. Ley, K.; Reutershan, J. Leucocyte-Endothelial Interactions in Health and Disease. *Handb. Exp. Pharmacol.* **2006**, *176*, 97–133. [CrossRef]
39. Vieira-De-Abreu, A.; Campbell, R.A.; Weyrich, A.S.; Zimmerman, G.A. Platelets: Versatile effector cells in hemostasis, inflammation, and the immune continuum. *Semin. Immunopathol.* **2011**, *34*, 5–30. [CrossRef]
40. Henn, V.; Slupsky, J.R.; Gräfe, M.; Anagnostopoulos, I.; Foörster, R.; Müller-Berghaus, G.; Kroczek, R.A. CD40 ligand on activated platelets triggers an inflammatory reaction of endothelial cells. *Nature* **1998**, *391*, 591–594. [CrossRef]
41. Zuchtriegel, G.; Uhl, B.; Puhr-Westerheide, D.; Pörnbacher, M.; Lauber, K.; Krombach, F.; Reichel, C.A. Platelets Guide Leukocytes to Their Sites of Extravasation. *PLoS Biol.* **2016**, *14*, e1002459. [CrossRef]
42. Halling, S.M.; Peterson-Burch, B.D.; Bricker, B.J.; Zuerner, R.L.; Qing, Z.; Li, L.-L.; Kapur, V.; Alt, D.P.; Olsen, S.C. Completion of the Genome Sequence of Brucella abortus and Comparison to the Highly Similar Genomes of Brucella melitensis and Brucella suis. *J. Bacteriol.* **2005**, *187*, 2715–2726. [CrossRef]
43. Tibor, A.; Decelle, B.; Letesson, J.-J. Outer Membrane Proteins Omp10, Omp16, and Omp19 of Brucella spp. Are Lipoproteins. *Infect. Immun.* **1999**, *67*, 4960–4962. [CrossRef] [PubMed]
44. Sun, Y.; Liu, W.-Z.; Liu, T.; Feng, X.; Yang, N.; Zhou, H.-F. Signaling pathway of MAPK/ERK in cell proliferation, differentiation, migration, senescence and apoptosis. *J. Recept. Signal Transduct.* **2015**, *35*, 600–604. [CrossRef] [PubMed]
45. Aramoto, H.; Breslin, J.W.; Pappas, P.J.; Hobson, R.W.; Durán, W.N. Vascular endothelial growth factor stimulates differential signaling pathways in in vivo microcirculation. *Am. J. Physiol. Circ. Physiol.* **2004**, *287*, H1590–H1598. [CrossRef] [PubMed]
46. Barone, F.C.; Irving, E.; Ray, A.; Lee, J.; Kassis, S.; Kumar, S.; Badger, A.; Legos, J.J.; A Erhardt, J.; Ohlstein, E.; et al. Inhibition of p38 mitogen-activated protein kinase provides neuroprotection in cerebral focal ischemia. *Med. Res. Rev.* **2001**, *21*, 129–145. [CrossRef]
47. Stins, M.F.; Gilles, F.; Kim, K.S. Selective expression of adhesion molecules on human brain microvascular endothelial cells. *J. Neuroimmunol.* **1997**, *76*, 81–90. [CrossRef]
48. Stins, M.F.; Badgera, J.; Kim, K.S. Bacterial invasion and transcytosis in transfected human brain microvascular endothelial cells. *Microb. Pathog.* **2001**, *30*, 19–28. [CrossRef]

49. Nizet, V.; Kim, K.S.; Stins, M.; Jonas, M.; Chi, E.Y.; Nguyen, D.; Rubens, C.E. Invasion of brain microvascular endothelial cells by group B streptococci. *Infect. Immun.* **1997**, *65*, 5074–5081. [CrossRef]
50. Ferrero, M.C.; Bregante, J.; Delpino, M.V.; Barrionuevo, P.; A Fossati, C.; Giambartolomei, G.H.; Baldi, P.C. Proinflammatory response of human endothelial cells to Brucella infection. *Microbes Infect.* **2011**, *13*, 852–861. [CrossRef]

# 5

# Impact of *Staphylococcus aureus* Small Colony Variants on Human Lung Epithelial Cells with Subsequent Influenza Virus Infection

**Janine J. Wilden [1], Eike R. Hrincius [1], Silke Niemann [2], Yvonne Boergeling [1], Bettina Löffler [3,4], Stephan Ludwig [1,5] and Christina Ehrhardt [6,*]**

1 Institute of Virology Muenster (IVM), Westfaelische Wilhelms-University Muenster, 48149 Muenster, Germany; wildenja@ukmuenster.de (J.J.W.); hrincius@uni-muenster.de (E.R.H.); borgelin@uni-muenster.de (Y.B.); ludwigs@uni-muenster.de (S.L.)
2 Institute of Medical Microbiology, Westfaelische Wilhelms-University Muenster, 48149 Muenster, Germany; Silke.Niemann@uni-muenster.de
3 Institute of Medical Microbiology, Jena University Hospital, 07747 Jena, Germany; bettina.loeffler@med.uni-jena.de
4 Cluster of Excellence EXC 2051 "Balance of the Microverse", FSU Jena, 07743 Jena, Germany
5 Cluster of Excellence EXC 1003 "Cells in Motion", WWU Muenster, 48149 Muenster, Germany
6 Section of Experimental Virology, Institute of Medical Microbiology, Jena University Hospital, 07745 Jena, Germany
* Correspondence: christina.ehrhardt@med.uni-jena.de

**Abstract:** Human beings are exposed to microorganisms every day. Among those, diverse commensals and potential pathogens including *Staphylococcus aureus* (*S. aureus*) compose a significant part of the respiratory tract microbiota. Remarkably, bacterial colonization is supposed to affect the outcome of viral respiratory tract infections, including those caused by influenza viruses (IV). Since 30% of the world's population is already colonized with *S. aureus* that can develop metabolically inactive dormant phenotypes and seasonal IV circulate every year, super-infections are likely to occur. Although IV and *S. aureus* super-infections are widely described in the literature, the interactions of these pathogens with each other and the host cell are only scarcely understood. Especially, the effect of quasi-dormant bacterial subpopulations on IV infections is barely investigated. In the present study, we aimed to investigate the impact of *S. aureus* small colony variants on the cell intrinsic immune response during a subsequent IV infection in vitro. In fact, we observed a significant impact on the regulation of pro-inflammatory factors, contributing to a synergistic effect on cell intrinsic innate immune response and induction of harmful cell death. Interestingly, the cytopathic effect, which was observed in presence of both pathogens, was not due to an increased pathogen load.

**Keywords:** *Staphylococcus aureus*; small colony variants; influenza virus; super-infection; pro-inflammatory response

---

## 1. Introduction

The respiratory tract is a major portal for microorganisms, through which virus infections can cause non-symptomatic, mild, and self-limiting but also severe diseases, sometimes with fatal outcomes [1]. A growing body of evidence shows that the human respiratory tract contains a highly adapted microbiota including commensal and opportunistic pathogens. Among those, *Staphylococcus aureus* (*S. aureus*) is of special importance, forming quasi-dormant subpopulations characterized by increased fitness compared to other phenotypes [2]. Colonization of *S. aureus* could either be persistent or non-persistent, whereby nasal colonization appears to be the most prominent localization [3]. *S. aureus* as a community-acquired pathogen is already colonized on approximately 30% of the human population, some without causing any symptoms [4]. During long-term colonization or infection, *S. aureus* can change phenotypes to so-called small colony variants (SCVs), which adapt in their metabolic and phenotypic characteristics, allowing them to evade the host's immune system. SCVs can be localized intracellularly and are characterized by a slow growth rate, non-pigmentation, less hemolytic activity, and decreased antibiotic susceptibility [5–7] but often enhanced surface presentation of adhesion molecules [8]. SCVs are often misdiagnosed [9]. Due to their slow growth, they often get overgrown by other bacteria, and an initially effective antibiotic treatment results in the development of resistances accompanied by chronic and relapsing infections [5,6,8,10,11]. The clinical relevance of colonizing SCVs gets obvious in patients with chronic respiratory diseases, such as chronic obstructive pulmonary disease (COPD) or cystic fibrosis (CF) [5]. Patients who are colonized with bacteria are more likely to suffer from recurring infections [12], as the phenotype can revert to the pathogenic phenotype.

Besides, simultaneous occurrence of different pathogens can induce or even exacerbate a pathological effect in the lung. Super-infections with influenza viruses (IV) and with the community-acquired *S. aureus* are known to be harmful and lead to increased inflammatory lung damage [13]. Due to their quick adaptation and genomic changes, both pathogens can evade the host's immune response, causing the tedious development of effective medications. Concerning super-infections, most studies describe infections with a primary viral infection that paves the path for a secondary bacterial infection [14–17]. However, there is evidence that primary bacterial colonization also occurs prior to viral infections [18].

However, the influence of colonizing *S. aureus* SCVs on subsequent IV infection is largely unexplored. Thus, in the present study, we aimed to investigate the effect of the bacterial strain *S. aureus* $3878_{SCV}$ on cell intrinsic immune responses to a subsequent IV infection, in vitro. Here, we observed that the response of anti-viral gene expression was barely changed. However, pro-inflammatory genes were highly upregulated upon super-infection, resulting in an induction of necrotic cell death. Thus, we were able to show that colonizing SCVs could enhance severity of subsequent viral infection.

## 2. Materials and Methods

### *2.1. Cell Lines, Virus Strains, and Bacteria Strain*

All cell lines were cultivated at 37 °C and 5% $CO_2$ under sterile conditions. Human lung epithelial cells A549 (American Type Culture Collection (ATCC), Wesel, Germany) were cultivated in Dulbeccos's modified eagle medium (DMEM; Sigma-Aldrich, St. Louis, MO, USA) and Madin-Darby canine kidney cells II (MDCKII) in minimum essential medium eagle (MEM; Sigma-Aldrich, St. Louis, MO, USA), supplemented with 10% fetal bovine serum (FBS; Biochrom, Berlin, Germany).

The human IV strains A/Puerto Rico/8/34 (H1N1, PR8-M) and A/Panama/2007/99 (H3N2, Panama) were taken from the virus stock of the Institute of Virology Muenster, 48149 Muenster, Germany, subcultured and passaged on MDCKII cells.

The persisting bacterial strain *S. aureus* $3878_{SCV}$, wildtype phenotype strain *S. aureus* $3878_{WT}$, and the human lung isolate of another SCV phenotype strain *S. aureus* $814_{SCV}$ (provided by Karsten Becker, Institute of Medical Microbiology, Muenster, Germany) were stored at −80 °C in a 30% glycerol/brain-heart infusion (BHI; Merck; Darmstadt, Germany) medium. *S. aureus* $3878_{SCV}$ and *S. aureus* $3878_{WT}$ were already characterized and described previously [10,19–21]. Before experiments, bacteria were plated on blood agar plates to take single clones, which were inoculated in BHI medium and incubated for 24 h at 37 °C and 5% $CO_2$. For bacterial infection, bacterial suspension was washed with phosphate buffered saline (PBS) (4000 rpm; 4 °C; 5 min) and adjusted to an optical density of $OD_{600nm} = 1$. Growth kinetics were performed to determine a colony forming unit (CFU) of $2 \times 10^8$ CFU/mL at $OD_{600nm} = 1$ for each bacterial strain used.

### 2.2. *Super-Infection Protocol*

Human lung epithelial cells were seeded in either 6-well plates ($0.5 \times 10^6$) or 12-well plates ($0.2 \times 10^6$) in 2 mL or 1 mL culture medium 24 h before infection. For bacterial infection, the overnight culture was set to $OD_{600nm} = 1$ to determine the multiplicity of infection (MOI). Cells were washed with PBS and infected with *S. aureus* $3878_{SCV}$ in invasion media ($DMEM_{INV}$: DMEM supplemented with 1% human serum albumin, 25 nmol/L HEPES) for 24 h with a MOI of 0.01. For viral infection, supernatant was aspirated, cells were washed with PBS and incubated with IV PR8-M (MOI = 0.1) or IV Panama (MOI = 0.01) in infection PBS ($PBS_{INF}$: PBS supplemented with 0.2% bovine serum albumin (BSA), 1 mM $MgCl_2$, 0.9 mM $CaCl_2$, 100 U/mL penicillin, 0.1 mg/mL streptomycin) for 30 min. Viral suspension was aspirated, and cells were washed with PBS and further incubated in infection media ($DMEM_{INF}$: DMEM supplemented with 0.2% bovine serum albumin (BSA), 1 mM $MgCl_2$, 0.9 mM $CaCl_2$) up to 8 hpvi, 24 hpvi, 32 hpvi, 44 hpvi, or 48 hpvi (hours post-viral infection).

### 2.3. *Transfection Protocol*

For transfection of the 3× NFκB reporter plasmid construct as described elsewhere [22] (0.1 μg/μL) A549 cells were seeded in 12-well plates as described above. Cells were transfected with 0.1 μg/μL of the indicated plasmid for 4 h with Lipofectamine® 2000 (Invitrogen, Carlsbad, CA, USA) corresponding to the manufacturer's protocol. Afterwards, cells were washed with PBS and further incubated in cell culture media up to 24 h. Afterwards transfected cells were infected up to 8 hpvi. Performance of luciferase assay was done as described elsewhere [23].

### 2.4. *Intra- and Extracellular Bacterial Titer Measurements*

Extracellular bacterial titers were determined by collecting the supernatant of infected cells including the washing with PBS. Cells were lysed via hypotonic shock with 2 mL $ddH_2O$ according to Tuchscherr et al. [7,8] (37 °C, 30 min) to determine intracellular bacterial titers, including adherent bacteria at the cells surface. Bacterial suspensions were centrifuged (4000 rpm, 4 °C, 10 min), pellets were resuspended in 1 mL PBS, and serial dilutions (1:10) were plated on BHI agar plates and incubated for 32 h at 37 °C.

### 2.5. *Standard Plaque Assay*

Infectious virus particles in the supernatant were titrated to determine viral titers. A standard plaque assay was performed as described earlier [24].

### 2.6. *Quantitative Real-Time PCR (qRT-PCR)*

RNA isolation was performed with RNeasy Kit (Qiagen, Hilden, Germany) according to the manufacturer's instructions. Reverse transcription was performed with 2 μg of total RNA with Revert AID H Minus Reverse Transciptase (Thermo Fisher Scientific, Karlsruhe, Germany) and oligo (dT) primers according to the manufacturer's protocol. qRT-PCR was performed using a Roche LightCycler 480 and Brilliant SYBRGreen Mastermix (Agilent, Santa Clara, CA, USA) according to the manufacturer's instructions. The following primers were used: GAPDH: fwd 5′GCAAATTCCATGGCACCGT3′, rev 5′GCCCCACTTGATTTGGAGG3′; IL-6: fwd 5′AACCTGAACCTTCCAAAGATGG3′, rev 5′TCTGGCTTGTTCCTCACTAGT3′; IL-8: fwd 5′CTTGTTCCACTGTGCCTTGGTT3′, rev 5′GCTTCCACATGTCCTCACAACAT3′; TNFα: fwd 5′-ATGAGCACTGAAAGCATGATC-3′, rev 5′-GAGGGCTGATTAGAGAGAGGT-3′; IL-1β: fwd 5′-CAGCTACGAATCTCCGACCAC-3′, rev 5′-GGCAGGGAACCAGCATCTTC-3′; IFNγ: fwd 5′AAACGAGATGACTTCGAAAAGCTG3′, rev 5′TGTTTAGCTGCTGGCGACAG3′; RIG-I: fwd 5′CCTACCTACATCCTGAGCTACAT3′, rev 5′TCTAGGGCATCCAAAAAGCCA3′; IFNβ: fwd 5′TCTGGCACAACAGGTAGTAGGC3′, rev 5′GAGAAGCACAACAGGAGAGCAA3′; MxA: fwd 5′GTTTCCGAAGTGGACATCGCA3′, rev 5′GAAGGGCAACTCCTGACAGT3′; OAS1: fwd

5′GATCTCAGAAATACCCCAGCCA3′, rev 5′AGCTACCTCGGAAGCACCTT3′. Relative changes in expression levels (n-fold) were calculated according to the $2^{-\Delta\Delta Ct}$ method [25].

Bacterial RNA was isolated with the RNeasy Protect Bacteria Mini Kit (Qiagen, Hilden, Germany), and cDNA synthesis was performed using QuantiTect Reverse Transcription Kit (Qiagen, Hilden, Germany) according to the manufacturer's instructions. qRT-PCR was performed using a Roche LightCycler 480 (Basel, Switzerland) and Brilliant SYBRGreen Mastermix (Agilent, Santa Clara, CA, USA) according to the manufacturer's instructions. The primers to determine the gene expression of *gyrB*, *aroE*, *arg*, *hla*, *sarA*, and *sigB* were already described elsewhere [7].

### 2.7. *RT$^2$ Profiler Array Analysis*

For pathway focused gene expression analysis, we used *RT$^2$ Profiler PCR Arrays* (Qiagen, Hilden, Germany). RNA isolation, cDNA synthesis and procedure were performed according to the manufacturer's protocol and instructions. Analysis of data was accomplished by using the GeneGlobe Data Analysis Center recommended by Qiagen [26].

### 2.8. *FACS Analysis*

Determination of secreted proteins in the supernatant was performed with BioLegend's LEGENDplex™ (San Diego, CA, USA) according to the manufacturer's protocol. The human anti-viral and pro-inflammatory chemokine panels were used. Results were analyzed by BioLegend's cloud-based LEGENDplex™ Data Analysis Software. To analyze apoptotic or necrotic cells, infection was performed as described above until 44 hpvi. Cells were treated with tumor necrosis factor related apoptosis inducing ligand (TRAIL; Enzo Life Sciences, Farmingdale, NY, USA) (150 ng/mL) 4.5 h before harvested and used as a positive control for apoptosis. The supernatant was collected for this purpose, and cells were detached from the wells with trypsin-EDTA and recombined with the supernatant. Cell suspension was centrifuged at 1000× *g* at room temperature (RT) for 5 min, and cells were washed with PBS supplemented with 5% FCS. Afterwards, cells were stained with annexin V FITC (20 µL) (ImmunoTool, Friesoythe, Germany) and 1:2000 eBioscience™ Fixable Viability Dye eFluor™ 660 (Thermo Fisher Scientific, Karlsruhe, Germany) in 100 µL 1× annexin V staining buffer (10× annexin V staining buffer: 0.1 M HEPES, 1.4 M NaCl, and 25 mM $CaCl_2$ (pH 7.5)) for 30 min at RT in the dark. Further 150 µL of staining buffer were added and the supernatant was removed after centrifugation. Cells were fixed with 500 µL PBS containing 4% formaldehyde and 1.25 mM $CaCl_2$ for 20 min at RT in the dark. Cells were finally resuspended in 150 µL staining buffer and stored at 4 °C until measurement with the FACSCalibur flow cytometer (BD Biosciences, Heidelberg, Germany), followed by the analysis with FlowJo software (v.10; Flow Jo, Ashland, OR, USA). Three gates were set as the following: annexin V positive cells (early apoptotic cells) and live/dead marker positive cells (cells with a membrane rupture tending to necrosis).

### 2.9. *Recording of Cytopathic Effect of Infected Cells*

To record the CPE at different time points, cells were visualized with Canon (EOS 500D) by light microscopy (Axiovert 40C, ZEISS, Jena, Germany) with a 10× magnification.

### 2.10. *SDS-PAGE and Western Blot Analysis*

Protein expressions were determined by separating proteins in a polyacrylamide gel and subsequent transfer on nitrocellulose membranes by western blot analysis as described earlier [27]. The following antibodies were used: pMLKL [(S353) #91689 Cell Signaling, Frankfurt, Germany], PARP (#611039 BD, Heidelberg, Germany) and ERK1/2 (#4696 Cell Signaling, Frankfurt, Germany).

### 2.11. Lactate Dehydrogenase (LDH) Assay

The lactate dehydrogenase assay (CellBiolabs, San Diego, CA, USA) was used to measure the cell cytotoxicity and was used according to the manufacturer's instructions. Cells were infected as described previously, and 90 μL of the supernatant was mixed with 10 μL of the LDH cytotoxicity reagent in a 96-well plate. This plate was incubated at 37 °C and 5% $CO_2$ for 30 min, and the $OD_{450nm}$ was measured on a Spectromax M2 Instrument (Molecular Devices, Munich, Germany). Triton X-100 used according to the manufacturer's instructions served as a positive control.

### 2.12. Quantification and Statistical Analysis

All data represent the means + standard deviation (SD) of three independent experiments. Statistical significances were determined by unpaired *t*-test (Figure S4A), one-way ANOVA followed by Tukey's, (Figures 4D,E, 5, 6A,B,D,E, Figure S1B,D,E, S2D,E, S3 and S4C–E) or two-way ANOVA followed by Sidak's (Figure 2, Figures S1C and S4B) or followed by Tukey's (Figure 4A–C,F–I and Figure S2A–C,F–I) multiple comparison test using GraphPad Prism software (v.7.03, GraphPad Prism, Inc., La Jolla, CA, USA).

## 3. Results

### 3.1. Primary S. aureus $3878_{SCV}$ Infection Provokes a Cytopathic Effect in Presence of IV

Cell death mechanisms induced by *S. aureus* or IV alone are very well investigated and described [28–33]. With respect to IV and *S. aureus* super-infection, we recently were able to show a *S. aureus*-mediated switch from IV-induced apoptosis to necrosis [27]. It is known that IV infection paves the path for secondary bacterial infection, resulting in enhanced pathogen-load [15,34], cytokine expression [35,36], and cell death [27]. Since *S. aureus* often persist in humans without any harm, we aimed to investigate the effects of colonizing *S. aureus* SCVs on secondary IV super-infection.

In a first set of experiments, we focused on the cell morphology of A549 human lung epithelial cells in absence and presence of *S. aureus* $3878_{SCV}$ and IV. For this reason, A549 human lung epithelial cells were infected with *S. aureus* $3878_{SCV}$, which is a well described SCV patient isolate [10,37], for 24 h followed by infection with IV strain A/Puerto Rico/8/34 (PR8-M; H1N1) for the indicated points in time. The morphology of single- and super-infected cells was analyzed by light microscopy in comparison to uninfected control (mock) (Figure 1). While the cell monolayer is still intact in un-, single-, and super-infected cells up to 32 hpvi (hours post-viral infection), first changes in the cell morphology were visible 48 hpvi in single virus-infected and super-infected cells. Pictures of virus-infected cells showed a less confluent cell monolayer compared to uninfected cells, and in super-infected samples a clear cytopathic effect was observed, indicated by cell monolayer disruption and floating cells. To be able to ascribe these findings to the SCV phenotype, we additionally specified the pathological difference between *S. aureus* wildtype phenotype and SCV phenotype (*S. aureus* $3878_{WT}$ and *S. aureus* $3878_{SCV}$) by infecting A549 human lung epithelial cells. Cell morphology was monitored by light-microscopy (Supplementary Figure S1A) and cell viability was quantified by lactate dehydrogenase assay (LDH) assay (Supplementary Figure S1B). Both assays indicate a massive destruction of the cell monolayer 8 h post bacterial infection (hpbi) with *S. aureus* $3878_{WT}$ in comparison to *S. aureus* $3878_{SCV}$. Further, the determination of the expression of distinct bacterial genes, which are involved in the virulence of the pathogens, verified the reduced virulence of *S. aureus* $3878_{SCV}$ in comparison to the *S. aureus* $3878_{WT}$ (Supplementary Figure S1C). Based on these results, *S. aureus* $3878_{WT}$ was not used in the following experiments. The analysis of cell viability at 32 hpvi and 48 hpvi confirmed the cell disturbance in presence of *S. aureus* $3878_{SCV}$ and IV infection (Supplementary Figure S1D,E).

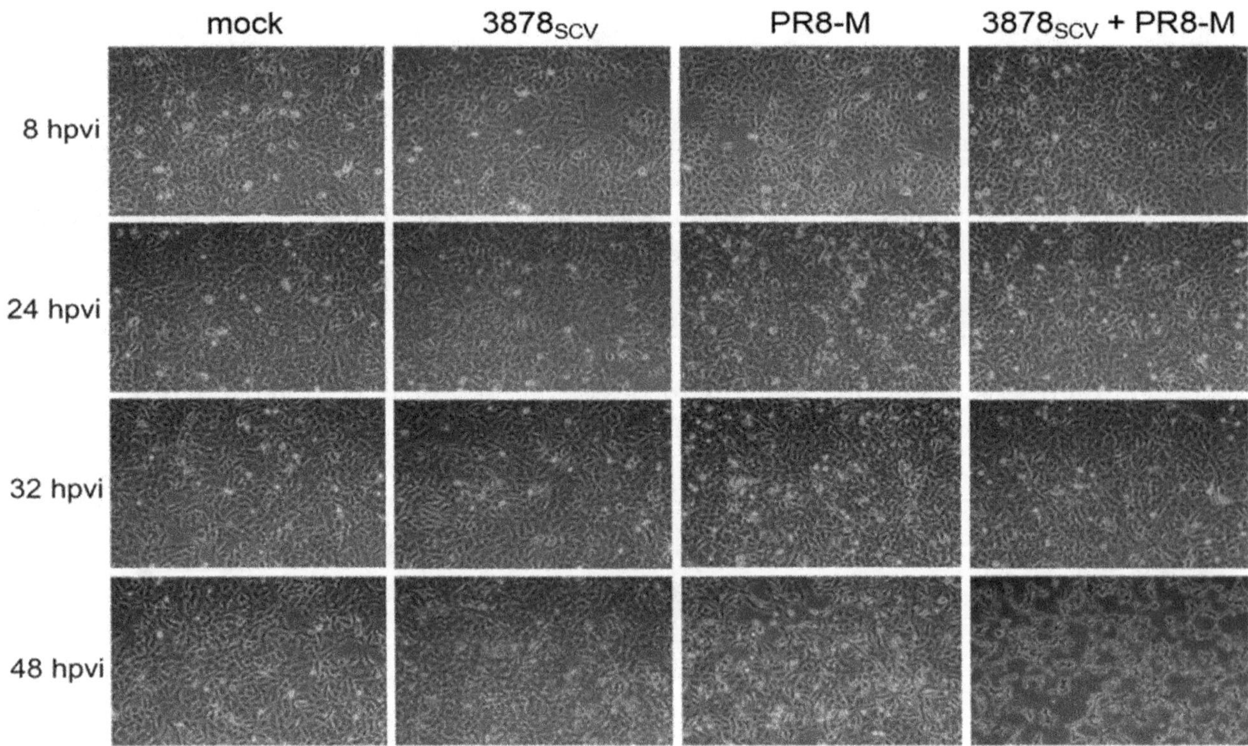


**Figure 1.** *S. aureus* $3878_{SCV}$ colonization and subsequent influenza virus infection provokes a cytopathic effect. A549 human lung epithelial cells were infected with *S. aureus* $3878_{SCV}$ (multiplicity of infection (MOI) = 0.01) for 24 h at 37 °C and 5% $CO_2$. Afterwards, cells were infected with influenza viruses (IV) Puerto Rico/8 (PR8)-M (MOI = 0.1) until the indicated points in time (hours post-viral infection, hpvi). Cells were visualized by light microscopy with a 10× magnification. Shown are representative images of three independent experiments ($n = 3$).

These results point to an altered cell culture environment and/or cellular signaling upon super-infection with *S. aureus* $3878_{SCV}$ and IV, which could be triggered by increased pathogen load or cell intrinsic signaling changes in presence of both pathogens.

### *3.2. Primary Infection with S. aureus $3878_{SCV}$ Followed by IV Infection Had No Impact on Bacterial or Viral Titers*

First, we analyzed whether the observed cytotoxicity of co-infected A549 human lung epithelial cells with *S. aureus* $3878_{SCV}$ and IV was due to increased pathogen load. For this, we infected A549 cells with *S. aureus* $3878_{SCV}$ for 24 h and super-infected with two different IV strains for the indicated points in time to determine the amount of plaque forming units (PFU) or colony forming units (CFU) of viruses or bacteria, respectively (Figure 2).

In general, titers of IV and SCVs increased with time, but neither viral (Figure 2A,B) nor bacterial titers (Figure 2C–F) were significantly changed upon super-infection compared to single-infected cells, a phenomenon independent of the virus strain used [PR8-M (H1N1), A/Panama/2007/99 (Panama; H3N2)].

Thus, these data indicate that the disruption of the cell monolayer upon super-infection is not induced by increased amounts of pathogens but by a different mechanism that is altered by the presence of both pathogens.

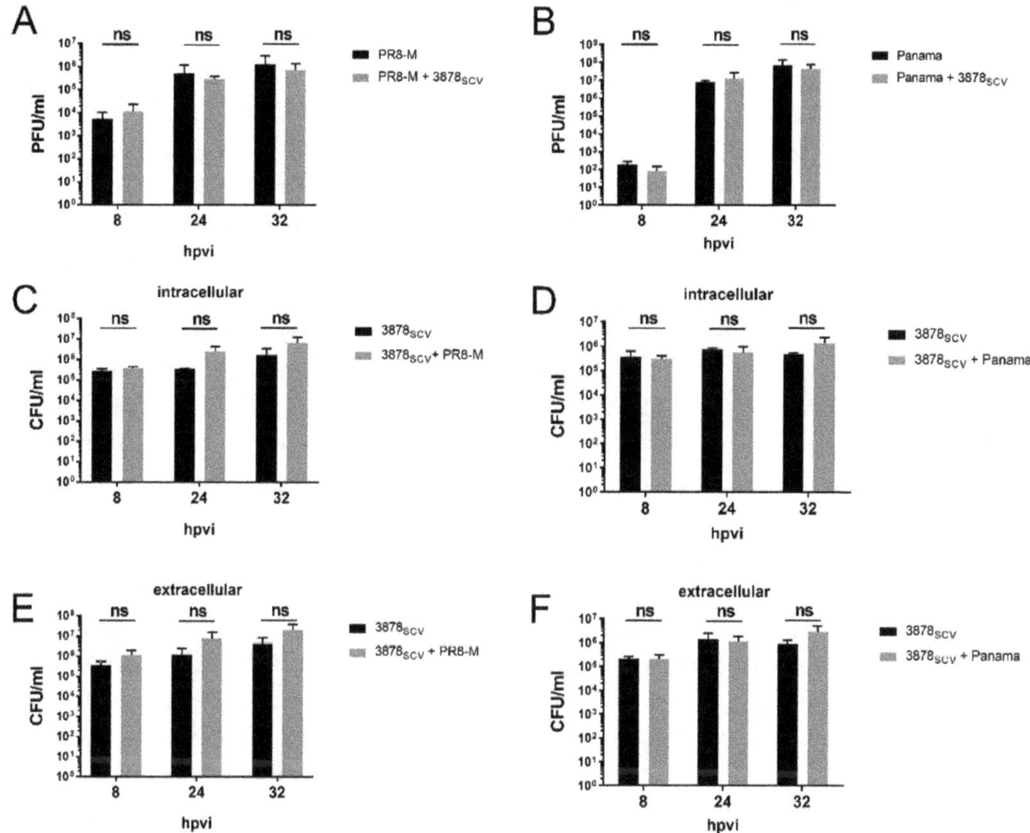


**Figure 2.** Pathogen load is not affected during *S. aureus* $3878_{SCV}$ colonization and subsequent influenza virus infection. A549 human lung epithelial cells were infected with *S. aureus* $3878_{SCV}$ (MOI = 0.01) for 24 h and/or super-infected with (**A**,**C**,**E**) IV PR8-M (H1N1; MOI = 0.1) or (**B**,**D**,**F**) IV Panama (H3N2; MOI = 0.01) for 8 hpvi, 24 hpvi, or 32 hpvi. At the indicated times post-viral infection, supernatants were collected to determine viral and extracellular bacterial titers. Afterwards, cells were lysed via hypotonic shock to analyze intracellular bacterial titers. Means + SD of three independent experiments with technical duplicates are shown ($n$ = 3). Statistical significance (compared to single-IV infection (**A**,**B**) or single-bacteria infection (**C**–**F**) was analyzed by a two-way ANOVA, followed by Sidak's multiple comparison test; (hpvi = hours post-viral infection; ns = not significant).

### 3.3. *Pro-Inflammatory Gene Expression Is Highly Upregulated after Super-Infection of* S. aureus *$3878_{SCV}$ and IV*

Given the observation that super-infection of *S. aureus* $3878_{SCV}$ and IV PR8-M induced a cytopathic effect (Figure 1), which was not caused by increased pathogen load (Figure 2), we aimed to elucidate if changes of cell intrinsic signaling and inflammatory gene expression might be responsible for this phenomenon. We analyzed the gene expression of 84 different genes, involved in different signaling cascades by use of a *RT*$^2$ *profiler Array* (Qiagen, Hilden, Germany) in a single experiment to gain a first insight in the complexity of cellular signaling (Figure 3). This enables a quick analysis of expression levels of different genes that are organized by their function to be able to limit the amount of genes, altering the cell intrinsic signaling. Here, we used the anti-viral immune response panel, including pattern recognition receptors (PRRs), cytokines, and chemokines involved in pathogen recognition and immune responses. The bioinformatic analysis is based on conventional ct-values and was performed with the recommended GeneGlobe online software [26]. A clustergram was generated to visually illustrate all up- and downregulated genes that were analyzed (Figure 3A). To further interpret the results of the $RT^2$ *profiler Array*, we did an in silico clustering of the upregulated genes of the array that were highly upregulated (difference of an n-fold of 2) in *S. aureus* $3878_{SCV}$ and PR8-M super-infected cells compared to single-infected cells (*APOBEC3G, CASP1, CASP10, CCL3, CCL5, CD40, CD80, CTSS, CXCL10, CXCL11, CYLD, IL1B, IL6, CXCL8, MEFV, TLR3, TNF, B2M*) (see Supplementary Table S1), with respect to their linkage to specific signaling pathways (Figure 3B) by using the Kyoto Encyclopedia of Genes and Genomes mapper (KEGG mapper). KEGG mapper is a database resource of collected information about pathways and the involved genes representing a pool of molecular interactions, reactions, and their relation to each other [38–40]. Down-regulated genes were excluded, as the gene expressions were negligible (Supplementary Table S1). Besides gene clusters connected to expected PRR pathways including TLR-(11 genes involved, out of the 18 highly upregulated genes comparing co- and single-infected cells identified (11/18), NLR- (7/18), TNFR- (6/18), RLR- (5/18), and NFκB- (5/18) signaling pathways (Figure 3B), we identified gene clusters belonging to two cell death mechanisms, necroptosis (5/18) and apoptosis (3/18). Furthermore, we identified genes involved in the IL-17 (5/18) and c-type lectin (5/18) signaling pathways. To further classify the activated genes leading to the observed cytopathic effect on human lung epithelial cells, we searched for a specific induction pattern in which super-infected cells led to upregulated genes. We, therefore, compared all upregulated genes of single-infected to super-infected samples in a Venn diagram (Figure 3C,D). We identified 11 genes that were induced in all three infection-scenarios compared to uninfected cells and 9 genes that were upregulated in super-infected cells only. We also compared the upregulated genes for super-infection with IV Panama. Here, all infection scenarios shared the induction of 12 genes, where 7 genes were exclusively induced by the super-infection of *S. aureus* $3878_{SCV}$ and IV Panama. The upregulated genes of the Venn diagram are listed in Table S2A,B. With respect to the mRNA expression levels shown in Supplementary Table S1 and the cytopathic effect observed in super-infected cells (Figure 1), an induction of pro-inflammatory immune response can be concluded, which was further visualized by graphs, exhibiting the gene expression of the highly upregulated genes (Figure 3E).

To confirm an increased pro-inflammatory status of the human lung epithelial cells upon super-infection, we analyzed the mRNA expression of different representative pro-inflammatory cytokines and chemokines (IL-6, IL-8, TNFα, IL-1β, and IFN-γ) in detail (Figure 4A–E). Furthermore, we analyzed the mRNA expression of molecules that are involved in the induction of the type-I-IFN signaling (RIG-I, IFN-β, MxA, and OAS1) (Figure 4F–I), since it was described that IV-induced type-I-IFN signaling had an impact on bacterial infections [41].

In *S. aureus* $3878_{SCV}$ colonized cells subsequently infected with IV PR8-M the mRNA expression of IL-6, IL-8, TNFα, and IL-1β 32 hpvi was induced if compared to uninfected cells or single-infected cells (Figure 4A–D) and IFN-γ showed the same tendency (Figure 4E). Single-infection of *S. aureus* $3878_{SCV}$ or IV PR8-M resulted in no significant induction of the mRNA expression 8 hpvi, 24 hpvi, or 32 hpvi, except for IL-8, which was significantly induced 8 hpvi in bacteria single-infected cells (Figure 4B). Nevertheless, this induction was abolished over time. Genes, encoding key proteins involved in the recognition, and induction of type-I-IFN signaling were upregulated in IV PR8-M infected cells 8 hpvi (IFN-β by tendency) (Figure 4G) or 24 hpvi (RIG-I, MxA and OAS1) (Figure 4F,H,I). Previous colonization with *S. aureus* $3878_{SCV}$ had no impact on IV-induced mRNA expression of factors linked to the type-I-IFN response, except for RIG-I at 32 hpvi, which was significantly decreased in super-infected cells. However, the enhanced RIG-I mRNA synthesis did not result in alterations of viral titers. Similar results were obtained upon super-infection with *S. aureus* $3878_{SCV}$ and IV Panama, indicating a virus-independent effect (Supplementary Figure S2A–I).

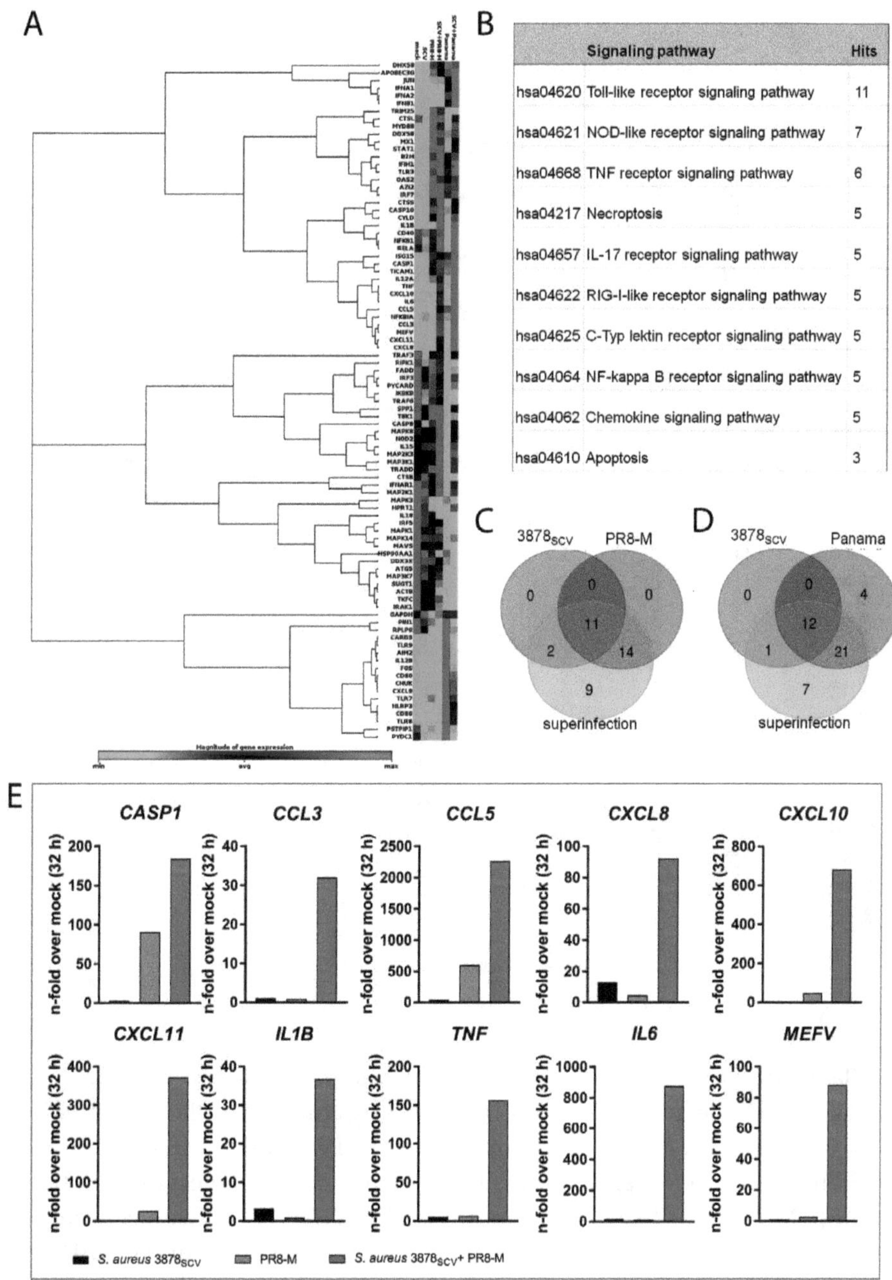


**Figure 3.** Gene expression analysis by *RT*[2] *Profiler Array*. (**A–E**) A549 lung epithelial cells were infected with *S. aureus* $3878_{SCV}$ (MOI = 0.01) for 24 h and/or super-infected with IV PR8-M (H1N1; MOI = 0.1) or IV Panama (H3N2; MOI = 0.01) for 32 h. Subsequently, RNA was isolated and further used to perform the *RT*[2] *Profiler Array* (Qiagen, Hilden, Germany). Ct-values were analyzed with the recommended QIAGEN web portal [26]. (**A**) A clustergram is shown, visualizing the up- and downregulated genes of the customized 84-gene array. (**B**) Ten potential signaling pathways are listed, which can be analyzed by the *RT*[2] *Profiler Array* with the correlating count of genes involved. The mapping was done by using the Kyoto Encyclopedia of Genes and Genomes (KEGG) mapper [38–40]. (**C**,**D**) A Venn diagram of the upregulated genes in a super-infection scenario with *S. aureus* $3878_{SCV}$ and IV PR8-M (**C**) or IV Panama (**D**) is shown. The analysis was performed by use of http://bioinformatics.psb.ugent.be/webtools/Venn/. (**E**) Gene expression of highly induced genes indicating an increased pro-inflammatory cytokine response. Values are shown as n-fold over mock (32 hpvi); ($n$ = 1); (hpvi = hours post-viral infection).

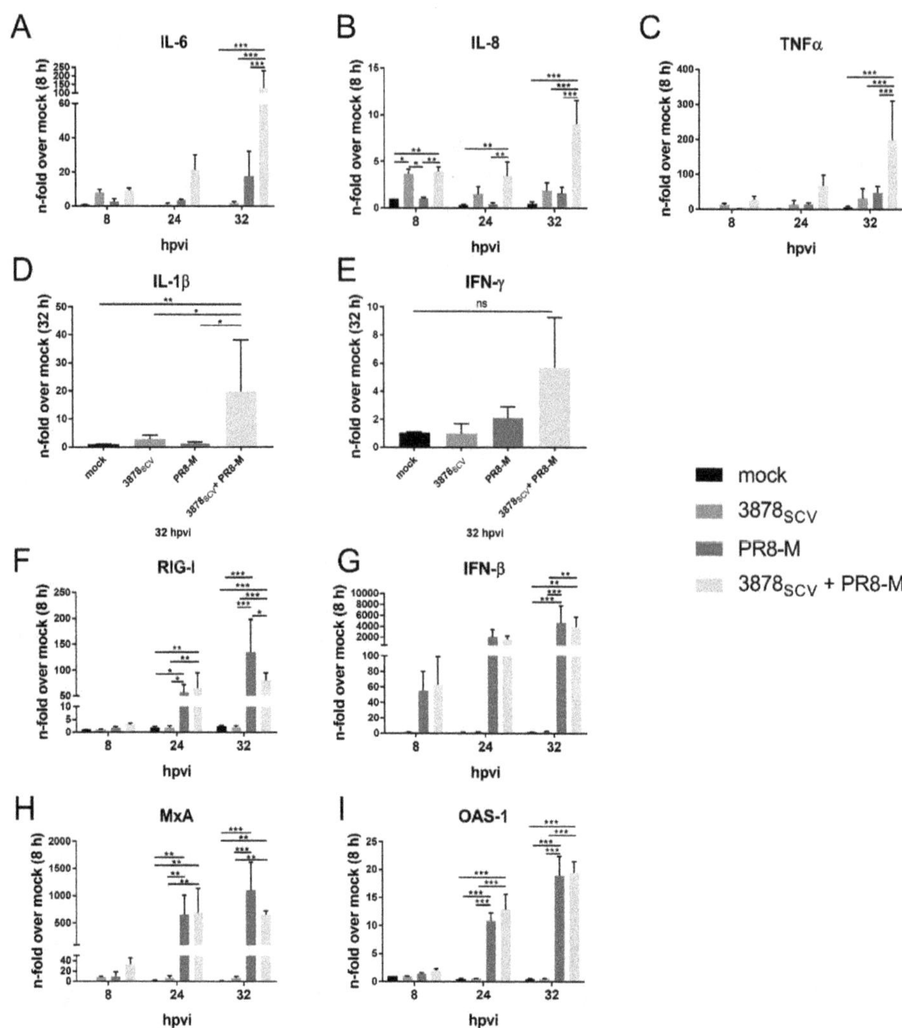


**Figure 4.** Pro-inflammatory cytokines and chemokines are enhanced after *S. aureus* $3878_{SCV}$ colonization and subsequent IV PR8-M infection. (**A–I**) A549 human lung epithelial cells were infected with *S. aureus* $3878_{SCV}$ (MOI = 0.01) for 24 h and/or super-infected with IV PR8-M (H1N1; MOI = 0.1) for 8 hpvi, 24 hpvi and/or 32 hpvi. Afterwards, RNA was isolated and mRNA levels of IL-6, IL-8, TNFα, IL-1β, IFN-γ, RIG-I, IFN-β, MxA, and OAS1 were determined by qRT-PCR. All values were correlated to the representative mock-control 8 hpvi (IL-6, IL-8, TNFα, RIG-I, IFNβ, MxA and OAS1) or 32 hpvi (IL-1β and IFN-γ). Means + SD of three independent experiments including technical duplicates are shown. Statistical significance was analyzed by a two-way (**A–C**), (**F–I**) or one-way (**D**,**E**) ANOVA, followed by Tukey's multiple comparison test (* $p < 0.05$, ** $p < 0.01$, *** $p < 0.001$); (hpvi = hours post-viral infection; ns = not significant).

To analyze whether the induction of mRNA synthesis of pro-inflammatory genes could also be detected on protein level and to get further insights into the cell intrinsic innate immune status of super-infected A549 cells, the protein expression of exemplary cytokines and chemokines was monitored by FACS analysis (Figure 5A–H). Remarkably, FACS analysis verified the increased pro-inflammatory response of A549 human lung epithelial cells for the secretion of representative factors. In super-infected cells, protein levels of IL-6, RANTES (CCL5), IP-10, and I-TAC were significantly induced compared to uninfected or single-infected cells with either *S. aureus* $3878_{SCV}$ or IV PR8-M (Figure 5A,C,E,F). TNFα was also significantly upregulated upon super-infection compared to uninfected and bacteria single-infected cells but not to IV PR8-M single-infected cells. Furthermore, IV PR8-M infection provoked TNFα protein expression 32 hpvi (Figure 5B). Representative IFN protein concentrations of IFN-γ and IFNβ (Figure 5G,H) showed no alteration in the amount of secreted proteins.

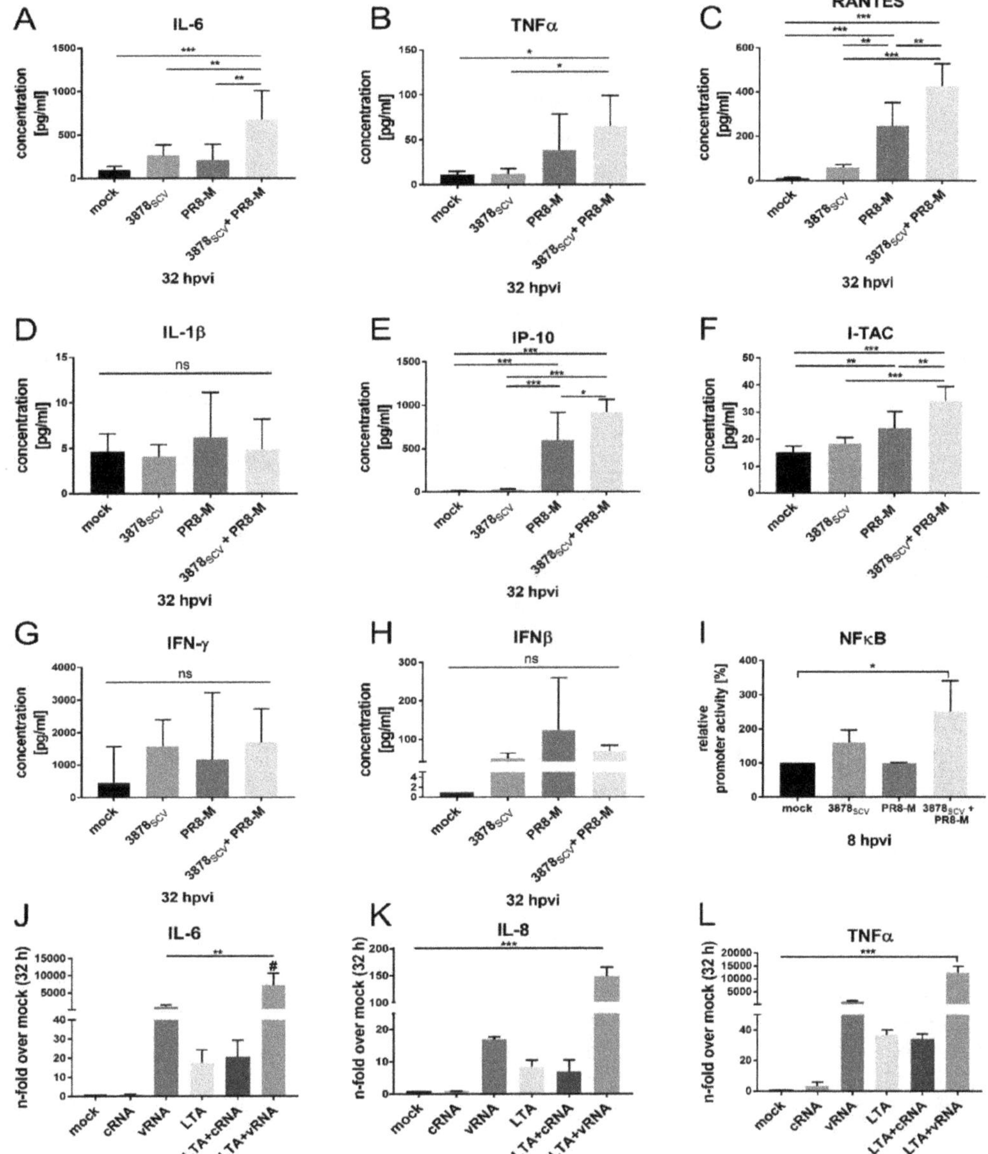


**Figure 5.** Secretion of the pro-inflammatory cytokines and chemokines are enhanced after *S. aureus* $3878_{SCV}$ colonization and subsequent IV PR8-M infection regulated by TLR2- and RIG-I-mediated NFκB promoter activation. (**A–H**) A549 human lung epithelial cells were infected with *S. aureus* $3878_{SCV}$ (MOI = 0.01) for 24 h and/or super-infected with IV PR8-M (H1N1; MOI = 0.1) for 32 h. Afterwards, supernatants were collected to measure the concentration of secreted proteins via FACS analysis. Means + SD of three independent experiments, including technical duplicates, are shown. (**I**) A549 human lung epithelial cells were transfected with 3× NFκB luciferase promoter reporter construct for 24 h prior to super-infection as described before. Afterwards cells were harvested and analyzed for luciferase activity. (**J–L**) A549 human lung epithelial cells were stimulated with LTA (100 ng/mL) for 24 h at 37 °C and 5% $CO_2$. Afterwards, cells were stimulated with cellular RNA (cRNA) or viral RNA (vRNA) (100 ng/mL) in the presence or absence of LTA for 4 h at 37 °C and 5% $CO_2$. Subsequently, RNA was isolated, and mRNA levels of IL-6, IL-8, and TNFα, were measured by qRT-PCR. All values are correlated to the respective mock-control ($n = 3$). Statistical significance was analyzed by one-way ANOVA followed by Tukey's multiple comparison test (* $p < 0.05$, ** $p < 0.01$, *** $p < 0.001$; # = *** $p < 0.001$ compared to LTA + vRNA, except for vRNA); (hpvi = hours post-viral infection; ns = not significant).

These results emphasize the enhanced cell intrinsic pro-inflammatory status of the super-infected cells. We could confirm these results by the use of IV strain Panama, verifying a viral strain independent effect (Supplementary Figure S3A–H). Based on these results and due to the fact that the induction of pro-inflammatory cytokines and chemokines is mainly driven by NFκB activation, we hypothesized an induction of the pro-inflammatory response via specific PRRs, resulting in the activation of the NFκB-signaling cascade [42]. To confirm the induction of NFκB-signaling we transfected A549 cells with an artificial NFκB promoter-dependent luciferase reporter plasmid prior to super-infection with *S. aureus* $3878_{SCV}$ and subsequent IV PR8-M infection. An increase of NFκB activation was observed in super-infected cells compared to uninfected and IV PR8-M-infected cells, while *S. aureus* $3878_{SCV}$ infection only resulted in an increase of NFκB activation by trend (Figure 5I). This induction pattern was also confirmed in cells super-infected with IV Panama (Supplementary Figure S3I).

The induction of pro-inflammatory responses via NFκB in epithelial cells after pathogen exposure can be triggered by both pathogens through different factors and their corresponding receptors [43–45], which were further analyzed. To exclude specific pathogen-mediated interference with cellular factors due to differences in protein expression, such as virulence factors or surface proteins, viral RNA (vRNA) and bacterial lipoteichonic acid (LTA, Invivogen, San Diego, CA, USA) were used as pathogen specific molecular stimuli. In human lung epithelial cells vRNA is mainly recognized by RIG-I, leading to a strong induction of the type-I-IFN-signaling cascade [46], while LTA is mainly recognized by TLR-2 [47]. We investigated mRNA expression of IL-6, IL-8, and TNFα after stimulating A549 cells with vRNA and LTA (Figure 5J–L). The results matched our findings obtained from super-infected cells, since significant enhancement of mRNA expression of IL-6, IL-8, and TNFα was observed in presence of both stimuli. Artificial effects caused by RNA transfection could be excluded due to equal cytokine mRNA expression induced by cellular RNA (cRNA) and cRNA + LTA stimulated cells. Stimulation with vRNA tended to induce the expression of IL-6, IL-8, and TNFα, which, however, was not significant compared to unstimulated cells.

Overall, these data suggest an induction of pro-inflammatory gene expression responses through the detection of bacterial and viral components via the pathogen-associated molecular pattern receptors (PAMP) RIG-I and TLR-2, followed by the induction of NFκB. To exclude bacterial strain-specific effects, another SCV strain (*S. aureus* $814_{SCV}$) was used to determine pathogen loads and pro-inflammatory gene expression in IV super-infection (Supplementary Figure S4). While neither viral titers nor intra- and extracellular bacterial load were increased in presence of both pathogens, pro-inflammatory cytokine expression was enhanced, verifying the former observations.

### 3.4. *S. aureus* $3878_{SCV}$ *Provoke Enhanced Necrotic Cell Death in Presence of IV Infection*

The observed disruption of the cell monolayer (Figure 1) could be induced by a variety of mechanisms. Besides the involvement of pro-inflammatory cytokines in the innate immune response, these factors are also involved in the induction of cell death mechanisms, like apoptosis and necrosis. As the results shown in Figure 3 indicate, an upregulation of pro-inflammatory cytokines, the cell death mechanisms might be triggered by TLRs or cell death receptors through PAMPs or cytokines, like TNFα, among others [48–50]. Therefore, we further investigated the induction of apoptosis and necrosis, correlating to the cell death mechanisms identified in the *$RT^2$ Profiler Array* analysis (Figure 3B).

As the results of the LDH assay led to the hypothesis of an induced necrotic cell death mechanism, we performed FACS analysis to determine the number of early apoptotic cells by detecting phosphatidylserine which switches to the cells' surface in early apoptotic cells and can be labeled with annexin V. Cells with a membrane rupture tending to necrosis were detected by using a viability marker comparable to 7-aminoactinomycin D and propidium iodide staining. Therefore, we performed the infection up to 44 hpvi to be able to still distinguish between early apoptosis and necrotic-like cells and stained the cells accordingly. The amount of necrotic cells significantly increased comparing un- or single-infected with super-infected cells, probably indicating necrosis (Figure 6A).

Furthermore, the amount of apoptotic cells was significantly higher in IV-infected cells compared to un-, bacteria-, or super-infected cells 44 hpvi (Figure 6B).

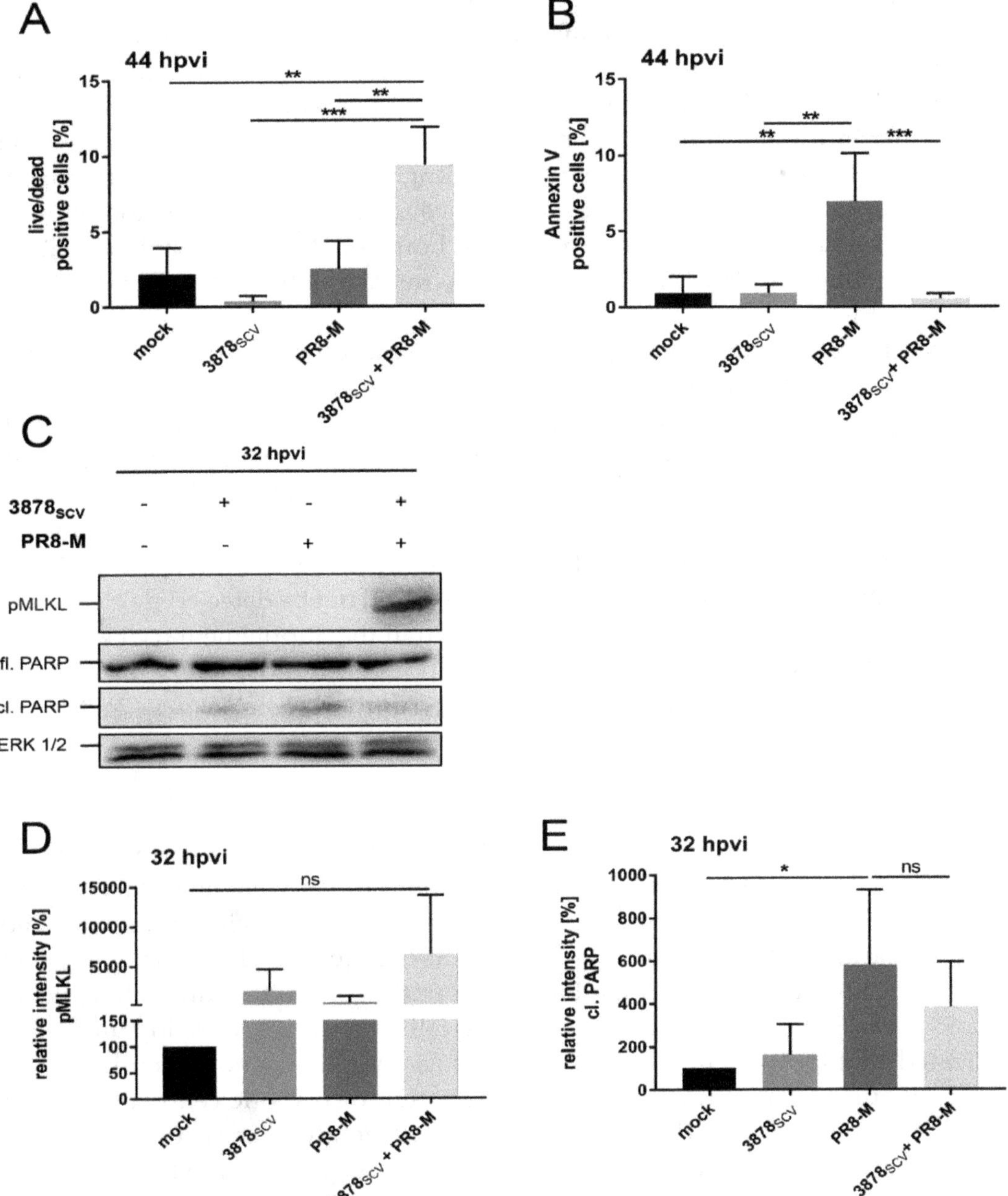


**Figure 6.** *S. aureus* $3878_{SCV}$ colonization and subsequent IV infection inhibits IV-induced apoptosis but results in the induction of necrosis. (**A–E**) A549 human lung epithelial cells were infected with *S. aureus* $3878_{SCV}$ (MOI = 0.01) for 24 h and/or super-infected with IV PR8-M (H1N1; MOI = 0.1) for 44 hpvi (**A**,**B**) or 32 hpvi (**C–E**). At the indicated times post-viral infection, total amount of cells was collected to perform FACS analysis to determine the relative amount of viability marker positive cells (**A**) or annexin V positive cells (**B**). Furthermore, whole cell lysates were subjected to western blot analysis (**C**). (**D**,**E**) Densitometrical analysis of three independent western blot experiments of cleaved pMLKL (**D**) and PARP (**E**) 32 hpvi are shown. Equal protein amounts were calculated by correlating the signal intensities to their corresponding ERK1/2 signals. Means + SD of three independent experiments are shown ($n = 3$). Statistical significance was analyzed by a one-way ANOVA, followed by Tukey's multiple comparison test (**A**,**B**,**D**,**E**); (* $p < 0.05$, ** $p < 0.01$, *** $p < 0.001$); (hpvi = hours post-viral infection; ns = not significant).

As cell death mechanisms like necrosis can be further defined in specific mechanisms and to compare our findings to previously described inductions of cell death mechanisms upon infection with SCV [51] or in co-infection scenarios [27], we performed western blot analysis to be able to differentiate between necroptosis and apoptosis.

Necroptosis is an inflammatory programmed form of necrosis, which was already described in a recent publication reporting its induction by *S. aureus* SCV in single-infected human primary keratinocytes [51]. Necroptosis is induced via a receptor-interacting protein (RIP) kinase-mediated activation, resulting in the phosphorylation and oligomerization of mixed lineage kinase domain like pseudokinase (MLKL) and pore formation, leading to the release of inflammatory cytokines. To distinguish necroptosis from apoptosis induction, we monitored the induction of phosphorylated MLKL and PARP cleavage, which are indications for both cell death mechanisms [52]. We infected A549 human lung epithelial cells with *S. aureus* $3878_{SCV}$ for 24 h, followed by IV infection with PR8-M for 32 h (Figure 6C–E).

In super-infected cells, induction of pMLKL 32 hpvi was observed in comparison to uninfected, bacteria-, or virus single-infected cells, respectively. PARP cleavage was more likely to be induced in IV PR8-M-infected cells, and was slightly decreased in super-infected samples 32 hpvi (Figure 6C). However, the densitometrical analysis of three independent experiments could only confirm a trend of activated MLKL due to induced phosphorylation in super-infected cells (Figure 6D), whereas an induction of apoptosis upon IV infection could be verified (Figure 6E). Additionally, pyroptosis as another form of regulated necrotic cell death mechanism, which is activated via the induction of the inflammasome resulting in the cleavage of gasdermin D, could not be detected by cleaved gasdermin D (Supplementary Figure S5). The original blots are shown in the Supplementary Figure S6.

Thus, our results indicate a necrotic cell death induction, most likely induced by increased pro-inflammatory gene expression response after super-infection with *S. aureus* $3878_{SCV}$, followed by secondary IV infection with PR8-M and Panama.

## 4. Discussion

The first occurrence of persisting bacteria or SCVs was already described about 100 years ago [53]. Even though they are known for such a long time, not many studies were undertaken to elucidate their impact on cellular responses or their impact on additional infections with other pathogens. Our aim was to investigate the interaction of *S. aureus* SCVs with a subsequent IV infection in respect to epithelial cell responses, which built the first cellular barrier for pathogens in the lung. Here, we demonstrate that invasive *S. aureus* SCVs do have an impact on the cell intrinsic response in human lung epithelial cells, as indicated by highly secreted pro-inflammatory cytokines and chemokines (Figure 5 and Supplementary Figure S3) and, furthermore, an induction of necrotic cell death of super-infected compared to single-infected cells (Figure 6). This was somehow surprising, since the majority of SCVs are not described to significantly induce cell intrinsic responses, due to decreased secretion of virulence factors [54], a feature that would match their dormant status. In particular, not much is known about the impact of SCVs on lung tissue responses and nothing so far about their impact on a secondary IV infection. In this study, we were able to show a cytopathic effect accompanied by increased pro-inflammatory cytokine and chemokine release and necrotic cell death through colonizing *S. aureus* $3878_{SCV}$ and different IV strains, such as PR8-M and Panama.

Typically, super-infections with pathogenic *S. aureus* strains and IV led to increased pathogen loads accompanied with the induction of pro-inflammatory responses [13,35]. However, this could not be confirmed within the present study in the SCV and IV super-infection scenario. It was shown previously that super-infection with pathogenic *S. aureus* leads to the enhancement of viral titers due to the inhibition of STAT1 and STAT2 dimerization, resulting in decreased production of anti-viral factors [15]. This inhibitory effect could be excluded since mRNA expression of RIG-I, IFN$\beta$, MxA, or OAS1 in super-infected cells compared to IV-infected cells were not altered (Figure 4 and Supplementary Figure S2). Based on these results, we could exclude an effect of the anti-viral response and the involvement of an altered pathogen load. Nevertheless, we identified a clear induction of pro-inflammatory cytokines and chemokines in human lung epithelial cells. Besides the attraction of immune cells and the induction of an anti-pathogen status of the cell, pro-inflammatory cytokines induce a stress response leading to the induction of cell death mechanisms via TLRs or death receptors [55–57].

To proof the impact of two main cell death mechanisms we performed FACS analysis to monitor early apoptotic and necrotic-like cells. Correlating to the LDH assays, we could confirm an increase in necrosis during super-infection (Figure 6A). Besides, the amount of apoptotic cells was decreased in super- compared to IV-infected cells. Further specifications of cell-death mechanisms by western blot analysis revealed the tendency for an increase of phosphorylated MLKL in super-infected cells, giving the hit of probably induced necroptosis. Concomitantly, IV-induced PARP cleavage was reduced in super-infected cells compared to IV PR8-M-infected cells by trend.

Interestingly, there are two different mechanisms described, how pathogenic *S. aureus* and *S. aureus* SCVs are able to induce necroptosis [27,51]. During the critical phase of *S. aureus* infection the virulence factor *agr* is induced [8], resulting in possible secretion of different toxins, which induces necroptosis [27,33]. In SCV-infected keratinocytes, necroptosis was driven by the activation of glycolysis [51]. *S. aureus* adopts its whole metabolism to persist within the host. The metabolic changes of *S. aureus* were already described elsewhere [58]. As the utilization of the tricarboxylic acid cycle for the host cell and the persisting bacteria is decreased, the glycolysis is stronger induced to generate adenosine triphosphate (ATP). As we observed a disruption of cell monolayer and a possible induction of phosphorylated MLKL upon super-infection, we linked our findings more to necroptosis (Figures 1 and 6). Even though *S. aureus*-induced necroptosis might be independent of TLR stimulation [59], our data indicate a synergistic effect of *S. aureus* $3878_{SCV}$ and IV inducing cell death, which can be related to TLR2- and RIG-I-mediated pro-inflammatory response induction. In addition, our data show that the superinfection could be imitated with the stimuli LTA and vRNA. This underlines that the initial induction of the pro-inflammatory response and the subsequent cell death must be different from that of pathogenic bacterial strains that induce cell death much more quickly. In case of SCV, this indicates a lower virulence probably due to the decreased secretion of virulence factors. Nonetheless, dormant SCVs can work synergistically and affect the virus-induced immune response. As we performed pure ligand experiments, inhibitory effects of molecules of this pro-inflammatory cell intrinsic response is supposed to trigger cellular stress in the form of reactive oxygen species [60].

So far, the impact of *S. aureus* SCVs with subsequent IV infection had not been investigated. Interestingly, we could give first insights in this super-infection scenario and unravel one extraordinary role of a SCV patients' isolate *S. aureus* $3878_{SCV}$ with subsequent IV infection. We observed an induction of pro-inflammatory cytokines and chemokines, which underlines the severity of the coincident occurrence of *S. aureus* SCVs and IV. These data point to a cross-interaction of necrotic cell death and pro-inflammatory cell intrinsic response, as the pathogens alone can induce an inflammatory response through PAMPs and secreted cell damage-associated molecular patterns (DAMPs). Upon necrotic cell death induction, further pro-inflammatory responses are induced via DAMP receptors [50], leading to an enhancement of pro-inflammatory cytokines and chemokines seen on transcriptional and translational level.

In summary, we were able to show that persistent *S. aureus* SCV and subsequent IV infection affects cell-internal immune response by inducing the release of pro-inflammatory cytokines and chemokines, resulting in cell death induction.

**Supplementary Materials:**

Table S1: List of ct-values analyzed by GeneGlobe online software of the *RT*$^2$ *Profiler Array* plate. Table S2: (A,B) Listed gene names of the venn diagrams shown in Figure 3. Figure S1: Wildtype phenotype *S. aureus* 3878 is more virulent compared to *S. aureus* $3878_{SCV}$, but *S. aureus* $3878_{SCV}$ induces LDH release upon super-infection with PR8-M. Figure S2: Pro-inflammatory cytokines and chemokines are enhanced after *S. aureus* $3878_{SCV}$ colonization and subsequent IV Panama infection. Figure S3: Secretion of the pro-inflammatory cytokines and chemokines are enhanced after *S. aureus* $3878_{SCV}$ colonization and subsequent IV Panama infection regulated by TLR2- and

RIG-I-mediated NFκB promoter activation. Figure S4: Pathogen load and pro-inflammatory cytokines and chemokines are enhanced after super-infection with the SCV strain *S. aureus* $814_{SCV}$. Figure S5: *S. aureus* $3878_{SCV}$ colonization and subsequent influenza virus infection has no effect on the induction of pyroptosis. Figure S6: Original western blots of Figure 6C and S5A.

**Author Contributions:** Conceptualization, J.J.W., E.R.H., S.L. and C.E.; methodology, J.J.W.; software, J.J.W.; validation, J.J.W., Y.B. and E.R.H.; formal analysis, J.J.W., E.R.H., Y.B., S.L., S.N. and C.E.; investigation, J.J.W.; resources, S.L. and C.E.; data curation, S.L. and C.E.; writing—original draft preparation, J.J.W. and C.E.; writing—review and editing, J.J.W., E.R.H., S.N., Y.B., B.L., S.L. and C.E.; visualization, J.J.W.; supervision, C.E.; project administration, C.E.; funding acquisition, J.J.W., S.L. and C.E. All authors have read and agreed to the published version of the manuscript.

**Acknowledgments:** We would like to thank Karsten Becker for providing us with the bacterial isolate.

## Abbreviations

| | |
|---|---|
| DAMP | damage-associated molecular pattern |
| hpbi | hours post bacterial infection |
| hpvi | hours post-viral infection |
| I-TAC | interferon-inducible T-cell alpha chemoattractant |
| IFN | interferon |
| IL | interleukin |
| IP-10 | interferon gamma-induced protein 10 |
| IV | influenza virus |
| LDH | Lactate dehydrogenase |
| MAPK | mitogen-activated protein kinase |
| MxA | interferon-induced GTP-binding protein MxA |
| NFκB | nuclear factor kappa-light-chain-enhancer of activated B-cells |
| NLR | NOD-like receptor |
| NOD | nucleotide-binding oligomerization domain |
| OAS1 | 2′-5′-oligoadenylate synthetase 1 |
| PAMP | pathogen associated molecular pattern |
| PRR | pattern recognition receptor |
| RANTES | CC-chemokine ligand 5 |
| RELA | nuclear factor NF-kappa-B p65 subunit |
| RIG-I | retinoic acid inducible gene I |
| RLR | RIG-I like receptor |
| *S. aureus* | *Staphylococcus aureus* |
| SCVs | small colony variants |
| TLR | toll-like receptor |
| TNFR | tumor necrosis factor receptor |
| TNFα | tumor necrosis factor alpha |

## References

1. Lynch, S.V. Viruses and Microbiome Alterations. *Ann. Am. Thorac. Soc.* **2014**, *11*, S57–S60. [CrossRef] [PubMed]
2. Lee, J.; Zilm, P.S.; Kidd, S.P. Novel Research Models for Staphylococcus aureus Small Colony Variants (SCV) Development: Co-pathogenesis and Growth Rate. *Front. Microbiol.* **2020**, *11*, 1–8. [CrossRef] [PubMed]

3. van Belkum, A.; Verkaik, N.J.; de Vogel, C.P.; Boelens, H.A.; Verveer, J.; Nouwen, J.L.; Verbrugh, H.A.; Wertheim, H.F.L. Reclassification of Staphylococcus aureus Nasal Carriage Types. *J. Infect. Dis.* **2009**, *199*, 1820–1826. [CrossRef] [PubMed]
4. Jenul, C.; Horswill, A.R. Regulation of Staphylococcus aureus Virulence. In *Gram-Positive Pathogens*; ASM Press: Washington, DC, USA, 2019; Volume 6, pp. 669–686.
5. Kahl, B.C.; Becker, K.; Löffler, B. Clinical Significance and Pathogenesis of Staphylococcal Small Colony Variants in Persistent Infections. *Clin. Microbiol. Rev.* **2016**, *29*, 401–427. [CrossRef] [PubMed]
6. Tuchscherr, L.; Löffler, B. Staphylococcus aureus dynamically adapts global regulators and virulence factor expression in the course from acute to chronic infection. *Curr. Genet.* **2016**, *62*, 15–17. [CrossRef] [PubMed]
7. Tuchscherr, L.; Bischoff, M.; Lattar, S.M.; Noto Llana, M.; Pförtner, H.; Niemann, S.; Geraci, J.; Van de Vyver, H.; Fraunholz, M.J.; Cheung, A.L.; et al. Sigma Factor SigB Is Crucial to Mediate Staphylococcus aureus Adaptation during Chronic Infections. *PLoS Pathog.* **2015**, *11*, 1–26. [CrossRef] [PubMed]
8. Tuchscherr, L.; Medina, E.; Hussain, M.; Völker, W.; Heitmann, V.; Niemann, S.; Holzinger, D.; Roth, J.; Proctor, R.A.; Becker, K.; et al. Staphylococcus aureus phenotype switching: An effective bacterial strategy to escape host immune response and establish a chronic infection. *EMBO Mol. Med.* **2011**, *3*, 129–141. [CrossRef]
9. Kahl, B.C.; Belling, G.; Reichelt, R.; Herrmann, M.; Proctor, R.A.; Peters, G. Thymidine-dependent small-colony variants of Staphylococcus aureus exhibit gross morphological and ultrastructural changes consistent with impaired cell separation. *J. Clin. Microbiol.* **2003**, *41*, 410–413. [CrossRef]
10. Kriegeskorte, A.; König, S.; Sander, G.; Pirkl, A.; Mahabir, E.; Proctor, R.A.; von Eiff, C.; Peters, G.; Becker, K. Small colony variants of Staphylococcus aureus reveal distinct protein profiles. *Proteomics* **2011**, *11*, 2476–2490. [CrossRef]
11. Proctor, R.A.; Balwit, J.M.; Vesga, O. Variant subpopulations of Staphylococcus aureus as cause of persistent and recurrent infections. *Infect. Agents Dis.* **1994**, *3*, 302–312.
12. Papi, A.; Bellettato, C.M.; Braccioni, F.; Romagnoli, M.; Casolari, P.; Caramori, G.; Fabbri, L.M.; Johnston, S.L. Infections and Airway Inflammation in Chronic Obstructive Pulmonary Disease Severe Exacerbations. *Am. J. Respir. Crit. Care Med.* **2006**, *173*, 1114–1121. [CrossRef] [PubMed]
13. Iverson, A.R.; Boyd, K.L.; McAuley, J.L.; Plano, L.R.; Hart, M.E.; McCullers, J.A. Influenza Virus Primes Mice for Pneumonia From Staphylococcus aureus. *J. Infect. Dis.* **2011**, *203*, 880–888. [CrossRef] [PubMed]
14. Bakaletz, L.O. Viral–bacterial co-infections in the respiratory tract. *Curr. Opin. Microbiol.* **2017**, *35*, 30–35. [CrossRef] [PubMed]
15. Warnking, K.; Klemm, C.; Löffler, B.; Niemann, S.; van Krüchten, A.; Peters, G.; Ludwig, S.; Ehrhardt, C. Super-infection with Staphylococcus aureus inhibits influenza virus-induced type I IFN signalling through impaired STAT1-STAT2 dimerization. *Cell. Microbiol.* **2015**, *17*, 303–317. [CrossRef]
16. LeMessurier, K.S.; Tiwary, M.; Morin, N.P.; Samarasinghe, A.E. Respiratory Barrier as a Safeguard and Regulator of Defense Against Influenza A Virus and Streptococcus pneumoniae. *Front. Immunol.* **2020**, *11*, 1–15. [CrossRef]
17. Morris, D.E.; Cleary, D.W.; Clarke, S.C. Secondary Bacterial Infections Associated with Influenza Pandemics. *Front. Microbiol.* **2017**, *8*, 1–17. [CrossRef]
18. Yu, D.; Wei, L.; Zhengxiu, L.; Jian, L.; Lijia, W.; Wei, L.; Xiqiang, Y.; Xiaodong, Z.; Zhou, F.; Enmei, L. Impact of bacterial colonization on the severity, and accompanying airway inflammation, of virus-induced wheezing in children. *Clin. Microbiol. Infect.* **2010**, *16*, 1399–1404. [CrossRef]
19. Von Eiff, C.; Becker, K.; Metze, D.; Lubritz, G.; Hockmann, J.; Schwarz, T.; Peters, G. Intracellular Persistence of Staphylococcus aureus Small-Colony Variants within Keratinocytes: A Cause for Antibiotic Treatment Failure in a Patient with Darier's Disease. *Clin. Infect. Dis.* **2001**, *32*, 1643–1647. [CrossRef]
20. von Eiff, C.; Bettin, D.; Proctor, R.A.; Rolauffs, B.; Lindner, N.; Winkelmann, W.; Peters, G. Recovery of Small Colony Variants of Staphylococcus aureus Following Gentamicin Bead Placement for Osteomyelitis. *Clin. Infect. Dis.* **1997**, *25*, 1250–1251. [CrossRef]
21. Abu-Qatouseh, L.F.; Chinni, S.V.; Seggewiß, J.; Proctor, R.A.; Brosius, J.; Rozhdestvensky, T.S.; Peters, G.; von Eiff, C.; Becker, K. Identification of differentially expressed small non-protein-coding RNAs in Staphylococcus aureus displaying both the normal and the small-colony variant phenotype. *J. Mol. Med.* **2010**, *88*, 565–575. [CrossRef]

22. Flory, E.; Kunz, M.; Scheller, C.; Jassoy, C.; Stauber, R.; Rapp, U.R.; Ludwig, S. Influenza Virus-induced NF-κB-dependent Gene Expression Is Mediated by Overexpression of Viral Proteins and Involves Oxidative Radicals and Activation of IκB Kinase. *J. Biol. Chem.* **2000**, *275*, 8307–8314. [CrossRef] [PubMed]
23. Ludwig, S.; Ehrhardt, C.; Neumeier, E.R.; Kracht, M.; Rapp, U.R.; Pleschka, S. Influenza Virus-induced AP-1-dependent Gene Expression Requires Activation of the JNK Signaling Pathway. *J. Biol. Chem.* **2001**, *276*, 10990–10998. [CrossRef]
24. Mazur, I.; Wurzer, W.J.; Ehrhardt, C.; Pleschka, S.; Puthavathana, P.; Silberzahn, T.; Wolff, T.; Planz, O.; Ludwig, S. Acetylsalicylic acid (ASA) blocks influenza virus propagation via its NF-kB-inhibiting activity. *Cell. Microbiol.* **2007**, *9*, 1683–1694. [CrossRef] [PubMed]
25. Livak, K.J.; Schmittgen, T.D. Analysis of Relative Gene Expression Data Using Real-Time Quantitative PCR and the $2-\Delta\Delta CT$ Method. *Methods* **2001**, *25*, 402–408. [CrossRef] [PubMed]
26. Qiagen GeneGlobe RT2 Profiler PCR Data Analysis. Available online: https://geneglobe.qiagen.com/de/analyze/ (accessed on 16 June 2020).
27. Van Krüchten, A.; Wilden, J.J.; Niemann, S.; Peters, G.; Löffler, B.; Ludwig, S.; Ehrhardt, C. Staphylococcus aureus triggers a shift from influenza virus-induced apoptosis to necrotic cell death. *FASEB J.* **2018**, *32*, 2779–2793. [CrossRef] [PubMed]
28. Wurzer, W.J.; Planz, O.; Ehrhardt, C.; Giner, M.; Silberzahn, T.; Pleschka, S.; Ludwig, S. Caspase 3 activation is essential for effcient infuenza virus propagation. *EMBO J.* **2003**, *22*, 2717–2728. [CrossRef]
29. Wurzer, W.J.; Ehrhardt, C.; Pleschka, S.; Berberich-Siebelt, F.; Wolff, T.; Walczak, H.; Planz, O.; Ludwig, S. NF-κB-dependent Induction of Tumor Necrosis Factor-related Apoptosis-inducing Ligand (TRAIL) and Fas/FasL Is Crucial for Efficient Influenza Virus Propagation. *J. Biol. Chem.* **2004**, *279*, 30931–30937. [CrossRef]
30. Ehrhardt, C.; Wolff, T.; Ludwig, S. Activation of phosphatidylinositol 3-kinase signaling by the nonstructural NS1 protein is not conserved among type A and B influenza viruses. *J. Virol.* **2007**, *81*, 12097–12100. [CrossRef]
31. Korea, C.G.; Balsamo, G.; Pezzicoli, A.; Merakou, C.; Tavarini, S.; Bagnoli, F.; Serruto, D.; Unnikrishnan, M. Staphylococcal Esx Proteins Modulate Apoptosis and Release of Intracellular Staphylococcus aureus during Infection in Epithelial Cells. *Infect. Immun.* **2014**, *82*, 4144–4153. [CrossRef]
32. Chi, C.-Y.; Lin, C.-C.; Liao, I.-C.; Yao, Y.-C.; Shen, F.-C.; Liu, C.-C.; Lin, C.-F. Panton-Valentine Leukocidin Facilitates the Escape of Staphylococcus aureus From Human Keratinocyte Endosomes and Induces Apoptosis. *J. Infect. Dis.* **2014**, *209*, 224–235. [CrossRef]
33. Kitur, K.; Parker, D.; Nieto, P.; Ahn, D.S.; Cohen, T.S.; Chung, S.; Wachtel, S.; Bueno, S.; Prince, A. Toxin-Induced Necroptosis Is a Major Mechanism of Staphylococcus aureus Lung Damage. *PLoS Pathog.* **2015**, *11*, e1004820. [CrossRef] [PubMed]
34. McCullers, J.A. Preventing and treating secondary bacterial infections with antiviral agents. *Antivir. Ther.* **2011**, *16*, 123–135. [CrossRef] [PubMed]
35. Klemm, C.; Bruchhagen, C.; van Krüchten, A.; Niemann, S.; Löffler, B.; Peters, G.; Ludwig, S.; Ehrhardt, C. Mitogen-activated protein kinases (MAPKs) regulate IL-6 over-production during concomitant influenza virus and Staphylococcus aureus infection. *Sci. Rep.* **2017**, *7*, 42473. [CrossRef] [PubMed]
36. Tse, L.V.; Whittaker, G.R. Modification of the hemagglutinin cleavage site allows indirect activation of avian influenza virus H9N2 by bacterial staphylokinase. *Virology* **2015**, *482*, 1–8. [CrossRef] [PubMed]
37. Kriegeskorte, A.; Grubmüller, S.; Huber, C.; Kahl, B.C.; von Eiff, C.; Proctor, R.A.; Peters, G.; Eisenreich, W.; Becker, K. Staphylococcus aureus small colony variants show common metabolic features in central metabolism irrespective of the underlying auxotrophism. *Front. Cell. Infect. Microbiol.* **2014**, *4*, 1–8. [CrossRef] [PubMed]
38. Kanehisa, M. Toward understanding the origin and evolution of cellular organisms. *Protein Sci.* **2019**, *28*, 1947–1951. [CrossRef]
39. Kanehisa, M.; Sato, Y.; Furumichi, M.; Morishima, K.; Tanabe, M. New approach for understanding genome variations in KEGG. *Nucleic Acids Res.* **2019**, *47*, D590–D595. [CrossRef]
40. Kanehisa, M. KEGG: Kyoto Encyclopedia of Genes and Genomes. *Nucleic Acids Res.* **2000**, *28*, 27–30. [CrossRef]

41. Li, W.; Moltedo, B.; Moran, T.M. Type I Interferon Induction during Influenza Virus Infection Increases Susceptibility to Secondary Streptococcus pneumoniae Infection by Negative Regulation of T Cells. *J. Virol.* **2012**, *86*, 12304–12312. [CrossRef]
42. Liu, T.; Zhang, L.; Joo, D.; Sun, S.-C. NF-κB signaling in inflammation. *Signal Transduct. Target. Ther.* **2017**, *2*, 17023. [CrossRef]
43. Claro, T.; Widaa, A.; McDonnell, C.; Foster, T.J.; O'Brien, F.J.; Kerrigan, S.W. Staphylococcus aureus protein A binding to osteoblast tumour necrosis factor receptor 1 results in activation of nuclear factor kappa B and release of interleukin-6 in bone infection. *Microbiology* **2013**, *159*, 147–154. [CrossRef] [PubMed]
44. Chiaretti, A.; Pulitanò, S.; Barone, G.; Ferrara, P.; Romano, V.; Capozzi, D.; Riccardi, R. IL-1 β and IL-6 Upregulation in Children with H1N1 Influenza Virus Infection. *Mediat. Inflamm.* **2013**, *2013*, 1–8. [CrossRef] [PubMed]
45. Chan, M.; Cheung, C.; Chui, W.; Tsao, S.; Nicholls, J.; Chan, Y.; Chan, R.; Long, H.; Poon, L.; Guan, Y.; et al. Proinflammatory cytokine responses induced by influenza A (H5N1) viruses in primary human alveolar and bronchial epithelial cells. *Respir. Res.* **2005**, *6*, 135. [CrossRef] [PubMed]
46. Yoneyama, M.; Fujita, T. RIG-I family RNA helicases: Cytoplasmic sensor for antiviral innate immunity. *Cytokine Growth Factor Rev.* **2007**, *18*, 545–551. [CrossRef]
47. Fournier, B.; Philpott, D.J. Recognition of Staphylococcus aureus by the Innate Immune System. *Clin. Microbiol. Rev.* **2005**, *18*, 521–540. [CrossRef]
48. Holler, N.; Zaru, R.; Micheau, O.; Thome, M.; Attinger, A.; Valitutti, S.; Bodmer, J.-L.; Schneider, P.; Seed, B.; Tschopp, J. Fas triggers an alternative, caspase-8–independent cell death pathway using the kinase RIP as effector molecule. *Nat. Immunol.* **2000**, *1*, 489–495. [CrossRef]
49. Berghe, T.V.; Linkermann, A.; Jouan-Lanhouet, S.; Walczak, H.; Vandenabeele, P. Regulated necrosis: The expanding network of non-apoptotic cell death pathways. *Nat. Rev. Mol. Cell Biol.* **2014**, *15*, 135–147. [CrossRef]
50. Kearney, C.J.; Martin, S.J. An Inflammatory Perspective on Necroptosis. *Mol. Cell* **2017**, *65*, 965–973. [CrossRef]
51. Wong Fok Lung, T.; Monk, I.R.; Acker, K.P.; Mu, A.; Wang, N.; Riquelme, S.A.; Pires, S.; Noguera, L.P.; Dach, F.; Gabryszewski, S.J.; et al. Staphylococcus aureus small colony variants impair host immunity by activating host cell glycolysis and inducing necroptosis. *Nat. Microbiol.* **2020**, *5*, 141–153. [CrossRef]
52. Tewari, M.; Quan, L.T.; O'Rourke, K.; Desnoyers, S.; Zeng, Z.; Beidler, D.R.; Poirier, G.G.; Salvesen, G.S.; Dixit, V.M. Yama/CPP32β, a mammalian homolog of CED-3, is a CrmA-inhibitable protease that cleaves the death substrate poly(ADP-ribose) polymerase. *Cell* **1995**, *81*, 801–809. [CrossRef]
53. Jacobsen, K.A. Mitteilungen über einen variablen Typhusstamm (Bacterium typhi mutabile), sowie über eine eigentümliche hemmende Wirkung des gewöhnlichen Agar, verursacht durch Autoklavierung. *Zentralbl. Bakteriol.* **1910**, *56*, 208–2016.
54. Tuchscherr, L.; Heitmann, V.; Hussain, M.; Viemann, D.; Roth, J.; von Eiff, C.; Peters, G.; Becker, K.; Löffler, B. Staphylococcus aureus Small-Colony Variants Are Adapted Phenotypes for Intracellular Persistence. *J. Infect. Dis.* **2010**, *202*, 1031–1040. [CrossRef] [PubMed]
55. Betáková, T.; Kostrábová, A.; Lachová, V.; Turianová, L. Cytokines Induced During Influenza Virus Infection. *Curr. Pharm. Des.* **2017**, *23*, 2616–2622. [CrossRef] [PubMed]
56. Demine, S.; Schiavo, A.A.; Marín-Cañas, S.; Marchetti, P.; Cnop, M.; Eizirik, D.L. Pro-inflammatory cytokines induce cell death, inflammatory responses, and endoplasmic reticulum stress in human iPSC-derived beta cells. *Stem Cell Res. Ther.* **2020**, *11*, 7. [CrossRef] [PubMed]
57. Vanlangenakker, N.; Bertrand, M.J.M.; Bogaert, P.; Vandenabeele, P.; Vanden Berghe, T. TNF-induced necroptosis in L929 cells is tightly regulated by multiple TNFR1 complex I and II members. *Cell Death Dis.* **2011**, *2*, e230. [CrossRef]
58. Riquelme, S.A.; Wong, T.F.L.; Prince, A. Pulmonary Pathogens Adapt to Immune Signaling Metabolites in the Airway. *Front. Immunol.* **2020**, *11*, 1–14. [CrossRef]
59. Surewaard, B.G.J.; de Haas, C.J.C.; Vervoort, F.; Rigby, K.M.; DeLeo, F.R.; Otto, M.; van Strijp, J.A.G.;

Nijland, R. Staphylococcal alpha-phenol soluble modulins contribute to neutrophil lysis after phagocytosis. *Cell. Microbiol.* **2013**, *15*, 1427–1437. [CrossRef]

60. Yang, D.; Elner, S.G.; Bian, Z.-M.; Till, G.O.; Petty, H.R.; Elner, V.M. Pro-inflammatory cytokines increase reactive oxygen species through mitochondria and NADPH oxidase in cultured RPE cells. *Exp. Eye Res.* **2007**, *85*, 462–472. [CrossRef]

6

# The Role of ST2 Receptor in the Regulation of *Brucella abortus* Oral Infection

**Raiany Santos [1], Priscila C. Campos [2], Marcella Rungue [2], Victor Rocha [2], David Santos [2], Viviani Mendes [2], Fabio V. Marinho [2], Flaviano Martins [3], Mayra F. Ricci [4], Diego C. dos Reis [4], Geovanni D. Cassali [4], José Carlos Alves-Filho [5], Angelica T. Vieira [2,†] and Sergio C. Oliveira [2,*,†]**

1. Department of Genetics, Institute of Biological Sciences, Federal University of Minas Gerais—Belo Horizonte, Minas Gerais 31270-901, Brazil; raianyas@yahoo.com.br
2. Department of Biochemistry and Immunology, Institute of Biological Sciences, Federal University of Minas Gerais—Belo Horizonte, Minas Gerais 31270-901, Brazil; pccampos78@gmail.com (P.C.C.); marcellarungue@hotmail.com (M.R.); victormelobio@gmail.com (V.R.); davidmartinsst@gmail.com (D.S.); mendesviviani.a@gmail.com (V.M.); fabiovitarelli@yahoo.com.br (F.V.M.); angelicathomaz@gmail.com (A.T.V.)
3. Department of Microbiology, Institute of Biological Sciences, Federal University of Minas Gerais—Belo Horizonte, Minas Gerais 31270-901, Brazil; flaviano@icb.ufmg.br
4. Department of General Pathology, Institute of Biological Sciences, Federal University of Minas Gerais—Belo Horizonte, Minas Gerais 31270-901, Brazil; riccimayra@gmail.com (M.F.R.); diegobiomed.reis@gmail.com (D.C.d.R.); geovanni.cassali@gmail.com (G.D.C.)
5. Department of Pharmacology, Ribeirao Preto Medical School, University of Sao Paulo, Ribeirao Preto 14049-900, Brazil; jcafilho@usp.br

* Correspondence: scozeus@icb.ufmg.br

† These authors contribute to this paper equally.

**Abstract:** The ST2 receptor plays an important role in the gut such as permeability regulation, epithelium regeneration, and promoting intestinal immune modulation. Here, we studied the role of ST2 receptor in a murine model of oral infection with *Brucella abortus*, its influence on gut homeostasis and control of bacterial replication. Balb/c (wild-type, WT) and ST2 deficient mice ($ST2^{-/-}$) were infected by oral gavage and the results were obtained at 3 and 14 days post infection (dpi). Our results suggest that $ST2^{-/-}$ are more resistant to *B. abortus* infection, as a lower bacterial colony-forming unit (CFU) was detected in the livers and spleens of knockout mice, when compared to WT. Additionally, we observed an increase in intestinal permeability in WT-infected mice, compared to $ST2^{-/-}$ animals. Breakage of the intestinal epithelial barrier and bacterial dissemination might be associated with the presence of the ST2 receptor; since, in the knockout mice no change in intestinal permeability was observed after infection. Together with enhanced resistance to infection, $ST2^{-/-}$ produced greater levels of IFN-$\gamma$ and TNF-$\alpha$ in the small intestine, compared to WT mice. Nevertheless, in the systemic model of infection ST2 plays no role in controlling *Brucella* replication in vivo. Our results suggest that the ST2 receptor is involved in the invasion process of *B. abortus* by the mucosa in the oral infection model.

**Keywords:** ST2 receptor; *Brucella abortus*; oral infection

---

## 1. Introduction

Brucellosis is a worldwide zoonotic disease caused by facultative intracellular pathogen of the genus *Brucella* [1]. Bacteria of the genus *Brucella* infect a wide variety of land and aquatic mammals, including pigs, cattle, goats, sheep, dogs, dolphins, whales, seals, and desert wooden mice. Traditionally, the genus *Brucella* consisted of six recognized species, grouped according to their primary host preferences, i.e.,

*B. abortus*, bovine; *B. melitensis*, sheep and goats; *B. suis*, pigs; *B. ovis*, sheep; *B. kennels*, dogs; and *B. neotomae*, desert wood mice. Recent new species were isolated from humans (*B. inopinata*), aquatic mammals (*B. pinnipedialis* and *B. ceti*), and from a common rat (*B. microti*), raising the current number to 10 species of the genus [2]. Human brucellosis can be mainly caused by *Brucellaabortus* and *Brucella melitensis*, leading not only to cases of morbidity but also severe economic losses caused mainly by abortions and infertility in infected animals [3]. The natural infection by *Brucella* occurs mainly by the oral and nasal routes through consumption of raw milk and unpasteurized dairy products from infected animals, inhalation of aerosols containing the pathogen, contact with infected animals and their secretions, and by the habit of cattle to lick and smell newborn animals or even aborted fetuses. In addition, there is laboratory and occupational contamination, affecting researchers, farmers, slaughterhouse workers, butchers, and veterinary doctors (many cases of accidental self-inoculation of the vaccine against animal brucellosis), and there are forms (although very unlikely) of human transmission such as contamination of plants by feces and urine from infected animals, and breastfeeding [4–7].

Brucellosis is a systemic disease in which any organ or tissue of the body might be involved. Affected individuals present nonspecific symptoms shared with several other diseases, which cause human brucellosis to present underestimated data of epidemiological distribution [8]. In humans, the main symptoms of the acute phase of the disease are undulating fever, headaches, fatigue, myalgia, and weight loss. In the chronic phase of the disease endocarditis, arthritis, osteomyelitis, and neurological complications can be observed [9]. In animals, brucellosis is a chronic infection that persists throughout life. In females, *Brucella* causes tropism through the bovine placental hormone, erythritol, leading to lesions in the uterine glands, while in males, the bacterium causes tropism through male hormones like testosterone, addressing the testicles. Thus, *Brucella* infection primarily affects the reproductive organs causing abortion and infertility [10].

*Brucella* spp. can resist death by neutrophils, and replicate within macrophages and dendritic cells, thus, maintaining a long lasting interaction with the host cells [11]. Therefore, innate immunity has developed important mechanisms for detecting and eliminating these bacteria. Toll-like receptors (TLRs) have already proven to be important in the control of *Brucella abortus* infection. The recognition of *B. abortus* molecules by TLR2 (external membrane proteins, Omp16 and Omp19), TLR4 (*Brucella* LPS and *Brucella* lumazine synthase), and TLR9 (*Brucella* DNA), activates intracellular signaling via MyD88, resulting in the activation of NF-κB, MAP kinases, and expression of pro-inflammatory cytokines [12–16]. TLR2 does not participate in the in vivo control of infection, contributing only to the production of pro-inflammatory cytokines [12,14]. However, TLR9 has played a prominent role in relation to in vivo and in vitro control of *B. abortus* infection [17]. In addition to the receptors mentioned above, *B. abortus* leads to activation of NLRP3 (through reactive mitochondrial oxygen species induced by bacteria) and AIM2 (recognition of bacterial DNA) inflammasomes, leading to activation of innate immunity and infection control [18]. The STING protein was also determined as an important adapter molecule required for resistance against this bacterium [19,20].

In the context of intestinal immunity, it is important to highlight the importance of these receptors in this microenvironment, since several of these innate immunity receptors such as the TLRs, NRLs, G protein-coupled receptors (GPCRs), and STING are also expressed in the intestinal mucosa, having an important function in the maintenance of host commensal microbiota and intestinal homeostasis [21–23]. Among these innate immunity receptors, the ST2 receptor and its ligand, cytokine IL-33 has been widely studied since its discovery [24]. The ST2 receptor of the IL-1 family, also called IL1rl1, tumorigenicity suppressor 2, growth stimulation expressed in gene 2 and serum stimulation 2, was classified as a receptor for IL-33 in 2005 [25,26].There are four isoforms encoded by the ST2 gene. The two most prominent isoforms include the ST2L transmembrane, which acts as a membrane receptor, responsible for binding the IL-33 and activating the signaling cascades to improve the functions of the cells that express this receptor, and the sST2, presented in a soluble form, which acts by sequestering the free IL-33, preventing its signaling. They are the consequence of a double system of promoters (sST2 proximal promoter and ST2L distal promoter) that results in the differential expression of mRNA.

ST2L, like other IL-1 receptors, consists of an extracellular domain, transmembrane domain and cytoplasmic domain (Toll/interleukin-1 receptor (TIR)), while sST2 does not have the transmembrane and cytoplasmic domains and therefore exists as a soluble protein. In addition, alternative splicing results in the formation of ST2V and ST2LV. ST2V shares the same extracellular and transmembrane domain as ST2L, but is remarkable for its unique hydrophobic tail and is particularly enriched in the gastrointestinal tract. Finally, ST2LV notably does not have the ST2L transmembrane domain, but maintains the intracellular domain [27,28].The ST2 receptor is expressed in a wide variety of immune cells, such as conventional T cells, particularly regulatory T cells (T regs) [29], innate type 2 lymphoid cells (ILC2) [30], polarized macrophages M2 [31], eosinophils [32], basophils [33], neutrophils [33], NK cells [34], iNKT cells [34], and several other immune cells and their soluble isoform.sST2 can be produced spontaneously by the small intestine.

As an alarmine, IL-33 is one of the first molecules that "sounds the alarm" to indicate that there has been a violation of the primary defenses of the intestinal epithelium against pathogens and other threats [35]. IL-33 is produced by a variety of stromal cells and organ parenchyma, such as smooth muscle cells, fibroblasts, myofibroblasts, endothelial cells, glia cells, osteoblasts, adipocytes, and by different cells of the immune system, such as macrophages, dendritic cells, and mast cells [36]. IL-33 acts on several cell types, including cells of hematopoietic origin and non-hematopoietic cells. The secretion of IL-33 has been described in monocyte lineage (THP-1 cells), in response to different stimuli—bacterial infection, lipopolysaccharide (LPS) with aluminum adjuvant, and isolated LPS. [29,32]. In order to maintain the integrity of this mucosal barrier, the intestinal epithelium undergoes rapid and continuous self-renewal to replace the damaged cells. Activation of the IL-33/ST2 pathway in epithelial progenitor cells leads to inhibition of Notch signaling and results in differentiation of stem cells towards a line of secretory intestinal cells [37], resulting in the production of mucin, an important barrier mechanism of intestinal immunity, decreasing the interaction of the intestinal epithelium and pathogenic bacteria [38]. Moreover, the activation of this axis is important to recruit and activate innate immune cells, inducing Th1 or Th2 responses, according to the required immune response [39]. Although brucellosis is a worldwide zoonosis, the mechanisms involved during the course and establishment of the natural oral infection by *Brucella abortus* are still poorly studied. With regard to the process of invasion of *Brucella* through mucosal barriers, there are few studies on the mechanisms involved in the ability of this pathogen to interact with the epithelial cells of the gastrointestinal (GI) tract with the host microbiota, and also with the subsequent immune and homeostatic response in the gastrointestinal tract. The intraperitoneal infection pathway is the most commonly used in studies using the murine model. This route favors the immediate systemic dissemination of *Brucella* and its proliferation in lymphoid tissues, especially in the spleen. However, considering that the oral route is the main route of natural infection in humans and animals, there is a need to understand the mechanisms of the establishment of oral infection, so new therapeutic strategies can be developed in order to control this disease. Since, the IL-33/ST2 axis is positioned to interact with the main components of the intestine, which include epithelial cells in response to cell damage and a microbiome composed of commensal bacteria and immune mucosal cells [24], we investigated the role of the ST2 receptor in the immune response against *Brucella abortus* oral infection.

## 2. Results

### 2.1. The Absence of the ST2 Receptor Confers Partial Resistance to Oral Infection

Considering that one of the main routes of the *Brucella* infection is through oral surfaces, we assessed the susceptibility of wild-type (WT) mice and animals deficient for the ST2 receptor ($ST2^{-/-}$) to oral infection with *Brucella abortus*, by determining the number of colony forming units in livers and spleens, 3to 14 days post-infection. We observed higher CFUs of *Brucella* in livers (Figure 1A) and spleens (Figure 1B) of WT mice compared to $ST2^{-/-}$ animals, 3 days after oral infection. Regarding the time of 14 days post-infection, we also observed reduced numbers of bacterial CFUs in livers of $ST2^{-/-}$

mice compared to WT, but not in the spleens of these animals. These findings suggest an enhanced resistance to *Brucella* infection in ST2 knockout mice compared to WT.

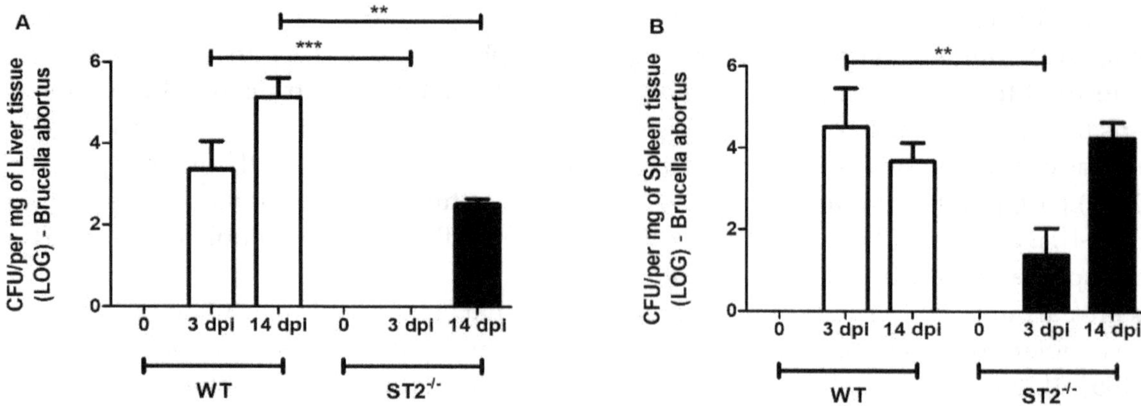


**Figure 1.** The absence of the ST2 receptor confers partial resistance to oral infection. Wild-type (WT) mice and ST2-deficient mice were orally infected by $1 \times 10^9$ colony-forming unit (CFU) of *Brucella abortus* and were sacrificed after 3 and 14 days of infection. The livers (**A**) and spleens (**B**) of the mice were collected and processed for evaluation of the number of viable bacteria through CFU counts. Results expressed as mean ± standard deviation ($n$ = 5–7). The data are representative of 3 independent experiments. ** $p < 0.01$, *** $p < 0.001$.

### 2.2. *Absence of ST2 Resulted in Change of Intestinal Architecture*

Intestinal epithelial cells produce antimicrobial effectors that play a central role in shaping the gut microbial community and protecting mucosal tissues from colonization and invasion of commensal microorganisms. To investigate the potential role of ST2 in intestine homeostasis, we analyzed histology sections of small intestine from WT and ST2$^{-/-}$ mice. First, we observed in H&E-stained sections that the villi, crypt, and mucosa thickness of small intestine in naive ST2$^{-/-}$ mice were shorter than in WT animals (Figure 2A–C), regardless of the infection. After oral infection by *Brucella abortus*, we did not observe a major alteration in the gut architecture within each mouse group. Representative photomicrographies of hematoxylin-and-eosin-stained duodenum sections from WT (Figure 2D) and ST2$^{-/-}$ (Figure 2E) are shown. Together, these data indicate that an intact ST2 signaling is important to maintain gut mucosa integrity.

### 2.3. *ST2 Receptor is Important in the Maintenance of the Intestinal Epithelial Barrier*

The role of the ST2 receptor in maintaining the integrity of the intestinal epithelial barrier following *Brucella infection* was evaluated by the FITC-labeled dextran flow method (Figure 3A). We observed that in the WT animals, *Brucella* infection led to an increased permeability of the epithelial barrier (Figure 3A) (observed by increase of FITC-dextran in the serum of animals). In contrast, ST2$^{-/-}$ mice intestinal permeability was not altered after infection (3 days). We also evaluated the regulation of amphiregulin (AREG) and mucin molecule 2 (MUC2) expression in WT and ST2$^{-/-}$ mice, after 3 days of oral infection with *Brucella*. Amphiregulin and MUC2 are two important components to protect the intestinal epithelium. Regarding the expression of amphiregulin, which is critical for intestinal epithelial regeneration after injury, *B. abortus* infection increased the expression of *AREG* in both WT and ST2$^{-/-}$ mice (Figure 3B). Additionally, we determined that *MUC2* expression following *B. abortus* infection requires ST2 (Figure 3C), suggesting the participation of ST2 in the transcriptional regulation of this molecule. Tight junctions (TJs) play an important role in intestinal function. TJs in intestinal epithelial cells are composed of different junctional molecules, such as claudins, zonula occludens (ZO-1, -2, and -3), among others. Therefore, we determined the role of ST2 in *ZO-1, -2,* and *-3* and *claudin-1* expression in intestinal tissue. Herein, we showed that animals lacking ST2 had reduced expression levels of *ZO-1* and to a less extent, that of *ZO-2* and *-3*, when compared to WT mice

(Figure 3D,E,F). Regarding *claudin-1* mRNA transcripts, the levels of this TJ remained similar between both mouse groups (Figure 3G).

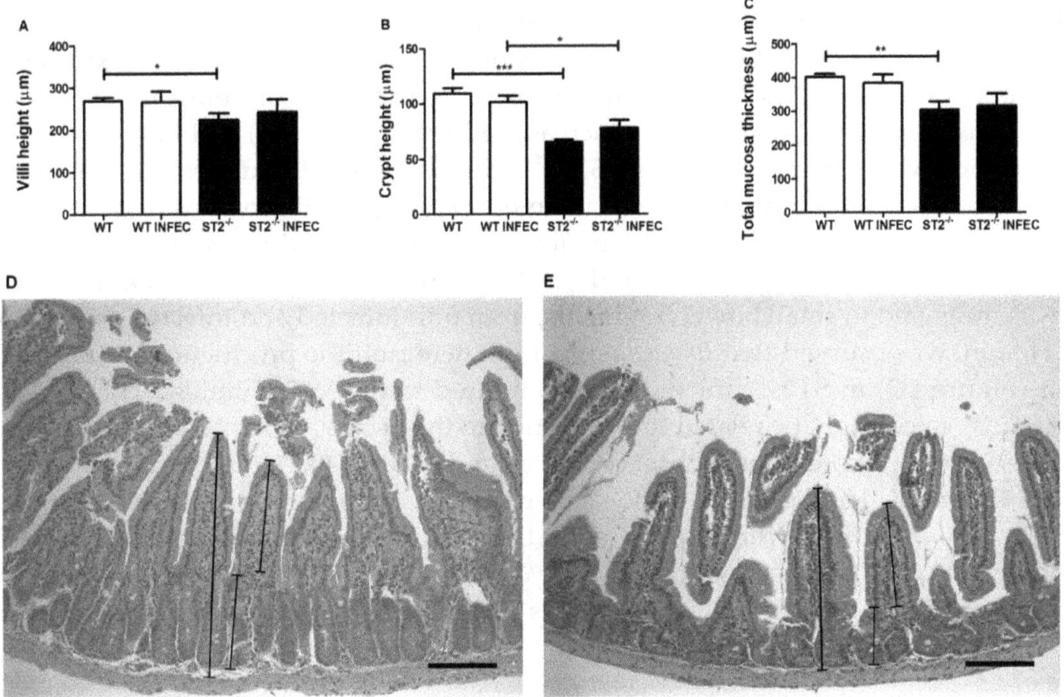


**Figure 2.** Alterations in mucosa structure in WT and ST2$^{-/-}$ mice during *Brucella* infection. Duodenum of wild-type (WT) and ST2$^{-/-}$ uninfected and infected mice were collected for analysis of (**A**) villi height, (**B**) crypt height and (**C**) total mucosa thickness. Representative photomicrographies of hematoxylin and eosin–stained duodenum sections from WT (**D**) and ST2$^{-/-}$ (**E**) mice evidencing total mucosa thickness, crypt and villi height. Bars represent 100 μm. * $p < 0.05$; ** $p < 0.01$; *** $p < 0.001$.

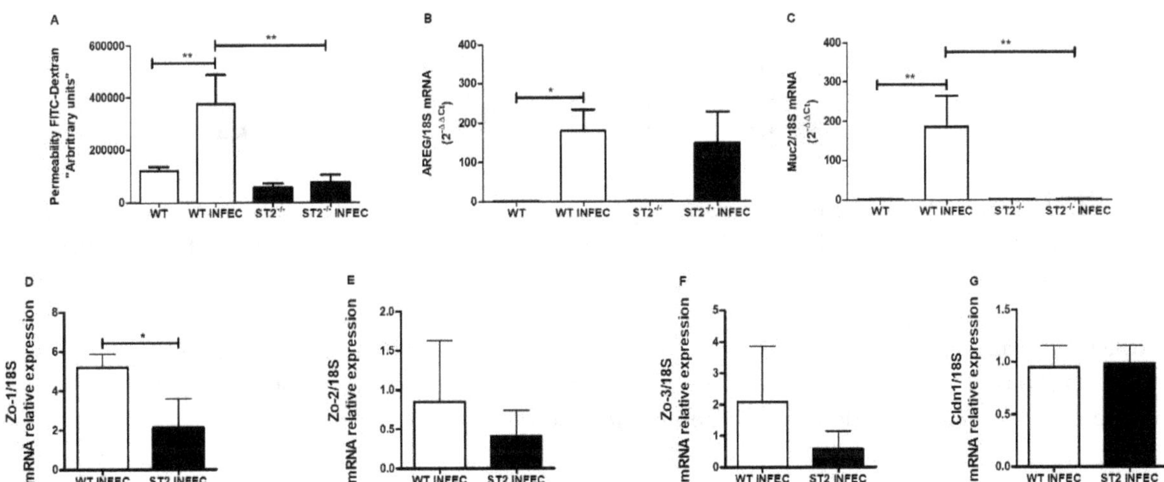


**Figure 3.** The ST2 receptor is important in the maintenance of the intestinal epithelial barrier and in the transcriptional regulation of Muc2 and ZO-1. WT mice and ST2-deficient mice were orally infected with $1 \times 10^9$ CFU of *B. abortus* and after 3 days of infection intestinal permeability was evaluated (**A**). Small bowel samples were also collected for transcriptional analysis of *AREG* (**B**), *Muc2* (**C**), *ZO-1* (**D**), *ZO-2* (**E**), *ZO-3* (**F**), and *claudin-1* (**G**) genes, using qPCR. Results expressed as mean ± standard deviation ($n$ = 5–7). The data are representative of two experiments. * $p < 0.05$; ** $p < 0.01$.

### 2.4. Lack of ST2 Receptor Modulates the Recruitment of Neutrophils and Eosinophils and Increases the Production of IFN-$\gamma$ and TNF-$\alpha$ in Small Intestine after Brucella abortus Infection

In order to investigate whether the inflammatory response could be involved in the resistance phenotype observed in ST2$^{-/-}$ after *B. abortus* infection, we determined myeloperoxidase (MPO) and eosinophilic peroxidase (EPO) activity as an indirect measurement of neutrophils and eosinophils influx. After *Brucella* infection, there was an increase in MPO (Figure 4A) and EPO (Figure 4B) activity in WT mice, which was not observed in the ST2$^{-/-}$-infected animals, suggesting that the absence of the ST2 receptor somehow modulated the recruitment of neutrophils and eosinophils after infection. Additionally, we also determined the participation of ST2 in the production of cytokines involved in the intestinal immune response, such as IFN-$\gamma$, TNF-$\alpha$, IL-10, IL-1$\beta$, and IL-33. The level of these cytokines was measured in small bowel fragments, from non-infected and infected mice after 3 days of infection. Herein, we observed that *Brucella* infection increased the production of IFN-$\gamma$ (Figure 4C) and TNF-$\alpha$ (Figure 4D) in ST2$^{-/-}$ mice, when compared to the WT animals. This Th1-like profile detected in ST2$^{-/-}$ mice might be related to a reduction in the bacterial load observed in the spleens and livers of these animals, as previously observed by us and others [40,41]. Regarding the production of IL-10 (Figure 4E) and IL-1$\beta$ (Figure 4F), there is no difference in the levels of these cytokines produced between ST2$^{-/-}$-infected mice when compared to the WT-infected animals. As for IL-33 (Figure 4G), an ST2-binding cytokine, we observed that in ST2$^{-/-}$ mice the production of this cytokine was already naturally decreased and after infection there was no change in this profile when compared to the WT mice.

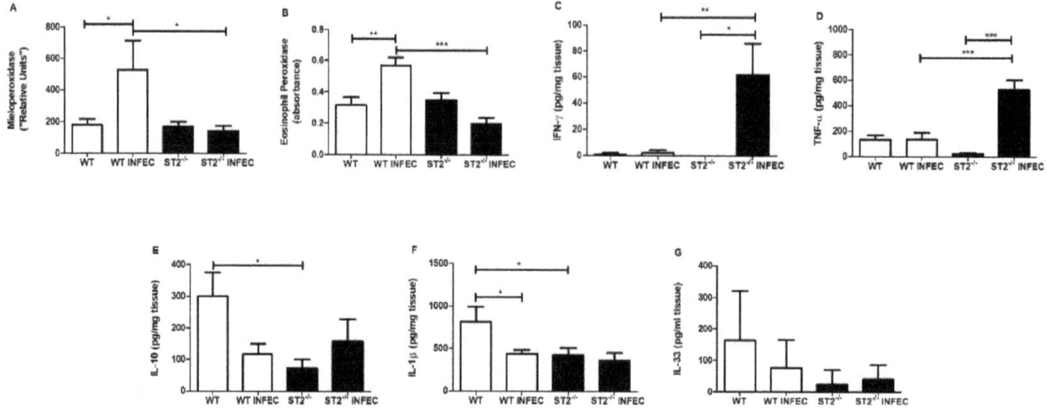

**Figure 4.** ST2 receptor deficiency modulates the recruitment of neutrophils and eosinophils and increases the production of IFN-$\gamma$ and TNF-$\alpha$ after *Brucella abortus* infection. WT and ST2$^{-/-}$ mice were infected orally with $1 \times 10^9$ CFU of *B. abortus* and after 3 days of infection, small intestine samples were collected for processing and evaluation of myeloperoxidase (**A**) and eosinophil peroxidase (**B**). Tissue samples were also assessed for cytokine production, such as IFN-$\gamma$ (**C**), TNF-$\alpha$ (**D**), IL-10 (**E**), IL-1$\beta$ (**F**), and IL-33 (**G**) by ELISA. Results are expressed as mean ± standard deviation ($n$ = 5–7). The data are representative of 3 independent experiments. * $p < 0.05$; ** $p < 0.01$; *** $p < 0.001$.

### 2.5. ST2 Receptor Does Not Play a Role in Systemic Infection Caused by Brucella abortus

The resistance or susceptibility phenotype to systemic infection by *Brucella abortus* was evaluated by determining the number of CFU in the livers and spleens of WT versus ST2$^{-/-}$mice, after 3 and 14 days of intraperitoneal (i.p.) infection. We observed that the bacterial load was similar in livers (Figure 5A) and spleens (Figure 5B) of WT and ST2$^{-/-}$ mice after infection. These findings suggest that lack of ST2 plays no role in *Brucella* control in vivo, after intraperitoneal infection.

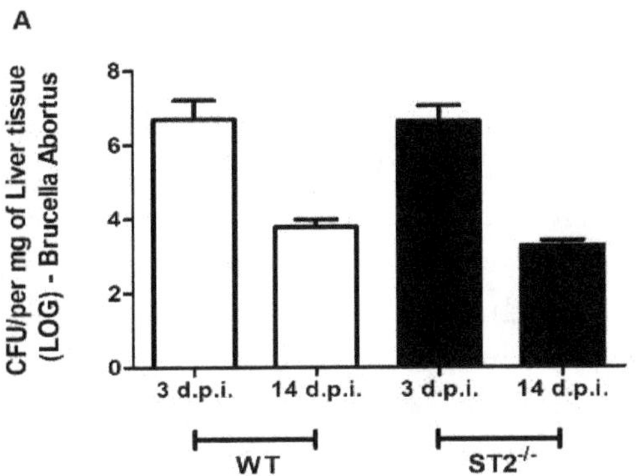

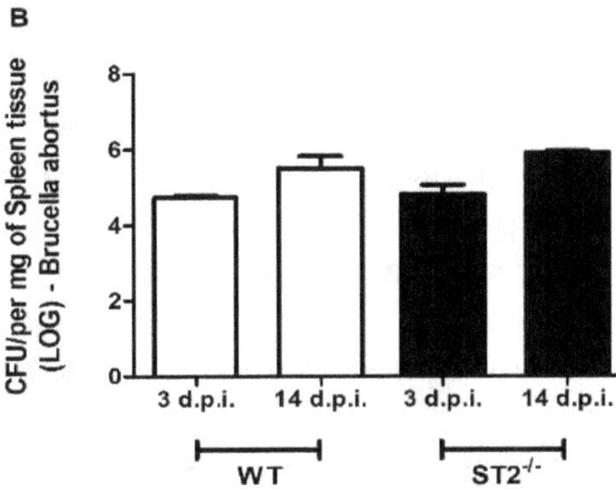


**Figure 5.** Lack of ST2 receptor does not influence systemic infection induced by *Brucella abortus*. WT mice and ST2-deficient mice were infected intraperitoneally with $1 \times 10^6$ CFU of *B. abortus* and sacrificed after 3 and 14 days of infection. The livers (**A**) and spleens (**B**) of these mice were collected and processed for evaluation of the number of viable bacteria through CFU count. Results expressed as mean ± standard deviation ($n$ = 5–7). The data are representative of 3 independent experiments.

### *2.6. The Absence of the ST2 Receptor Does Not alter the Production of Nitric Oxide by Macrophages*

To evaluate the nitric oxide (NO) production in WT and ST2$^{-/-}$ macrophages, and to correlate it with the potential microbicide activity, nitrite, a stable metabolite of NO, was measured using Griess reagent on macrophage supernatants. We observed that the production of NO in the macrophages of both mouse strains when stimulated with *Brucella* or LPS is similar, in the presence or absence of IFN-γ (Figure 6). Therefore, our findings suggest that ST2 deficiency does not influence the ability of *Brucella* infected macrophages to produce NO.

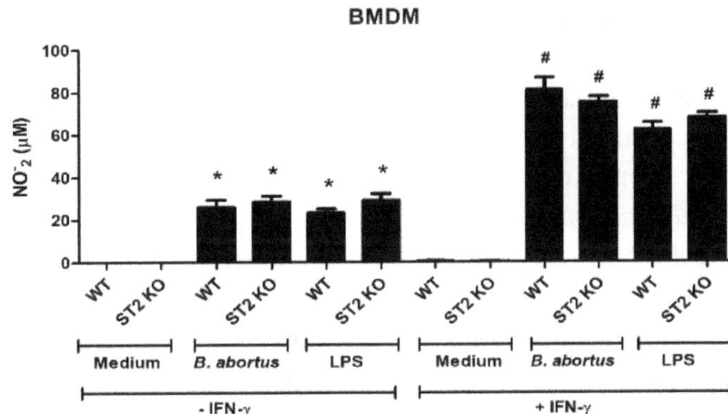


**Figure 6.** The absence of the ST2 receptor did not alter the production of nitric oxide by macrophages. Macrophage was derived from WT and ST2-deficient mice bone marrow and stimulation with *Brucella abortus* or lipopolysaccharide (LPS) was performed in the presence or absence of IFN-γ. The supernatant was collected to perform the Griess assay, as already described. * $p < 0.001$ when compared to the medium. # $p < 0.001$ when compared to the cells with no IFN-γ. Results expressed as mean ± standard deviation ($n$ = 5–7). The data are representative of two independent experiments.

## 3. Discussion

Infections caused by the bacteria of the genus *Brucella* were mainly transmitted orally to human and animals. *Brucella* has a rapid capacity for infectivity in the oral infection model (by gavage or inoculation in the oral cavity), where after one hour of infection, bacteria were already found in the lumen and in the epithelium of the duodenum [42]. Few virulence factors of *Brucella* that are important for the establishment of infection through the oral route have been described, such as urease [43], which confers resistance to gastric acidity, cholylglycine hydrolase (CGH) [44] which induces resistance to bile salts, and the *Brucella* protease inhibitor Omp19 [42] which induces resistance to the action of proteases.

When gastrointestinal tract defense cells fail to capture microorganisms, they are drained mainly through the portal vein into the liver [45]. Previous studies have shown that in bacterial infections, higher concentrations of LPS are detected in the portal vein when compared to other hepatic or peripheral veins and, interestingly, bacteria can be cultivated even from healthy liver explants [46,47]. The phenotype of resistance exhibited by ST2-receptor deficient mice during oral infection was lost when intraperitoneal infection was performed. Considering that the portal vein might be the main route for systemic dissemination of *Brucella*, we first speculated that the liver from ST2-deficient mice might be mounting its own immune response and consequently, decreasing the number of viable bacteria and their ability to spread systemically. However, when we measured IFN-$\gamma$ and TNF-$\alpha$ production by liver cells, ST2 knockout and WT mice produced similar levels of these cytokines (Figure S1 found in the Supplementary Materials). Therefore, we suggest that other mechanisms might be involved in reduced bacterial counts observed in ST2$^{-/-}$ livers.

The gut-associated lymphoid tissue (GALT), such as the Peyer's patches (PPs) along with the intestinal mucosal epithelium, act as a sentinel for recognition and initiation of immune responses against pathogenic bacteria [48]. The process of invasion of *Brucella* into the gastrointestinal tract occurs through its ability to translocate via M cells, which occurs by interaction with the prionic protein PrP$^c$ that are highly expressed on the apical surface of these cells [49]; however, this process does not lead to the rupture of the cell–cell junctions [50]. Another mechanism related to the invasion process is through the intestinal epithelial cells [51], but this mechanism has not yet been fully clarified. Tight junctions (TJs) play an important role in intestinal function. TJs in intestinal epithelial cells are composed of different junctional molecules, such as claudins, zonula occludens (ZO-1, -2, and -3), and occluding, among others. In this study, we determined the role of ST2 in *ZO-1*, *-2*, and *-3* and *claudin-1* expression in intestinal tissue. Herein, we showed that animals lacking ST2 had reduced expression levels of *ZO-1* and to a less extent that of *ZO-2* and *ZO-3*, when compared to WT mice. Regarding *claudin-1* mRNA transcripts, the levels of this TJ remained similar between both mouse groups. The reduced expression of zonula occludens (ZO) molecules might not have a direct relationship to intestinal permeability in this model since ST2$^{-/-}$ mice had reduced intestinal permeability, compared to WT animals, as measured by the FITC-dextran method. Rather, diminished expression of these tight junction gene products might correlate with enhanced IFN-$\gamma$ production observed in ST2$^{-/-}$ animals, as recently demonstrated in the *Salmonella enteritis* infection model [52]. Breaking the epithelial barrier after oral infection resulted in increased intestinal permeability observed in WT mice and could be one important mechanism that facilitates the entry and spread of this pathogen. Studies using a model of ex vivo infection in the ileal bowel loop showed that the migration of *Brucella* through the intestinal epithelium occurs via endocytosis by the follicle-associated epithelium (FAE) in Peyer's patches or by its uptake by the penetrating dendritic cells of the FAE [49,53]. Additionally, Rosseti and collaborators (2013) [54] observed through microarray analysis that two pathways related to the intestinal epithelial barrier were repressed during the initial phase of *Brucella* infection, suggesting the subversion of the barrier function and facilitating transepithelial migration. Thus, a variety of pathogens use molecules involved in cell adhesion and invasion, such as the *Helicobacter pylori*, whose type IV secretion system injects one of its effectors (CagA) into the host cell, modifying several processes and culminating in the rupture of the epithelial barrier and invasion of the bacteria [55]. Although, we did not explore the infection of *Brucella abortus* in intestinal epithelial cells in vitro, the breaking of the epithelial barrier in vivo might be associated with the presence of the ST2 receptor, since no increase in intestinal permeability was observed in ST2 knockout mice after infection, when compared to WT animals.

Another mechanism related to the process of maintaining the epithelial barrier, involves amphiregulin (AREG), which plays a role in intestinal epithelial regeneration after injury [56] and in cellular proliferation [57]. The level of expression of this molecule after infection was similar in both animals analyzed, suggesting that the change in intestinal permeability observed in WT mice is not mediated by the participation of ST2 in transcriptional regulation of *AREG*. The intestinal mucus is one of the main components of defense against invasion of pathogens and protects the epithelium from physical damage. Muc2 mucin is produced and secreted by intestinal goblet cells. We believe that the increased expression of MUC2 in WT mice might be linked to augmented intestinal permeability. Recently, a higher expression level of the mucin glycoprotein Muc2 in enteroids following *Shigella flexneri* infections was reported [58]. These findings suggest that mucus production might not be an important factor involved in the phenotype of decreased intestinal permeability observed in ST2 knockout-infected mice, and that other mechanisms are involved in the intestinal barrier of $ST2^{-/-}$ animals.

During infections, depending on the organ involved, IL-33/ST2 signaling might induce the necessary immune response to control the infectious foci, which might be a Th1 or Th2 type of response [38]. The increase of IFN-$\gamma$ and TNF-$\alpha$ after the infection observed in the knockout mice might be associated with a greater defense of the intestine against the invasion of *B. abortus*, which might be contributing to the phenotype of resistance observed in these animals. Previous studies have demonstrated the requirement of Th1-type cytokine profile to induce protection against *Brucella* infection [40,41]. IL-1$\beta$ plays an important role in the defense against pathogens and also in the maintenance of the intestinal homeostatic balance and in the regeneration of the epithelium [59,60]. High levels of this cytokine are found in the intestinal mucosa in a normal state (steady-state), implying its importance in maintaining the mucosal barrier and in immune monitoring. The decrease in IL-1$\beta$ levels observed in WT-infected mice corroborates the data of altered intestinal permeability through infection, suggesting that ST2 might have a regulatory role of this cytokine and consequently a function in the maintenance of intestinal permeability. Several studies have proposed that cell injury or death are the dominant mechanisms through which the IL-33 reaches the extracellular environment. Therefore, in a steady state, the IL-33 is not actively secreted by cells [35,61], being an important tool for the immune system, when there is a violation in the integrity of the mucosa, secondary to damage to the epithelial cells [61]. In this study, we observed that production of this cytokine in the small intestine was naturally higher in WT mice, compared to the knockout animals, suggesting that the change in intestinal permeability induced by oral infection was not through tissue damage, but via other mechanisms that need to be investigated.

The increase in MPO and EPO as an indirect measurement of neutrophils and eosinophils in the intestine of WT mice might contribute to the establishment of infection. Since *Brucella abortus* is an intracellular pathogen, and it is already described in the literature that neutrophils infected with *Brucella* are readily phagocytized by macrophages and replicate extensively within these cells, neutrophils then end up serving as "Trojan horse" vehicles for efficient bacterial dispersion, intracellular replication, and establishment of chronic infections [62]. In ST2 knockout mice, MPO and EPO were decreased after infection when compared to WT animals, which might contribute to the initial resistance profile exhibited by these mice, since there are less infected granulocytes that can carry the pathogen to spread into other host cells and organs. The absence of ST2 increased the bactericidal activity of neutrophils and macrophages against *Staphylococcus aureus* in a sepsis model [63], by increasing the production of nitric oxide of these cells. Thus, we sought to investigate whether, in an in vitro scenario, macrophages would show higher production of NO against *Brucella* infection. We observed that nitric oxide production rate is similarly influenced in WT and $ST2^{-/-}$ macrophages, either through stimulation with *B. abortus* or LPS, in the presence or absence of IFN-$\gamma$, suggesting that bone marrow-derived macrophages have the same microbicidal potential, and that ST2 in the context of *B. abortus* infection is not involved in the regulation of NO production by these cells.

In summary, we observed that lack of ST2 is important in the model of *Brucella* oral infection but not when the animals are infected by the intraperitoneal route. In this study, we revealed that the oral infection by *Brucella abortus* alters the intestinal homeostasis in favor of its invasion and establishment of systemic infection, and the mechanisms involved in this process were partially dependent on the ST2 receptor. The ST2 receptor proved to be important in maintaining the epithelial barrier and in the negative regulation of the inflammatory immune response to oral infection through *B. abortus*.

## 4. Materials and Methods

### 4.1. Mice

Wild-type Balb/C (WT) mice were purchased from the Federal University of Minas Gerais (UFMG), and ST2 KO (kindly provided by Dr. José Carlos Alves-Filho, Department of Pharmacology, Ribeirao Preto Medical School, University of Sao Paulo, Brazil). Genetically deficient and control mice were maintained at our facilities and used at 6–8 weeks of age. Mice were housed in filter-top cages and provided with sterile water and food ad libitum. Groups of 5 to 7 animals were used to perform all experiments. The procedures for animal experimentation were approved by the Ethics Committee for the Use of Animals of the Federal University of Minas Gerais—CEUA/UFMG under protocol number 273/2017.

### 4.2. Bacteria

*Brucella abortus* smooth virulent strain 2308 was obtained from our laboratory collection. Frozen stocks were prepared from isolated colonies previously grown in *Brucella* broth medium (BB) + 1.5% agar for 3 days. One day prior to infection, *B. abortus* was grown in liquid BB and the OD was measured in a spectrophotometer. In all experiments performed in this study, $OD_{600}$ 1 = 3 × $10^9$ CFU/mL.

### 4.3. Bacterial Counting in B. abortus Infected Mice

Five to seven mice from each group (Balb/c or $ST2^{-/-}$) were infected orally by intragastric gavage with 1 × $10^9$ or intraperitoneally (i.p.) with 1 × $10^6$ virulent *B. abortus* S2308 in 100 μL of PBS. After 3 or 14 days post-infection, mice were sacrificed and liver and spleens were used to determine the number of bacteria through CFU counting. All organs harvested from each animal were weighed and macerated in saline (NaCl 0.9%). To determine bacterial burden, livers and spleens were serially diluted in saline and plated in duplicates on BB agar. Plates were incubated for 3 days at 37 °C and the CFU number was determined.

### 4.4. Intestinal Permeability Assay

The in vivo intestinal permeability assay to verify the barrier function was performed using the FITC-labeled Dextran method with minor modifications [64]. Briefly, food and water were removed and, after 3 h, mice were weighed and received intragastric inoculation of FITC-Dextran (0.6 mg/g body weight, PM 4000; Sigma-Aldrich, St. Louis, MO, USA). Four hours after gavage, the animals were anesthetized with ketamine/xylazine (Syntec, São Paulo, Brazil) (0.6 mL of ketamine at the concentration of 100 mg/mL, 0.4 mL of xylazine at the concentration of 20 mg/mL, and 4 mL of saline), blood was taken by cardiac puncture and was subsequently euthanized. Blood was centrifugedat 10,000 rpm for 3 min at 4 °C and serum collected was pipetted in the volume of 100 μL/well, in a plate of 96 wells (Nunc, Thermo Fisher Scientific, Norcross, GA, USA). The measurement of the fluorescence intensity of each sample (excitation, 492 nm; emission 525 nm; Synergy2, Bio Tek Instruments, Inc., Winooski, VT, USA) was performed. The measurement of intestinal permeability was expressed as the mean of the fluorescence unit. Increased fluorescence in the serum indicated increased intestinal permeability.

### 4.5. Measurement of Myeloperoxidase (MPO) and Eosinophilic Peroxidase Activity (EPO)Activity

The evaluation of the MPO and EPO enzyme activity was used as an indirect index of neutrophil and eosinophil recruitment in the tissues, respectively. The protocol for dosage of this enzyme in homogenized tissues was performed with some modifications [65]. In brief, fragments of small intestine (100 mg) of the animals were removed and frozen at −80 °C. After thawing, the tissue was homogenized in 4.7 pH buffer (0.1 M NaCl, 0.02 M $NaH_2PO_4.1H_2O$, 0.015 M $Na_2$-EDTA) (100 mg of tissue in 1.0 mL buffer), using a tissue homogenizer, centrifuged at 10,000 rpm for 15 min at 4 °C and the precipitate was submitted to hypotonic lysis (500 μL of 0.2% NaCl solution followed by addition of equal volume of solution containing 1.6% NaCl and 5% glucose, 30 s after) for RBC lysis. After further centrifugation, the precipitate was resuspended in 0.05 M $NaH_2PO_4$ buffer (pH 5.4) containing 0.5% hexadeciltrimethylammonium bromide (HTAB) (Sigma) and was re-homogenized. Aliquots of 1 mL of the suspension were transferred to microcentrifuge tubes of 1.5 mL and submitted to three freezing/thawing cycles using liquid nitrogen. These samples were again centrifuged for 15 min at 10,000 rpm. The supernatant was collected and MPO activity was calculated by measuring the changes in optical density (OD) at 450 nm, using tetramethylbenzidine (TMB) (1.6 mM) (Sigma) and $H_2O_2$ (0.5 mM). The supernatant was also used to quantify the peroxidase activity. The assay was performed in 96-well plates, 75 μL per sample or blank well (PBS/HTAB 0.5%) was incubated with 75 μL of substrate (o-phenylenediamine (OPD) (Sigma) 1.5 mM, in Tris-HCl buffer—0.075 μM, pH 8, supplemented with $H_2O_2$ 6.6 mM). The plate was incubated at 20 °C in the dark for approximately 30 min and the reaction was interrupted by the addition of 50 μL of $H_2SO_4$ 1M. The reaction was measuredin a microplate reader (Multiskan FC Thermo Scientific, Norcross, GA, USA) with a 492 nm filter.

### *4.6. Measurement of Cytokine Concentrations*

To evaluate the production of cytokines, fragments of the small intestine with approximately 100 mg were homogenized using a tissue homogenizer (T10 Basic ULTRA-TURRAX®, IKA, Königswinter, Germany) in 1 mL of cytokine extraction solution—PBS containing antiprotease cocktail (0.1 mM PMSF, 0.1 mM benzethonium chloride, 10 mM EDTA, and 20 KI aprotinin A) and 0.05% Tween-20. Then, the homogenates were centrifuged at 4 °C for 10 min at 10,000 rpm. The supernatants were immediately collected and stored at −80 °C for subsequent measurement. The concentrations of IL-1β, IL-33, IFN-γ, TNF-α, and IL-10 was performed through the ELISA method, using kits purchased from R&D Systems (DuoSet) (R&D Systems, Minneapolis, MN, USA) according to manufacturers' recommendations.

### *4.7. Real-Time PCR (RT–PCR)*

RNA was extracted from small intestine with TRIzol reagent (Invitrogen, Thermo Fisher Scientific, Norcross, GA, USA) according to the manufacturer's instructions. cDNA was synthesized by reverse transcription (RT) from 1 μg of total RNA and was used to perform RT–PCR in a final volume of 10 μL containing SYBR green PCR Master Mix (Applied Biosystems, Carlsbad, CA, USA) and 20 μM of primers. RT–PCR was performed in triplicates, on an ABI 7900 Real-time PCR system (Applied Biosystems). The primers used for gene amplification were as follows: 18S forward 5′-CGTTCCACCAACTAAGAACG-3′, reverse 5′- CTCAACACGGGAAACCTCAC-3′; MUC2 forward 5′- CACCAACACGTCAAAAATCG -3′,reverse 5′- CGCAGAACTCCCAGTAGCA -3′; Amphiregulin forward 5′- GCCATTATGCAGCTGCTTTGGAGC -3′, reverse 5′- TGTTTTTCTTGGGCTTAATCACCT -3′; ZO-1 forward 5′-TGAACGCTCTCATAAGCTTCGTAA-3′, reverse 5′-ACCGTACCAACCATCATTC ATTG-3′; ZO-2 forward 5′-CCATGGGCGCGGACTATCTGA-3′, reverse 5′-CTGTGGCGGGGAGGTT TGACTTG-3′, ZO-3 forward 5′-AAGCACGCAATCCTGGATGTCACC-3′, reverse 5′-GTCGCG CCTGCTGTTGCTGTATTA-3′; claudin-1 forward 5′-AGCCAGGAGCCTCCCCCGCAGCTGCA-3′, reverse 5′-CGGGTTGCCTGCAAAGT-3′. The levels of mRNAs are presented as relative expression units after normalization to 18S transcripts.

### *4.8. Generation of BMDMs*

Bone-marrow cells were obtained from femur and tibiae of ST2 KO and WT mice and they were differentiated into BMDMs using a previously described protocol, with some modifications [66]. In brief, cells were seeded on 24-well plates at $5 \times 10^5$ cell/mL (day 0) and maintained in DMEM medium containing 10% FBS, 100 U/mL penicillin, 100 μg/mL streptomycin, and 20% LCCM (L929-conditioned medium), at 37 °C in a 5% $CO_2$ atmosphere for 7 days. On day 4 of incubation, the medium was fully replaced. Four hours before stimulation or infection, BMDMs were maintained only in the DMEM medium containing 1% FBS.

### *4.9. Nitrite Measurement by Griess Reagent*

The nitric oxide assay was performed as described previously [15]. The concentration of nitrite ($NO_2^-$), a stable metabolite of NO, was measured using Griess reagent (1% sulfanilamide and 0.1% naphthylethylenediaminedihydrochloride in 2.5% phosphoric acid). In brief, 50 μL of cell culture supernatants was mixed with 50 μL of Griess reagent. Subsequently, the mixture was incubated, protected from light at room temperature for 5 min, and the absorbance at 550 nm was measured in a microplate reader. Fresh culture medium (DMEM + 1% FBS) was used as a blank in every experiment. The quantity of nitrite was determined from a sodium nitrite ($NaNO_2$) standard curve.

### *4.10. Gut Pathology*

The small intestine of the animals was removed soon after the sacrifice, and the duodenum was separated for histological analysis. The tissues were extended in contact with the filter paper and opened by removing all their contents without damaging the mucosa. The fragments were transferred to a container containing 10% formaldehyde solution for a short period for pre-fixing. The prefixed material was placed on a flat surface and wound in a spiral with the mucosa facing inwards to form rolls. The rolls were tied with line and fixed by immersion in 10% formalin solution in PBS, pH 7.4 for 48 h, and embedded in paraffin. One 4-μm-thick sections were obtained and stained with hematoxylin-and-eosin (H&E) and examined under light microscopy by two pathologists blinded to the experiment. Measurement of villus heights, crypt, and total mucosa thickness depth was performed using the ImageJ software. Fifteen intact and well-oriented villi, crypts, and total mucosa thickness were measured from each animal of each mouse group ($n = 5$).

### *4.11. Statistical Analysis*

The experiments were repeated at least twice with similar results. Graphs and data analysis were performed using GraphPad Prism 5 (GraphPad Software, San Diego, CA, USA), using one-way ANOVA followed by a post-test of Student-Newman-Keuls.

**Author Contributions:** Conceptualization, R.S., A.T.V. and S.C.O.; Data curation, P.C.C., M.R., D.S. and F.V.M.; Formal analysis, P.C.C., J.C.A.-F., A.T.V. and S.C.O.; Funding acquisition, S.C.O.; Investigation, R.S., P.C.C., M.R., V.R., D.S., V.M., F.V.M., F.M., M.F.R., D.C.d.R., G.D.C. and J.C.A.-F.; Methodology, R.S., P.C.C., M.R., V.R., D.S., V.M., F.V.M., M.F.R., J.C.A.-F., and A.T.V.; Project administration, A.T.V. and S.C.O.; Resources, S.C.O., F.M., M.F.R. and J.C.A.-F.; Supervision, A.T.V. and S.C.O.; Validation, J.C.A.-F. and A.T.V.; Writing—original draft, R.S., A.T.V. and S.C.O. All authors have read and agreed to the published version of the manuscript.

## References


1. Moreno, E.; Moriyon, I. *Brucella melitensis*: A nasty bug with hidden credentials for virulence. *Proc. Natl. Acad. Sci. USA* **2002**, *99*, 1–3. [CrossRef] [PubMed]
2. De Figueiredo, P.; Ficht, T.A.; Rice-Ficht, A.; Rossetti, C.A.; Adams, L.G. Pathogenesis and immunobiology of brucellosis: Review of Brucella-host interactions. *Am. J. Pathol.* **2015**, *185*, 1505–1517. [CrossRef] [PubMed]
3. Moreno, E. Retrospective and prospective perspectives on zoonotic brucellosis. *Front. Microbiol.* **2014**, *5*, 213. [CrossRef] [PubMed]
4. Yoo, J.R.; Heo, S.T.; Lee, K.H.; Kim, Y.R.; Yoo, S.J. Foodborne outbreak of human brucellosis caused by ingested raw materials of fetal calf on Jeju Island. *Am. J. Trop. Med. Hyg.* **2015**, *92*, 267–269. [CrossRef]
5. Seleem, M.N.; Boyle, S.M.; Sriranganathan, N. Brucellosis: A re-emerging zoonosis. *Vet. Microbiol.* **2010**, *140*, 392–398. [CrossRef]
6. Silva, F.L.; Paixao, T.A.; Borges, A.M. Brucelose Bovina. *Cad. Tec. Vet. Zoot.* **2005**, *47*, 1–12.
7. Secretaria de Estado da Saúde do Paraná (Ed.) *Protocolo de Manejo Clínico e Vigilância em Saúde para Brucelose Humana No Estado do Paraná*; Secretaria de Estado da Saúde do Paraná: Curitiba, Brazil, 2015; p. 70.

8. World Health Organization (WHO). *Brucellosis in Humans and Animals;* World Health Organization: Geneva, Switzerland, 2006; pp. 1–102.
9. Solera, J. Update on brucellosis: Therapeutic challenges. *Int. J. Antimicrob. Agents* **2010**, *36* (Suppl. 1), S18–S20. [CrossRef]
10. Young, E.J. Brucellosis: A model zoonosis in developing countries. *Apmis. Suppl.* **1988**, *3*, 17–20.
11. Dornand, J.; Gross, A.; Lafont, V.; Liautard, J.; Oliaro, J.; Liautard, J.P. The innate immune response against Brucella in humans. *Vet. Microbiol.* **2002**, *90*, 383–394. [CrossRef]
12. Campos, M.A.; Rosinha, G.M.; Almeida, I.C.; Salgueiro, X.S.; Jarvis, B.W.; Splitter, G.A.; Qureshi, N.; Bruna-Romero, O.; Gazzinelli, R.T.; Oliveira, S.C. Role of Toll-like receptor 4 in induction of cell-mediated immunity and resistance to *Brucella abortus* infection in mice. *Infect. Immun.* **2004**, *72*, 176–186. [CrossRef]
13. Huang, L.Y.; Ishii, K.J.; Akira, S.; Aliberti, J.; Golding, B. Th1-like cytokine induction by heat-killed *Brucella abortus* is dependent on triggering of TLR9. *J. Immunol.* **2005**, *175*, 3964–3970. [CrossRef] [PubMed]
14. Weiss, D.S.; Takeda, K.; Akira, S.; Zychlinsky, A.; Moreno, E. MyD88, but not toll-like receptors 4 and 2, is required for efficient clearance of *Brucella abortus*. *Infect. Immun.* **2005**, *73*, 5137–5143. [CrossRef] [PubMed]
15. Macedo, G.C.; Magnani, D.M.; Carvalho, N.B.; Bruna-Romero, O.; Gazzinelli, R.T.; Oliveira, S.C. Central role of MyD88-dependent dendritic cell maturation and proinflammatory cytokine production to control *Brucella abortus* infection. *J. Immunol.* **2008**, *180*, 1080–1087. [CrossRef] [PubMed]
16. Oliveira, S.C.; De Oliveira, F.S.; Macedo, G.C.; De Almeida, L.A.; Carvalho, N.B. The role of innate immune receptors in the control of *Brucella abortus* infection: Toll-like receptors and beyond. *Microbes Infect.* **2008**, *10*, 1005–1009. [CrossRef]
17. Gomes, M.T.; Campos, P.C.; Pereira Gde, S.; Bartholomeu, D.C.; Splitter, G.; Oliveira, S.C. TLR9 is required for MAPK/NF-kappaB activation but does not cooperate with TLR2 or TLR6 to induce host resistance to *Brucella abortus*. *J. Leukoc. Biol.* **2016**, *99*, 771–780. [CrossRef]
18. Gomes, M.T.; Campos, P.C.; Oliveira, F.S.; Corsetti, P.P.; Bortoluci, K.R.; Cunha, L.D.; Zamboni, D.S.; Oliveira, S.C. Critical role of ASC inflammasomes and bacterial type IV secretion system in caspase-1 activation and host innate resistance to *Brucella abortus* infection. *J. Immunol.* **2013**, *190*, 3629–3638. [CrossRef]
19. Marim, F.M.; Franco, M.M.C.; Gomes, M.T.R.; Miraglia, M.C.; Giambartolomei, G.H.; Oliveira, S.C. The role of NLRP3 and AIM2 in inflammasome activation during *Brucella abortus* infection. *Semin. Immunopathol.* **2017**, *39*, 215–223. [CrossRef]
20. Costa Franco, M.M.; Marim, F.; Guimaraes, E.S.; Assis, N.R.G.; Cerqueira, D.M.; Alves-Silva, J.; Harms, J.; Splitter, G.; Smith, J.; Kanneganti, T.D.; et al. *Brucella abortus* Triggers a cGAS-Independent STING Pathway To Induce Host Protection That Involves Guanylate-Binding Proteins and Inflammasome Activation. *J. Immunol.* **2018**, *200*, 607–622. [CrossRef]
21. Rakoff-Nahoum, S.; Paglino, J.; Eslami-Varzaneh, F.; Edberg, S.; Medzhitov, R. Recognition of commensal microflora by toll-like receptors is required for intestinal homeostasis. *Cell* **2004**, *118*, 229–241. [CrossRef]
22. Artis, D. Epithelial-cell recognition of commensal bacteria and maintenance of immune homeostasis in the gut. *Nat. Rev. Immunol.* **2008**, *8*, 411–420. [CrossRef]
23. Barber, G.N. STING: Infection, inflammation and cancer. *Nat. Rev. Immunol.* **2015**, *15*, 760–770. [CrossRef]
24. Hodzic, Z.; Schill, E.M.; Bolock, A.M.; Good, M. IL-33 and the intestine: The good, the bad, and the inflammatory. *Cytokine* **2017**, *100*, 1–10. [CrossRef] [PubMed]
25. Yanagisawa, K.; Takagi, T.; Tsukamoto, T.; Tetsuka, T.; Tominaga, S. Presence of a novel primary response gene ST2L, encoding a product highly similar to the interleukin 1 receptor type 1. *FEBS Lett.* **1993**, *318*, 83–87. [CrossRef]
26. Schmitz, J.; Owyang, A.; Oldham, E.; Song, Y.; Murphy, E.; McClanahan, T.K.; Zurawski, G.; Moshrefi, M.; Qin, J.; Li, X.; et al. IL-33, an interleukin-1-like cytokine that signals via the IL-1 receptor-related protein ST2 and induces T helper type 2-associated cytokines. *Immunity* **2005**, *23*, 479–490. [CrossRef] [PubMed]
27. Bergers, G.; Reikerstorfer, A.; Braselmann, S.; Graninger, P.; Busslinger, M. Alternative promoter usage of the Fos-responsive gene Fit-1 generates mRNA isoforms coding for either secreted or membrane-bound proteins related to the IL-1 receptor. *EMBO J.* **1994**, *13*, 1176–1188. [CrossRef] [PubMed]
28. Iwahana, H.; Yanagisawa, K.; Ito-Kosaka, A.; Kuroiwa, K.; Tago, K.; Komatsu, N.; Katashima, R.; Itakura, M.; Tominaga, S. Different promoter usage and multiple transcription initiation sites of the interleukin-1

receptor-related human ST2 gene in UT-7 and TM12 cells. *Eur. J. Biochem.* **1999**, *264*, 397–406. [CrossRef] [PubMed]
29. Lohning, M.; Stroehmann, A.; Coyle, A.J.; Grogan, J.L.; Lin, S.; Gutierrez-Ramos, J.C.; Levinson, D.; Radbruch, A.; Kamradt, T. T1/ST2 is preferentially expressed on murine Th2 cells, independent of interleukin 4, interleukin 5, and interleukin 10, and important for Th2 effector function. *Proc. Natl. Acad. Sci. USA* **1998**, *95*, 6930–6935. [CrossRef] [PubMed]
30. Neill, D.R.; Wong, S.H.; Bellosi, A.; Flynn, R.J.; Daly, M.; Langford, T.K.A.; Bucks, C.; Kane, C.M.; Fallon, P.G.; Pannell, R.; et al. Nuocytes represent a new innate effector leukocyte that mediates type-2 immunity. *Nature* **2010**, *464*, U1367–U1369. [CrossRef]
31. Kurowska-Stolarska, M.; Stolarski, B.; Kewin, P.; Murphy, G.; Corrigan, C.J.; Ying, S.; Pitman, N.; Mirchandani, A.; Rana, B.; Van Rooijen, N.; et al. IL-33 Amplifies the Polarization of Alternatively Activated Macrophages That Contribute to Airway Inflammation. *J. Immunol.* **2009**, *183*, 6469–6477. [CrossRef]
32. Cherry, W.B.; Yoon, J.; Barternes, K.R.; Iijima, K.; Kita, H. A novel IL-1 family cytokine, IL-33, potently activates human eosinophils. *J. Allergy Clin. Immunol.* **2008**, *121*, 1484–1490. [CrossRef]
33. Suzukawa, M.; Iikura, M.; Koketsu, R.; Nagase, H.; Tamura, C.; Komiya, A.; Nakae, S.; Matsushima, K.; Ohta, K.; Yamamoto, K.; et al. An IL-1 Cytokine Member, IL-33, Induces Human Basophil Activation via Its ST2 Receptor. *J. Immunol.* **2008**, *181*, 5981–5989. [CrossRef] [PubMed]
34. Smithgall, M.D.; Comeau, M.R.; Yoon, B.R.P.; Kaufman, D.; Armitage, R.; Smith, D.E. IL-33 amplifies both T(h)1-and T(h)2-type responses through its activity on human basophils, allergen-reactive T(h)2 cells, iNKT and NK Cells. *Int. Immunol.* **2008**, *20*, 1019–1030. [CrossRef] [PubMed]
35. Martin, N.T.; Martin, M.U. Interleukin 33 is a guardian of barriers and a local alarmin. *Nat. Immunol.* **2016**, *17*, 122–131. [CrossRef] [PubMed]
36. Le, H.; Kim, W.; Kim, J.; Cho, H.R.; Kwon, B. Interleukin-33: A mediator of inflammation targeting hematopoietic stem and progenitor cells and their progenies. *Front. Immunol.* **2013**, *4*, 104. [CrossRef] [PubMed]
37. Mahapatro, M.; Foersch, S.; Hefele, M.; He, G.W.; Giner-Ventura, E.; McHedlidze, T.; Kindermann, M.; Vetrano, S.; Danese, S.; Gunther, C.; et al. Programming of Intestinal Epithelial Differentiation by IL-33 Derived from Pericryptal Fibroblasts in Response to Systemic Infection. *Cell Rep.* **2016**, *15*, 1743–1756. [CrossRef] [PubMed]
38. Liew, F.Y.; Girard, J.P.; Turnquist, H.R. Interleukin-33 in health and disease. *Nat. Rev. Immunol.* **2016**, *16*, 676–689. [CrossRef] [PubMed]
39. Hodzic, E.; Granov, N. Gigantic Thrombus of the Left Atrium in Mitral Stenosis. *Med. Arch.* **2017**, *71*, 449–452. [CrossRef]
40. Pham, O.H.; O'Donnell, H.; Al-Shamkhani, A.; Kerrinnes, T.; Tsolis, R.M.; McSorley, S.J. T cell expression of IL-18R and DR3 is essential for non-cognate stimulation of Th1 cells and optimal clearance of intracellular bacteria. *PLoS Pathog.* **2017**, *13*, e1006566. [CrossRef]
41. Brandao, A.P.; Oliveira, F.S.; Carvalho, N.B.; Vieira, L.Q.; Azevedo, V.; Macedo, G.C.; Oliveira, S.C. Host susceptibility to *Brucella abortus* infection is more pronounced in IFN-gamma knockout than IL-12/beta2-microglobulin double-deficient mice. *Clin. Dev. Immunol.* **2012**, *2012*, 589494. [CrossRef]
42. Pasquevich, K.A.; Carabajal, M.V.; Guaimas, F.F.; Bruno, L.; Roset, M.S.; Coria, L.M.; Rey Serrantes, D.A.; Comerci, D.J.; Cassataro, J. Omp19 Enables *Brucella abortus* to Evade the Antimicrobial Activity from Host's Proteolytic Defense System. *Front. Immunol.* **2019**, *10*, 1436. [CrossRef]
43. Sangari, F.J.; Seoane, A.; Rodriguez, M.C.; Aguero, J.; Garcia Lobo, J.M. Characterization of the urease operon of *Brucella abortus* and assessment of its role in virulence of the bacterium. *Infect. Immun.* **2007**, *75*, 774–780. [CrossRef] [PubMed]
44. Delpino, M.V.; Marchesini, M.I.; Estein, S.M.; Comerci, D.J.; Cassataro, J.; Fossati, C.A.; Baldi, P.C. A bile salt hydrolase of *Brucella abortus* contributes to the establishment of a successful infection through the oral route in mice. *Infect. Immun.* **2007**, *75*, 299–305. [CrossRef] [PubMed]
45. Brenchley, J.M.; Douek, D.C. Microbial translocation across the GI tract. *Annu. Rev. Immunol.* **2012**, *30*, 149–173. [CrossRef] [PubMed]
46. Lumsden, A.B.; Henderson, J.M.; Kutner, M.H. Endotoxin levels measured by a chromogenic assay in portal, hepatic and peripheral venous blood in patients with cirrhosis. *Hepatology* **1988**, *8*, 232–236. [CrossRef] [PubMed]

47. Singh, R.; Bullard, J.; Kalra, M.; Assefa, S.; Kaul, A.K.; Vonfeldt, K.; Strom, S.C.; Conrad, R.S.; Sharp, H.L.; Kaul, R. Status of bacterial colonization, Toll-like receptor expression and nuclear factor-kappa B activation in normal and diseased human livers. *Clin. Immunol.* **2011**, *138*, 41–49. [CrossRef]
48. Neutra, M.R.; Mantis, N.J.; Kraehenbuhl, J.P. Collaboration of epithelial cells with organized mucosal lymphoid tissues. *Nat. Immunol.* **2001**, *2*, 1004–1009. [CrossRef]
49. Nakato, G.; Hase, K.; Suzuki, M.; Kimura, M.; Ato, M.; Hanazato, M.; Tobiume, M.; Horiuchi, M.; Atarashi, R.; Nishida, N.; et al. Cutting Edge: *Brucella abortus* exploits a cellular prion protein on intestinal M cells as an invasive receptor. *J. Immunol.* **2012**, *189*, 1540–1544. [CrossRef]
50. Paixao, T.A.; Roux, C.M.; Den Hartigh, A.B.; Sankaran-Walters, S.; Dandekar, S.; Santos, R.L.; Tsolis, R.M. Establishment of systemic *Brucella melitensis* infection through the digestive tract requires urease, the type IV secretion system, and lipopolysaccharide O antigen. *Infect. Immun.* **2009**, *77*, 4197–4208. [CrossRef]
51. Zhao, X.; Yang, J.; Ju, Z.; Wu, J.; Wang, L.; Lin, H.; Sun, S. *Clostridium butyricum* Ameliorates Salmonella Enteritis Induced Inflammation by Enhancing and Improving Immunity of the Intestinal Epithelial Barrier at the Intestinal Mucosal Level. *Front. Microbiol.* **2020**, *11*, 299. [CrossRef]
52. Ferrero, M.C.; Fossati, C.A.; Rumbo, M.; Baldi, P.C. Brucella invasion of human intestinal epithelial cells elicits a weak proinflammatory response but a significant CCL20 secretion. *FEMS Immunol. Med. Microbiol.* **2012**, *66*, 45–57. [CrossRef]
53. Ackermann, M.R.; Cheville, N.F.; Deyoe, B.L. Bovine ileal dome lymphoepithelial cells: Endocytosis and transport of *Brucella abortus* strain 19. *Vet. Pathol.* **1988**, *25*, 28–35. [CrossRef] [PubMed]
54. Rossetti, C.A.; Drake, K.L.; Siddavatam, P.; Lawhon, S.D.; Nunes, J.E.; Gull, T.; Khare, S.; Everts, R.E.; Lewin, H.A.; Adams, L.G. Systems biology analysis of Brucella infected Peyer's patch reveals rapid invasion with modest transient perturbations of the host transcriptome. *PLoS ONE* **2013**, *8*, e81719. [CrossRef]
55. Churin, Y.; Al-Ghoul, L.; Kepp, O.; Meyer, T.F.; Birchmeier, W.; Naumann, M. *Helicobacter pylori* CagA protein targets the c-Met receptor and enhances the motogenic response. *J. Cell Biol.* **2003**, *161*, 249–255. [CrossRef] [PubMed]
56. Shao, J.; Sheng, H. Amphiregulin promotes intestinal epithelial regeneration: Roles of intestinal subepithelial myofibroblasts. *Endocrinology* **2010**, *151*, 3728–3737. [CrossRef] [PubMed]
57. Shoyab, M.; McDonald, V.L.; Bradley, J.G.; Todaro, G.J. Amphiregulin: A bifunctional growth-modulating glycoprotein produced by the phorbol 12-myristate 13-acetate-treated human breast adenocarcinoma cell line MCF-7. *Proc. Natl. Acad. Sci. USA* **1988**, *85*, 6528–6532. [CrossRef] [PubMed]
58. Ranganathan, S.; Doucet, M.; Grassel, C.L.; Delaine-Elias, B.; Zachos, N.C.; Barry, E.M. Evaluating *Shigella flexneri* Pathogenesis in the Human Enteroid Model. *Infect. Immun.* **2019**, 87. [CrossRef]
59. Rathinam, V.A.K.; Chan, F.K. Inflammasome, Inflammation, and Tissue Homeostasis. *Trends Mol. Med.* **2018**, *24*, 304–318. [CrossRef]
60. Elinav, E.; Thaiss, C.A.; Flavell, R.A. Analysis of microbiota alterations in inflammasome-deficient mice. *Methods Mol. Biol.* **2013**, *1040*, 185–194. [CrossRef]
61. Camilleri, M.; Madsen, K.; Spiller, R.; Van Meerveld, B.G.; Verne, G.N. Intestinal barrier function in health and gastrointestinal disease (vol 24, pg 503, 2012). *Neurogastroenterol. Motil.* **2012**, *24*, 976. [CrossRef]
62. Gutierrez-Jimenez, C.; Mora-Cartin, R.; Altamirano-Silva, P.; Chacon-Diaz, C.; Chaves-Olarte, E.; Moreno, E.; Barquero-Calvo, E. Neutrophils as Trojan Horse Vehicles for *Brucella abortus* Macrophage Infection. *Front. Immunol.* **2019**, *10*, 1012. [CrossRef]
63. Staurengo-Ferrari, L.; Trevelin, S.C.; Fattori, V.; Nascimento, D.C.; De Lima, K.A.; Pelayo, J.S.; Figueiredo, F.; Casagrande, R.; Fukada, S.Y.; Teixeira, M.M.; et al. Interleukin-33 Receptor (ST2) Deficiency Improves the Outcome of *Staphylococcus aureus*-Induced Septic Arthritis. *Front. Immunol.* **2018**, *9*, 962. [CrossRef] [PubMed]
64. Yan, Y.; Kolachala, V.; Dalmasso, G.; Nguyen, H.; Laroui, H.; Sitaraman, S.V.; Merlin, D. Temporal and spatial analysis of clinical and molecular parameters in dextran sodium sulfate induced colitis. *PLoS ONE* **2009**, *4*, e6073. [CrossRef] [PubMed]
65. Vieira, A.T.; Fagundes, C.T.; Alessandri, A.L.; Castor, M.G.; Guabiraba, R.; Borges, V.O.; Silveira, K.D.; Vieira, E.L.; Goncalves, J.L.; Silva, T.A.; et al. Treatment with a novel chemokine-binding protein or eosinophil lineage-ablation protects mice from experimental colitis. *Am. J. Pathol.* **2009**, *175*, 2382–2391. [CrossRef] [PubMed]
66. Weischenfeldt, J.; Porse, B. Bone Marrow-Derived Macrophages (BMM): Isolation and Applications. *Cold Spring Harb. Protoc.* **2008**, *2008*, pdb-prot5080. [CrossRef] [PubMed]

# 7

# ASC-Mediated Inflammation and Pyroptosis Attenuates *Brucella abortus* Pathogenesis Following the Recognition of gDNA

**Juselyn D. Tupik [1], Sheryl L. Coutermarsh-Ott [1], Angela H. Benton [1], Kellie A. King [1], Hanna D. Kiryluk [1], Clayton C. Caswell [1] and Irving C. Allen [1,2,*]**

[1] Department of Biomedical Sciences and Pathobiology, Virginia-Maryland College of Veterinary Medicine, Virginia Tech, Blacksburg, VA 24061, USA; jdtupik@vt.edu (J.D.T.); slc2003@vt.edu (S.L.C.-O.); ahbenton@vt.edu (A.H.B.); kellieking@vt.edu (K.A.K.); hanna98@vt.edu (H.D.K.); caswellc@vt.edu (C.C.C.)

[2] Department of Basic Science Education, Virginia Tech Carilion School of Medicine, Roanoke, VA 24016, USA

* Correspondence: icallen@vt.edu

**Abstract:** *Brucella abortus* is a zoonotic pathogen that causes brucellosis. Because of *Brucella's* unique LPS layer and intracellular localization predominately within macrophages, it can often evade immune detection. However, pattern recognition receptors are capable of sensing *Brucella* pathogen-associated molecular patterns (PAMPS). For example, NOD-like receptors (NLRs) can form a multi-protein inflammasome complex to attenuate *Brucella* pathogenesis. The inflammasome activates IL-1β and IL-18 to drive immune cell recruitment. Alternatively, inflammasome activation also initiates inflammatory cell death, termed pyroptosis, which augments bacteria clearance. In this report, we assess canonical and non-canonical inflammasome activation following *B. abortus* infection. We conducted in vivo studies using $Asc^{-/-}$ mice and observed decreased mouse survival, immune cell recruitment, and increased bacteria load. We also conducted studies with $Caspase\text{-}11^{-/-}$ mice and did not observe any significant impact on *B. abortus* pathogenesis. Through mechanistic studies using $Asc^{-/-}$ macrophages, our data suggests that the protective role of ASC may result from the induction of pyroptosis through a gasdermin D-dependent mechanism in macrophages. Additionally, we show that the recognition of *Brucella* is facilitated by sensing the PAMP gDNA rather than the less immunogenic LPS. Together, these results refine our understanding of the role that inflammasome activation and pyroptosis plays during brucellosis.

**Keywords:** brucellosis; canonical inflammasome; non-canonical inflammasome; NLR; pyroptosis; ASC; caspase-11; caspase-1; IL-1β; gDNA

---

## 1. Introduction

Brucellosis is a zoonotic bacterial disease that exhibits pathogenesis consistent with inflammation. Transmitted through *Brucella* spp. primarily from agricultural animals to humans in unpasteurized dairy products, brucellosis symptoms in humans often include inflammatory or influenza-like characteristics such as arthritis, undulant fevers, and neurological manifestations [1,2]. Because there is currently no human vaccine for brucellosis and effective antibiotic regiments for the disease require long treatment durations, these symptoms often persist throughout the infected individual's lifetime due to the well-adapted ability of *Brucella* to evade immune recognition [3]. Unlike the classical lipopolysaccharide (LPS) layer of Gram-negative bacteria such as *Escherichia coli* that contain a glucosamine backbone with short acyl groups, *Brucella* spp. contain a modified lipid A layer that consists of a diaminoglucose backbone with long branching acyl groups [2]. This deviation from a consistent molecular structure has the potential to subvert immune recognition by the innate immune system through complement interference and decreased cytokine production, leading to enhanced *Brucella* replication and pathogenesis. Despite its mechanisms of immune avoidance, there are some aspects of *Brucella* spp. that are recognized by the innate immune system, making the understanding of these mechanisms essential for targeting future treatments for brucellosis.

As hypothesized by Janeway (1989), the innate immune system has evolved over time to recognize consistent molecular structures in pathogens known as Pathogen or Damage-Associated Molecular Patterns (PAMPs or DAMPs). These PAMPs and DAMPs are recognized by protein structures known as pattern recognition receptors (PRRs) [4]. From previous studies, *Brucella* genomic DNA (gDNA) is known to be recognized by the PRR absent in melanoma 2 (AIM2) and subsequently promotes inflammation, making it an excellent PAMP for immune recognition [5–7]. PRRs include membrane-bound receptors, which consist of Toll-like receptors (TLRs) and C-type Lectin receptors (CLRs), as well as cytosolic receptors made up of a Nucleotide-Binding Domain and Leucine-Rich Repeat Containing receptors (NLRs), Aim-2-Like receptors (ALRs), Rig-I-Like Helicase receptors (RLRs), and the X-LR class of uncategorized receptors [8,9]. After the recognition of a PAMP or DAMP, PRRs generally serve as scaffolding proteins to promote the initiation or inhibition of immune signaling pathways [9]. Of the PRRs that have been described in brucellosis, the best characterized have been the TLRs. From previous literature, many TLRs have been implicated with *Brucella* detection, which plays a role in bacterial signaling, host resistance, and dendritic cell activation [10–16]. TLRs also play important roles in the transcriptional generation of inactive inflammatory cytokines in response to *Brucella* infections that can be activated by NLR or ALR immune signaling complexes [17,18]. This indicates that multiple PRRs work in tandem to attenuate brucellosis pathogenesis.

Inflammasome-forming NLRs and AIM2 have also been reported to play a role in *Brucella* sensing [5–7]. After recognition, the NLR or ALR is able to bind the apoptosis-associated speck-like protein containing a caspase activation recruitment domain (CARD) (ASC) and procaspase-1 to form the canonical inflammasome [8]. The inflammasome then cleaves caspase-1, which subsequently cleaves the cytokines pro-IL-1β and pro-IL-18, produced through TLR signaling, to their active forms to promote inflammation [8,9,19–23]. Inflammasome signaling can also lead to a form of inflammatory cell death known as pyroptosis. Pyroptosis occurs when activated capase-1 cleaves the protein gasdermin D, releasing the gasdermin N subunit [24]. This subunit binds with phosphoinositides on the cell membrane and oligomerizes, creating membrane pores that lead to an osmotic imbalance in the cell that eventually leads to cell lysis [24]. Recently, the formation of a non-canonical inflammasome has been described that utilizes caspase-11 to mediate the cleavage of gasdermin D to initiate pyroptosis.

Previous studies evaluating inflammasome activation in response to *Brucella* have predominately focused on characterizing the activation of inflammatory cytokine signaling associated with canonical inflammasome activation. The best described inflammasomes involved in *Brucella* infections are NLR Family Pyrin Domain Containing 3 (NLRP3) and AIM2. In mouse models, the NLRP3 inflammasome promotes survival and decreased bacterial load through enhanced cytokine secretion, in addition to sensing mitochondrial reactive oxygen species (ROS) generated from *Brucella* [6,7]. Looking at the *Brucella* PAMPs, AIM2, as a known sensor of bacterial DNA, becomes activated from *Brucella* gDNA recognition and initiates inflammatory cytokine signaling and pyroptosis [5,6,25]. These inflammasomes are ASC-dependent, as shown in the formation of punctate ASC structures during infection [6], indicating that ASC-dependent inflammasomes are important in *Brucella* recognition and targeting through inflammatory cytokine signaling. Despite these advancements in studying inflammasome-mediated inflammatory cytokine signaling, pyroptosis and the role of gasdermin D have not been extensively evaluated in response to *Brucella*.

In this study, we used *Asc*$^{-/-}$ and *Caspase-11*$^{-/-}$ mice to further elucidate the role of the canonical and non-canonical inflammasomes following *B. abortus* infection. We sought to assess survival, histopathology, bacterial load, and cell death to provide a more holistic view of cytokine responses and pyroptosis, both in vivo and *in vitro*. Additionally, we reassessed *Brucella* PAMPs using *B. abortus* gDNA and LPS to better define the mechanisms associated with pathogen recognition. Ultimately, we found that ASC functions to attenuate *B. abortus* pathogenesis through the modulation of inflammation and pyroptosis, requiring gasdermin D through a mechanism independent of caspase-11. Additionally, we determined that *Brucella* gDNA, rather than LPS, provoked an elevated inflammasome response that augmented pyroptosis. This report contributes to the current literature and provides some additional novel insights into potential mechanisms of inflammasome activation during brucellosis.

## 2. Results

### 2.1. ASC Attenuates B. abortus Pathogenesis and Is Critical for Host Survival

To explore canonical and non-canonical inflammasome activation to *Brucella abortus*, we used mice that lack either the ASC adaptor protein (*Asc*$^{-/-}$) or the non-canonical inflammasome-associated caspase, caspase-11 (*Caspase-11*$^{-/-}$). After intraperitoneally injecting mice with $1 \times 10^5$ colony forming units (CFUs) of *B. abortus*, we monitored mortality in all mouse groups for a 24-day period (Figure 1A). At Day 7, there was a 26.3% decrease in the survival of the *Asc*$^{-/-}$ mice. In these animals, the weight loss in 5 of the 19 mice exceeded 20%, and several of the *Asc*$^{-/-}$ mice developed clinical parameters associated with disease progression, such as decreased body condition, that required euthanasia (Figure 1B). However, there was no decrease in survival for the wildtype (WT) and *Caspase-11*$^{-/-}$ mice. These mortality data suggest that the canonical inflammasome plays a more critical role in host survival compared to the non-canonical inflammasome and caspase-11.

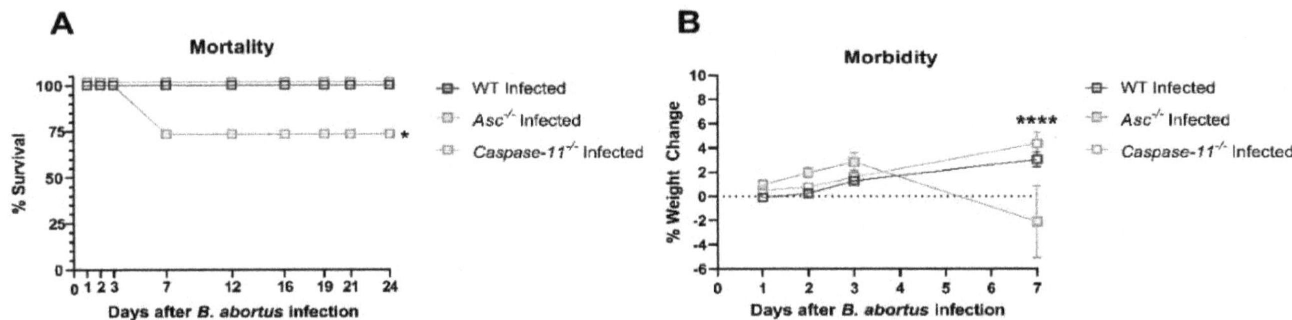


**Figure 1.** *Asc*$^{-/-}$ and *Caspase-11*$^{-/-}$ mortality and morbidity. *Asc*$^{-/-}$ ($n$ = 19), *Caspase-11*$^{-/-}$ ($n$ = 19), and C57BL/6 WT ($n$ = 26) mice were injected intraperitoneally with $1 \times 10^5$ *B. abortus* CFUs and assessed daily for excessive weight loss (>20%) warranting euthanasia according to the Institutional Animal Care and Use Committee (IACUC). (**A**) Mortality graph based on (**B**) a morbidity assessment of *Asc*$^{-/-}$, *Capsase-11*$^{-/-}$, and WT mice. All the mice were weighed for a 24-day period, with morbidity warranting euthanasia in *Asc*$^{-/-}$ mice only at Day 7. * $p < 0.05$, **** $p < 0.0001$.

### 2.2. ASC Contributes to Inflammatory Pathogenesis during B. abortus Infection

To further elucidate the role of ASC and caspase-11 activation *in vivo*, we conducted a histopathological analysis on the liver and spleen from wildtype, *Asc*$^{-/-}$, and *Caspase-11*$^{-/-}$ mice three days post-infection. Histopathology indicated that all the infected mouse groups exhibited elevated extramedullary hematopoiesis (EMH) and inflammation in the liver and spleen (Figure 2A,B).

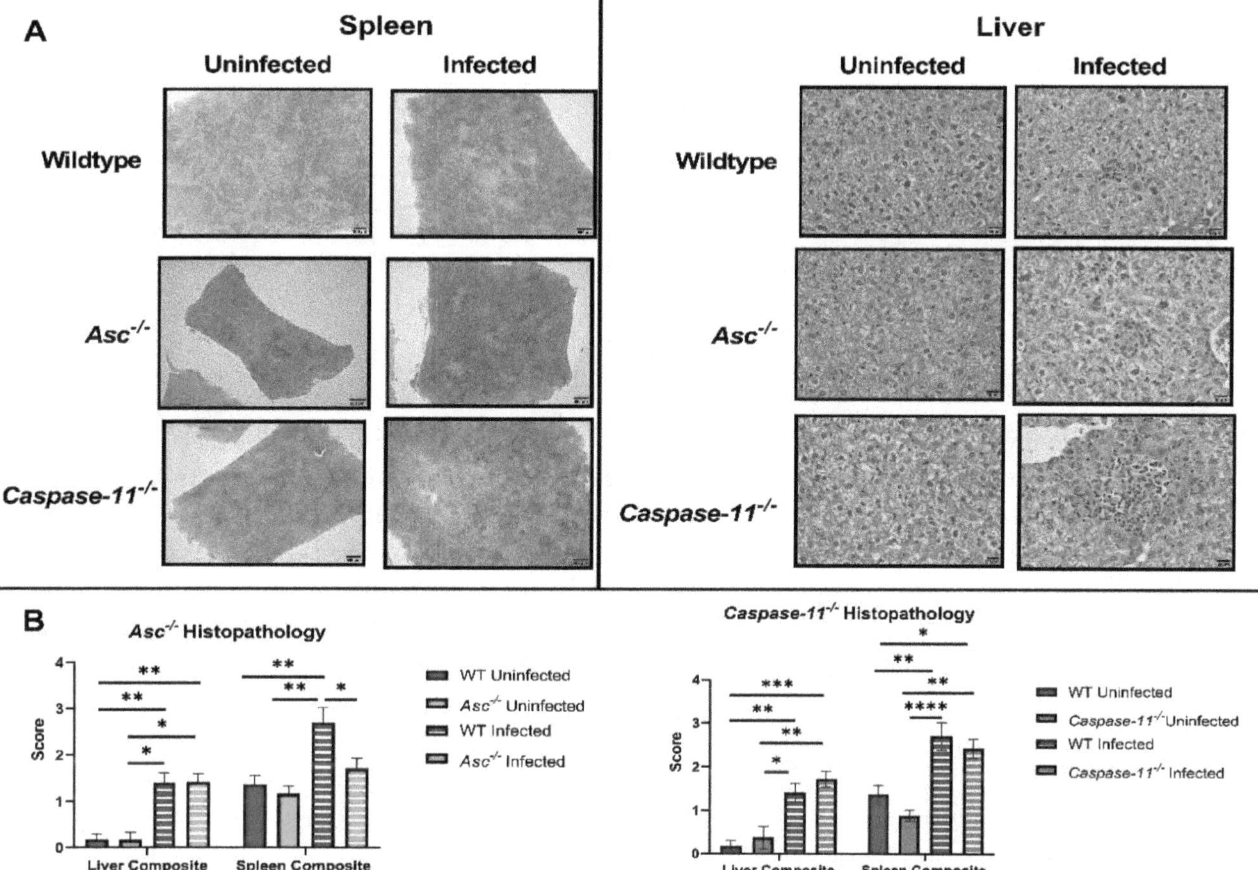


**Figure 2.** Inflammation in the liver and spleen. A total of 20 *Asc*$^{-/-}$ ($n = 6$ uninfected [U], 14 infected [I]), 27 *Caspase-11*$^{-/-}$ ($n = 8$ U, 19 I), and 34 C57BL/6 WT ($n = 11$ U, 23 I) livers and spleens were evaluated by histopathology 3 d.p.i. (**A**) H&E stained histological slides of the liver and spleen from WT, *Asc*$^{-/-}$, and *Caspase-11*$^{-/-}$ mice. All spleen images were taken at 4× power and all liver images were taken at 40× power. Inflammation and extramedullary hematopoiesis (EMH) were the dominant features observed in the histopathology evaluation. (**B**) Bar graphs of WT, *Asc*$^{-/-}$, and *Caspase-11*$^{-/-}$ histopathology composite scores were generated based on inflammation and EMH in the liver and spleen. * $p < 0.05$, ** $p < 0.01$, *** $p < 0.001$, **** $p < 0.0001$.

Wildtype mice exhibited higher EMH and inflammation scores compared to the *Asc*$^{-/-}$ animals in the spleen. This trend was not observed in the *Caspase-11*$^{-/-}$ mice in either the liver or the spleen. Together, these data suggest that ASC augments splenic inflammation and plays a key role in *B. abortus*-mediated pathology in the spleen, which is a target organ in this mouse model. The failure to mount a vigorous immune response to *B. abortus* in the *Asc*$^{-/-}$ mice is likely associated with the increased morbidity and mortality.

### 2.3. *Bacterial Load Is Significantly Increased in the Absence of ASC and Decreased in the Absence of Caspase-11*

Bacterial loads were determined in the spleen and liver of *B. abortus*-infected animals 3 d.p.i (Figure 3). Between wildtype, *Asc*$^{-/-}$, and *Caspase-11*$^{-/-}$ mice, there was no significant difference in the weight of the spleens used for analysis.

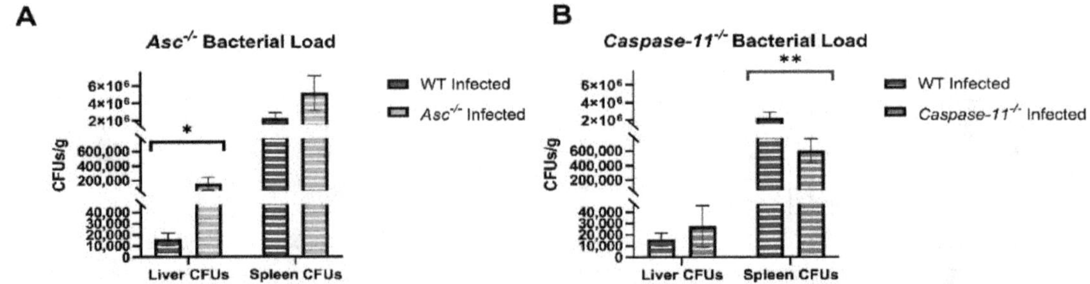


**Figure 3.** *Brucella* CFUs in the liver and spleen of $Asc^{-/-}$ and $Caspase\text{-}11^{-/-}$ mice. Three d.p.i. liver and spleens from infected C57BL/6 WT ($n = 23$), **(A)** $Asc^{-/-}$ ($n = 14$), and **(B)** $Caspase\text{-}11^{-/-}$ ($n = 19$) mice were homogenized and counted for CFUs/gram. * $p < 0.05$, ** $p < 0.01$.

In $Asc^{-/-}$ mice, we observed significantly elevated *B. abortus* CFUs in the liver and a trending increase in the spleen. In the $Caspase\text{-}11^{-/-}$ mice, we observed a similar trending increase in CFUs in the liver. However, in the spleen $Caspase\text{-}11^{-/-}$ mice had significantly decreased bacterial CFUs compared to the wildtype. The increased bacteria load in $Asc^{-/-}$ mice is consistent with the increased morbidity and reduced inflammation in Figures 1 and 2, and further illustrates the critical role of ASC in the host response to *B. abortus*. Likewise, these $Caspase\text{-}11^{-/-}$ data suggest a significant, but variable, role in controlling the *B. abortus* bacteria burden that does not appear to impact the overall host morbidity or inflammation.

## 2.4. *B. abortus Initiates a Weak Inflammasome-Mediated Inflammatory Cytokine Response*

Inflammasome activation results in the cleavage and processing of IL-1β and IL-18. To evaluate the generation of pro-IL-1β, so-called "Signal 1", at the transcript level, we conducted quantitative real-time PCR in liver and spleen homogenates (Figure 4). We observed significantly increased fold changes in *Il1β* in the livers of $Asc^{-/-}$ and wildtype mice infected with *B. abortus* versus the uninfected mice (Figure 4A). Within infected mouse groups in the liver, we also found that infected $Asc^{-/-}$ mice had a significantly decreased fold change of *Il1β* compared to wildtype mice (Figure 4A). However, when assessing the total IL-1β (uncleaved and cleaved) protein, we did not see a significant difference between the infected wildtype and $Asc^{-/-}$ mice.

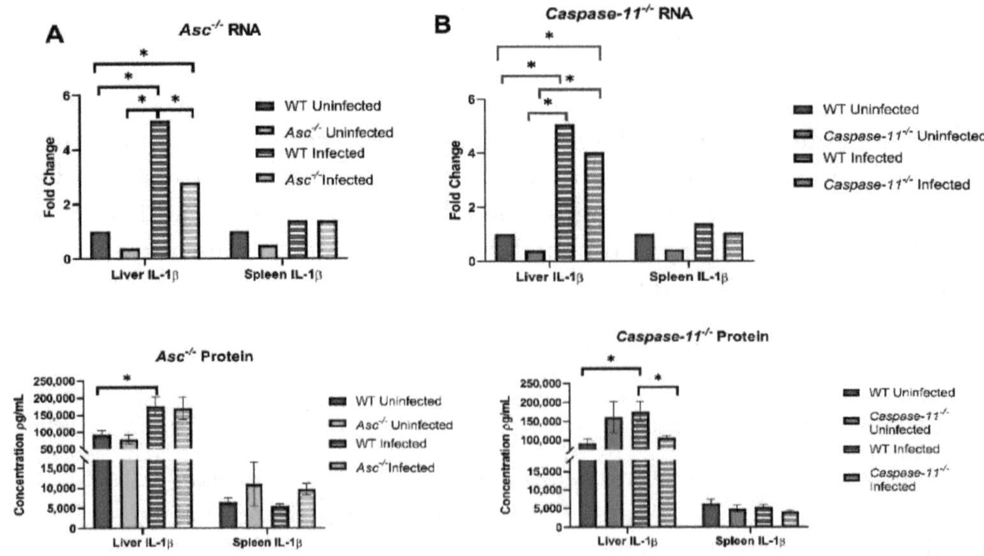


**Figure 4.** Inflammatory signaling in the liver and spleen of $Asc^{-/-}$ and $Caspase\text{-}11^{-/-}$ mice. Three d.p.i. livers and spleens of 34 C57BL/6 WT ($n = 11$ U, 23 I), **(A)** 20 $Asc^{-/-}$ ($n = 6$ U, 14 I), and **(B)** 27 $Caspase\text{-}11^{-/-}$ ($n = 8$ U, 19 I) mice were homogenized and analyzed for their *Il1β* RNA fold change using RT-PCR and IL-1β protein concentration through ELISA. * $p < 0.05$.

In experiments with *Caspase-11*$^{-/-}$ mice, we also found an elevated RNA fold change between the infected and uninfected groups (Figure 4B), but there was no significant difference between the infected wildtype and *Caspase-11*$^{-/-}$ mice for *Il1β* transcription (Figure 4B). However, there was a minimal, but statistically significant, decrease in IL-1β protein in the livers from *Caspase-11*$^{-/-}$ mice compared to the wildtype animals.

### *2.5. B. abortus Infection Attenuated IL-1β and Induced a Strong ASC-Dependent Pyroptosis Response in Macrophages*

Due to the significant phenotype observed in the *Asc*$^{-/-}$ mice, we next sought to better define the underlying mechanism using ex vivo bone marrow-derived macrophages (BMDMs). Intracellular bacterial replication and survival was evaluated over a 48 h period in *B. abortus*-infected *Asc*$^{-/-}$ and wildtype BMDMs. Under these conditions, we observed a significant decrease in *B. abortus* growth in the *Asc*$^{-/-}$ macrophages compared with the wildtype BMDMs after 24 h (Figure 5A). By 48 h, the *B. abortus* replication and survival was no longer detectable, while increasing in the wildtype BMDMs. These results were unexpected based on our in vivo findings and suggest a possible disconnect between pathogen clearance, inflammasome function, and pyroptosis in BMDMs.

We next measured the RNA fold change and protein concentration of IL-1β (Figure 5B,C). Within 2 h of *B. abortus* exposure, we observed elevated *Il1β* transcription in the infected wildtype and *Asc*$^{-/-}$ macrophages. Transcription was statistically significant, but only slightly higher, in the *Asc*$^{-/-}$ cells compared to the wildtype BMDMs (Figure 5B). At 24 h and 48 h, we observed a significant decrease in *Il1β* transcription in both groups of infected mice, with significantly more *Il1β* in the *Asc*$^{-/-}$ BMDMs at 24 h compared to the wildtype (Figure 5B). Complementing the transcription data, we also evaluated the IL-1β protein levels in the cell supernatant using ELISA (Figure 5C). These levels were significantly attenuated in the *B. abortus*-infected cells (Figure 5C). We also observed a significant decrease in IL-1β in the *Asc*$^{-/-}$ BMDMs under all conditions (Figure 5C), emphasizing that IL-1β processing is ASC-dependent.

To further expand upon our cell death findings, we conducted a LDH assay in our *Asc*$^{-/-}$ and wildtype macrophages to quantify the lactate dehydrogenase enzyme released from dead cells (Figure 5D). We found that, starting at 24 h, there was significantly higher cell death occurring in our wildtype cells compared to the *Asc*$^{-/-}$ macrophages. To further define the mechanism of cell death, we evaluated pyroptosis by determining gasdermin D cleavage using Western blot. We found a significant increase in cleaved gasdermin D 24 h post-infection in the wildtype BMDMs compared to the significantly reduced levels observed in the *Asc*$^{-/-}$ macrophages (Figure 5E). This was confirmed using densitometry (Figure 5E).

### *2.6. B. abortus gDNA is a Potent PAMP Associated with ASC-Dependent Canonical Inflammasome Signaling*

To further define inflammasome activation following *B. abortus* infection, we next evaluated the potential pathogen-associated molecular patterns (PAMPs), focusing on bacterial gDNA. We challenged macrophages with 1 μg of gDNA (2 μg/mL) both externally, by adding gDNA to the cell media, and internally through the Lipofectamine 3000 reagent (Figure 6A). We also added 300 μM of ATP to augment the IL-1β release. *B. abortus* gDNA induced a significant increase in the *Il1β* gene transcription 24 h post challenge, following either extracellular or intracellular challenge (Figure 6A). We observed significant increases in *Il1β* transcription under several different conditions in gDNA-challenged *Asc*$^{-/-}$ macrophages compared to the wildtype cells. ELISA assessments revealed that IL-1β protein was only released into the supernatant following internal gDNA stimulation in wildtype cells (Figure 6B). This was highly dependent on ASC. The *B. abortus* gDNA challenge resulted in significant increases in IL-1β in the wildtype cells, whereas the levels were below the level of detection in the $Asc^{-/-}$ cells (Figure 6B). Together, these data confirm *B. abortus* gDNA as a potent PAMP and suggest that its recognition by the canonical inflammasome, in an ASC-dependent mechanism, underlies host defense. In addition to gDNA, we also evaluated the ability of the canonical inflammasome to recognize *B. abortus* LPS (1 μg/mL externally). The fold change in *Il1β* RNA at 8 h post challenge indicated that there was elevated transcription in wildtype macrophages over $Asc^{-/-}$ cells (Figure 6C). However, there was no IL-1β protein signaling for *Brucella* LPS (Figure 6D).

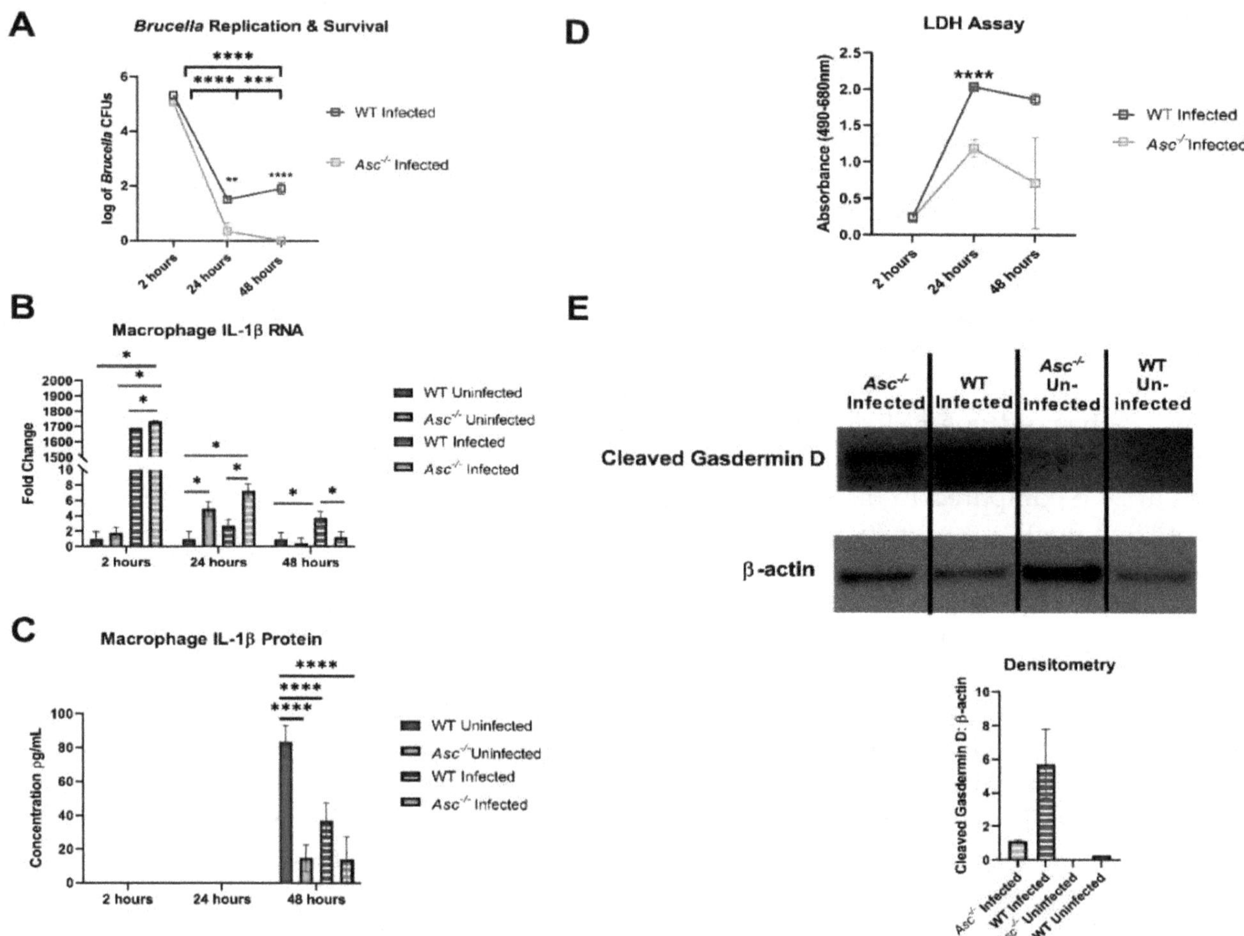


**Figure 5.** ASC-dependent IL-1β generation and pyroptosis following *B. abortus* infection in BMDMs. Bone marrow-derived macrophages (BMDMs) (500,000 cells per well, $n = 3$ per group) were harvested from C57BL/6 WT and $Asc^{-/-}$ mice and challenged with a MOI 100:1 ($10^7$ CFUs) of *Brucella abortus*. (**A**) BMDMs were measured for CFUs for a 48 h period post-challenge. (**B**) BMDMs were measured for the *Il1β* fold change in RNA through RT-PCR. (**C**) BMDMs were analyzed for IL-1β protein in the supernatant using an ELISA. (**D**) Lactate Dehydrogenase (LDH) was measured through spectrophotometry through a 48 h time period post-challenge. (**E**) BMDMs were lysed at 24 h post-challenge and used for Western blot analysis. BMDM protein (20μg) was probed for cleaved gasdermin D and β-actin as a control for the protein amount. Invitrogen IBright Analysis was used to determine the density ratio between the cleaved gasdermin D over β-actin. * $p < 0.05$, *** $p < 0.001$, **** $p < 0.0001$.

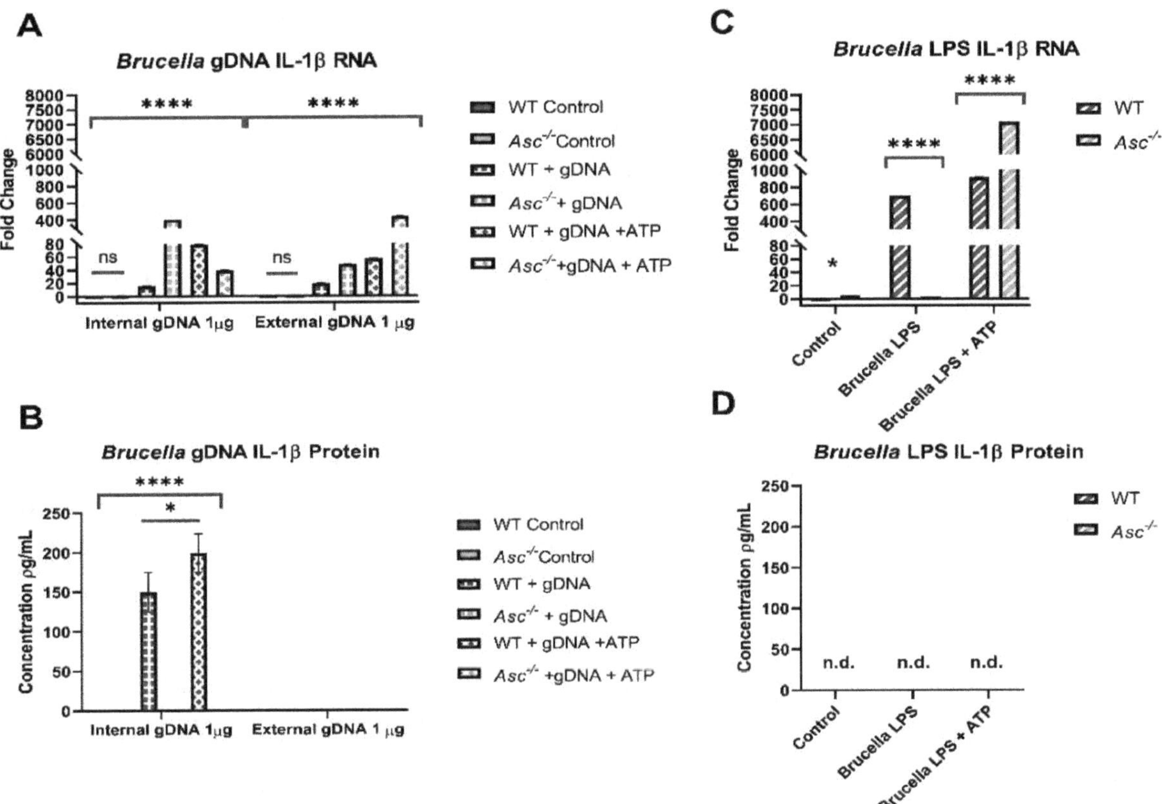


**Figure 6.** *B. abortus* gDNA is a potent PAMP and induces ASC-dependent IL-1β production. Bone marrow-derived macrophages (BMDMs) (500,000 cells per well, $n = 2$ per group) were harvested from C57BL/6 WT and $Asc^{-/-}$ mice. (**A**,**B**) BMDMs were challenged with 1 μg of gDNA (2μg/mL) externally in media or transfected internally with 300 μM of ATP and harvested after 24 h. BMDMs were analyzed for the IL-1β (**A**) RNA fold change through RT-PCR and (**B**) protein concentration through ELISA. (**C**,**D**) BMDMs were challenged with 1 μg/mL of *Brucella* LPS externally to the media with 300 μM of ATP and harvested after 8 h. BMDMs were analyzed for the IL-1β (**C**) RNA fold change through RT-PCR and (**D**) protein concentration through ELISA. Note that there was no detectable (*n*.d.) protein concentration in (**D**). * $p < 0.05$, **** $p < 0.0001$.

## 3. Discussion

In this report, we assessed inflammasome activation following *Brucella abortus* infection. During canonical inflammasome activation, a pathogen is sensed by a NLR or ALR pattern recognition receptor and forms a multi-protein complex with the binding protein ASC and caspase-1. This process can initiate the cleavage of IL-1β and IL-18 in addition to pyroptosis [8]. While highly related to the canonical inflammasome pathway, non-canonical inflammasome activation is often more closely associated with the promotion of pyroptosis and the activation of caspase-11 [26]. Together, our in vivo data reveal a more robust phenotype in the $Asc^{-/-}$ mice compared to the *Caspase-11*$^{-/-}$ animals, suggesting that the canonical inflammasome plays a significantly greater role in host pathogen defense following *B. abortus* infection. ASC and, by extension, the canonical inflammasome promotes survival, augments inflammation, and attenuates bacterial load *in vivo*. This result is consistent with others in the field that have used myriad of other inflammasome knockout models [5–7,27]. However, there are certainly exceptions to these findings. For example, Gomes et al. (2103) recently reported no inflammasome knockout mice ($n = 5$ per group) exhibited mortality under their experimental conditions [6]. One difference to note is that our study utilized a much larger sample size per group ($n = 19$). Because many of our animals recovered, it is certainly possible that greater power through that increased sample size is necessary to better reflect the mortality data for the $Asc^{-/-}$ animals. Ultimately, our comparison between $Asc^{-/-}$ and *Caspase-11*$^{-/-}$ mice expands upon the findings of many of these prior studies and provides a more direct assessment of the pathobiological effects of the canonical and non-canonical inflammasome in *B. abortus* host defense.

Although non-canonical signaling through the *Brucella* LPS activation of caspase-11-mediated pyroptosis has been previously indicated [28], we found inconsistent results in inflammasome activation under our conditions. In *Caspase-11*$^{-/-}$ mice, we found no loss in survival or morbidity, no significant inflammation through histopathology scoring, and no protective role in promoting inflammatory signaling. Previous studies using *Caspase-11*$^{-/-}$ mice mimicked our non-significant results in bacterial load at 3 days post-infection and only found significant bacterial load 1–2 weeks after infection [28]. Although we did not assess the bacterial load or inflammatory signaling after 3 d.p.i in this report, we did assess the morbidity and mortality of *Caspase-11*$^{-/-}$ mice over 3 weeks, in which knockouts exhibited no decrease in morbidity or loss in survival. This indicates that caspase-11 may play a small but relatively insignificant role in promoting pyroptosis that perhaps may be slightly amplified 1–2 weeks after *Brucella* infection. Additionally, these studies utilized immune cell priming in macrophages with PAMPs, such as *E. coli* LPS and Pam3CSK4, and subsequently observed high cytokine signaling in their challenge [28]. Contrasting this data, our results indicated no IL-1β protein response to *Brucella* LPS stimulation in unprimed macrophages. These data suggest that the immune adjuvants directly impact the activation of caspase-11 and that the role of non-canonical inflammasome in host defense against *B. abortus* is minimal in the absence of macrophage priming.

Previously, pyroptosis had only been attributed to non-canonical inflammasome activation by *Brucella* spp. [6,28]. This is further confirmed in this report through the presence of cleaved gasdermin D bands in *Asc*$^{-/-}$ macrophages, indicating that ASC-independent pyroptosis occurs in response to *Brucella*. However, our research demonstrates that the removal of ASC-dependent inflammasome activation significantly decreases the activation of gasdermin D to form pyroptotic pores. As described in the literature, ASC specks serve as recruitment factors for procaspase-1 through the polymerization of its caspase activation recruitment domain (CARD). Caspase-1 only becomes activated through this process during ASC-dependent inflammasome formation [29]. In turn, caspase-1 cleaves gasdermin D, which has been identified as the most significant gene initiating caspase-1 induced pyroptosis, and initiates inflammatory cell death [30]. Our results are consistent with this ASC-mediated pathway of the capsase-1 activation of pyroptosis that is dependent on gasdermin D cleavage. To date, we know that *Brucella* initiates the caspase-1 and -11 activation of pyroptosis in joints of animal models [31]. Additionally, pyroptosis is activated by gDNA in dendritic cells [5]. Our data supports a model where both caspase-1 and -11 promote pyroptosis, and where gDNA from *B. abortus* functions as a robust PAMP that specifically activates the canonical inflammasome, driving ASC-dependent inflammation and pyroptosis.

Previous literature indicates that the role of pyroptosis in brucellosis serves to restrict *Brucella* growth in macrophages of the joints and control infection [31]. Our findings are consistent with this previous study. Under our conditions, the ASC-mediated initiation of pyroptosis appears to ensure mouse survival, immune cell recruitment, and inflammatory signaling. However, we should also point out the bacteria clearance and IL-1β data in the BMDM studies (Figure 5A,C). In Figure 5A, these data would suggest that the lack of ASC and canonical inflammasome signaling actually improved the bacteria clearance from these BMDMs, despite having reduced IL-1β and pyroptosis. While these data seem to conflict with each other, several recent studies have reported similar findings for other bacterial pathogens. For example, *Citrobacter rodentium* infection results in significant osmotic changes in targeted cells that can augment inflammasome signaling [32]. However, the clearance of the pathogen itself appears to be independent of the inflammasome and the ASC modulation of inflammation and pyroptosis [33]. Thus, it is possible that a similar mechanism is associated here with *B. abortus*. Looking at the IL-1β graph in Figure 5C, it suggests that *B. abortus* infection in macrophages suppresses IL-1β production from wildtype cells. It is possible that the attenuation of total IL-1β in this figure may be due to the subversion of TLR signaling generating pro- IL-1β. Many studies have shown *Brucella* subversion of TLR signaling through proteins such as Tcbp, which leads to decreased proinflammatory cytokine expression [10]. Additionally, *Brucella* microRNAs can lead to the downregulation of the mRNA and protein expression of innate immune PRRs [34]. These mechanisms likely did not occur in our gDNA studies as there was no inclusion of these immunosuppressive proteins or production of inhibitory microRNAs. Therefore, it is possible that the full *Brucella* bacterium may be utilizing these methods of immunosuppression to contribute to decreased total IL-1β.

Our data suggests that ASC and the canonical inflammasome contribute to host defense in response to *B. abortus*. These results are consistent with several others in the field and provide additional insight into host defense against this highly intriguing pathogen. Currently, brucellosis is having a significantly negative impact on a growing number of human populations worldwide. Therefore, it is essential that we expand our understanding of the underlying disease mechanisms and host immune response to *B. abortus*. This finding that the canonical inflammasome plays a dominate role in driving the host innate immune response and pyroptosis following the sensing of gDNA is an encouraging discovery that may contribute to the development of future therapeutics or strategic approaches to combat this disease and its underlying pathogen.

## 4. Materials and Methods

### *4.1. Bacterial Strains and Growth Conditions*

*Brucella abortus* 2308 was routinely grown on Schaedler blood agar (SBA), which is composed of Schaedler agar (BD, Franklin Lakes, NJ, USA) containing 5% defibrinated bovine blood (Quad Five, Ryegate, MT, USA). All work with live *Brucella* strains was performed in a biosafety level 3 (BSL3) facility. Animal work conducted in ABSL3 conditions was conducted **under IACUC protocol # 14-055** at Virginia Tech following the ethical standards of animal use in research.

### *4.2. In Vivo Brucella Studies*

C57BL/6 WT mice ($n$ = 26), *Asc*$^{-/-}$ mice (C57BL/6 background; $n$ = 19), and *Caspase-11*$^{-/-}$ (C57BL/6 background; $n$ = 19) mice were inoculated intraperitoneally with $1 \times 10^5$ CFU of *Brucella abortus*. The percent weight change was measured each day post-*Brucella* inoculation to determine morbidity warranting euthanasia (>20%) following the IACUC protocol.

Additionally, 20 canonical knockout (*Asc*$^{-/-}$, $n$ = 6 U, 14 I), 27 non-canonical (*Caspase-11*$^{-/-}$, $n$ = 8 U, 19 I), and 34 C57BL/6 WT ($n$ = 11 U, 23 I) mice were euthanized at 3 days post infection and harvested for the liver and spleen. Both organs were sectioned into three equal parts. The first section of both organs was taken for histopathology analysis to determine the scoring of extramedullary hematopoiesis (EMH) and inflammation. The remaining sections were homogenized in 1×PBS and analyzed for the number of CFUs per gram and the RNA and protein concentrations of IL-1β. RNA was isolated from the liver and spleen with TRIzol reagents (Invitrogen) followed by ethanol precipitation. Genomic DNA was removed with DNase I, and samples were cleaned using phenol-chloroform extractions and precipitated with ethanol. RNA samples were then resuspended in nuclease-free $H_2O$, and the purity of samples was checked with a NanoDrop 1000 spectrophotometer (ThermoFisher).

After isolation, RNA was converted into 1 μg cDNA through a High-Capacity cDNA Reverse Transcription Kit (ThermoFisher). This cDNA was analyzed for IL-1β by RT-qPCR (40× cycles) using Taqman Fast MasterMix (ThermoFisher). Protein was determined through a sandwich ELISA kit (R&D systems).

### *4.3. Bone Marrow-Derived Macrophage (BMDM) Isolation*

Bone marrow-derived macrophages (BMDMs) were derived from the bone marrow of C57BL/6 WT and *Asc*$^{-/-}$ mice under the **IACUC protocol #18-104** at Virginia Tech. Two adult mice from each group were sacrificed using $CO_2$ fixation followed by cervical dislocation. Bone marrow was extracted from the tibias and femora of the mice. Cells were cultured in non-TC culture dishes in our formulation of culture media (Dulbecco's Modified Eagle Medium (DMEM) (ThermoFisher) containing 10% Fetal Bovine Serum (FBS), 1% penicillin/streptomycin, 1% nonessential amino acids, and 20% L929 conditioned media [35]). These culture dishes were incubated at 37 °C with 5% $CO_2$. After 6–7 days of culture, these cells had differentiated into macrophages. BMDMs were collected from the bottom of the plates through a cold 1× PBS solution containing 5 mM of EDTA. After 1 h on ice, we checked for macrophage detachment via microscope. Macrophages were collected and seeded at 500,000 cells/well in 24-well plates in media without antibiotics and left to adhere overnight in culture media.

### 4.4. Live Brucella Challenge in BMDMs

Macrophages were infected with a MOI 100:1 ($10^7$ CFUs/5 × $10^5$ BMDMs) with *B. abortus* 2308. At the 2, 24, and 48 h time points, intracellular *B. abortus* was determined through a gentamicin protection assay [36]. At 2 h, the infected macrophages were treated with gentamicin (50 µg/mL) for 1 h. Macrophages were then lysed with 0.1% deoxycholate in PBS, and serial dilutions were plated on Schaedler blood agar (SBA) containing 5% bovine blood (Quad Five). For the 24 and 48 h time points, macrophages were washed with PBS and fresh cell culture medium containing gentamicin (20 µg/mL) was added to the macrophages. At the indicated time points, macrophages were lysed, and serial dilutions were plated on SBA in triplicates.

The supernatant of BMDMs was sterile-filtered for later IL-1β protein analysis through a sandwich ELISA kit (R&D systems). Macrophages at each time point were lysed with 0.1% deoxycholate and isolated for RNA. After isolation, RNA was converted into 0.5µg cDNA through a High-Capacity cDNA Reverse Transcription Kit (ThermoFisher). This cDNA was analyzed for IL-1β by RT-qPCR (40× cycles) using Taqman Fast MasterMix (ThermoFisher).

### 4.5. LDH Assay

Macrophages were infected with a MOI 100:1 ($10^7$ CFUs/5 × $10^5$ BMDMs) with *B. abortus*. At the 2, 24, and 48 h time points, extracellular *B. abortus* was killed through gentamicin (50 µg/mL). The supernatant of BMDMs was collected and centrifuged at 1000× *g* for 10 min to remove cell debris. The remaining supernatant (50 µL) was used for the CyQUANT™ LDH Cytotoxicity Assay kit (Invitrogen) and read at a corrected absorbance of 490–680 nm.

### 4.6. Brucella PAMP Isolation

#### 4.6.1. *Brucella* gDNA

*B. abortus* gDNA was isolated from the 2308 strain by phenol:chloroform extraction. Approximately 3 mL of an overnight culture of *B. abortus* was pelleted by centrifugation. The pellet was resuspended in 200 µL of 0.04 M sodium acetate, 200 µL of 10% SDS, and 600 µL of TRIzol. A total of 250 µL of chloroform was added to this mixture in a Phase Lock tube (5PRIME) and centrifuged at 20,000× *g* (max speed) for 2 min. A second chloroform wash was performed in a Phase Lock tube. The resulting aqueous layer is then removed and added to 1 mL of 100% ethanol. DNA precipitation is carried out from this point.

#### 4.6.2. *Brucella* LPS

*B. abortus* LPS was isolated from the *B. abortus* 2308 using hot-phenol extraction, as described previously [31]. Bacteria were killed with ethanol: acetone, and the cells were recovered by centrifugation. The pellet was suspended in deionized water at 66 °C and then mixed with 90% phenol *w/v* that was heated to 66 °C. After stirring for 20 min, the suspension was chilled on ice. The solution was then subjected to centrifugation (15 min at 13,000× *g*). The phenol layer was aspirated, filtered through a Whatman #1 filter, and the LPS was precipitated with methanol containing 1% methanol saturated with sodium acetate. Following incubation at 4 °C for 1 h, the mixture was subjected to centrifugation (10,000× *g* for 10 min). The precipitate was stirred with deionized water for 12 h at 4 °C.

Following centrifugation (10,000× *g* for 10 min), the supernatant was precipitated with trichloroacetic acid, and the resulting supernatant following centrifugation (10,000× *g* for 10 min) was dialyzed with deionized water and stored at −20 °C. The concentration of LPS was determined through the Pierce™ Chromogenic Endotoxin Quant Kit (ThermoFisher).

### 4.7. *Brucella PAMP Challenge*

After PAMP isolation, *B. abortus* gDNA (2 µg/mL) and LPS (1 µg/mL) were introduced to BMDMs. gDNA was introduced both intracellularly, through the Lipofectamine 3000 Transfection Reagent (Invitrogen), and extracellularly in media. LPS was only introduced extracellularly. Timepoints for this challenge included 24 h for gDNA and 8 h for LPS post-challenge. Samples were run with and without the priming of 300 µM of ATP 45 min before each time point to stimulate IL-1β protein release after transcription. At each time point, supernatant was collected from each well and centrifuged at 1000× *g* for 10 min to remove cell debris. The supernatant was later used for protein quantification using an IL-1β sandwich ELISA (R&D Systems). Macrophages were lysed with 200 µL of TRIzol Reagent (Invitrogen) and followed the TRIzol RNA isolation protocol. After isolation, RNA was converted into 1 µg of cDNA through a High-Capacity cDNA Reverse Transcription Kit (ThermoFisher). This cDNA was analyzed for IL-1β by RT-qPCR (40× cycles) using the Taqman Fast MasterMix (ThermoFisher).

### 4.8. *Western Blot*

Macrophages from live *Brucella* challenge were lysed using a sodium lysis buffer (0.3% SDS, 200 mM dithiothreitol, 22 mM Tris-base, and 28 mM Tris-HCl pH 8.0). Samples were then boiled for a period of 1 h, vortexing the samples every 10 min. Samples were frozen at −20 °C until use. Prior to running the gel, samples were sonicated for 10 s each. Protein quantification was determined using the Pierce™ Detergent Compatible Bradford Assay Kit (ThermoFisher). Samples (20 µg of protein) were heated at 97 °C with a reducing buffer for 7 min and run on pre-cast Bolt™ 4 to 12%, Bis-Tris, 1.0 mm, Mini Protein Gel, 10-well gels (Invitrogen) for 45 min at 165 V with a 1× Micro Extraction Packet Sorbent (MEPS) buffer (ThermoFisher). Gel was transferred onto a polyvinylidene difluoride (PVDF) membrane using a transfer chamber with transfer buffer (20% methanol in 1× Tris Glycine (TGE)). The membrane was then blocked in 5% milk for 1 h, and then incubated overnight with a cleaved gasdermin D rabbit antibody (diluted 1:1000, Cell Signaling #36425S). Membrane was washed in Tris-buffered saline with Tween 20 (1× TBST) 4× for 15 min each and then blocked with goat anti-rabbit IgG antibody conjugated with horseradish peroxidase (HRP) (diluted 1:2000, Cell Signaling #7074) for 1 h. Membrane was washed again with 1× TBST 4× for 15 min each and then imaged using West Pico Substrate for imaging (ThermoFisher). Gels were imaged using an IBright CL1500 imaging machine. Sample bands were normalized using the β-actin rabbit antibody (Cell Signaling #4970) using the same protocol above. The density of bands was calculated using the IBright Analysis software (ThermoFisher) to calculate a cleaved gasdermin D/β-actin ratio.

### 4.9. *Graphing and Statistical Analyses*

All the figures and statistical analyses were generated in GraphPad 8.4.3 (Prism). Statistical tests included two-way ANOVAs, using the Tukey or Sidak post-hoc tests, and two sample *t*-tests when appropriate. All the data are contained within the article.

**Author Contributions:** The authors contributed to the following project aspects: Conceptualization: I.C.A., C.C.C., J.D.T.; methodology: all authors; software: J.D.T.; validation: all authors; formal analysis: J.D.T. and S.L.C.-O.; investigation: J.D.T., S.L.C.-O., A.H.B., K.A.K., H.D.K.; resources: C.C.C. and I.C.A.; data curation: J.D.T. and S.L.C.-O.; writing—original draft preparation: all authors; writing—review and editing: all authors; visualization: J.D.T.; supervision: C.C.C. and I.C.A.; project administration: C.C.C. and I.C.A.; funding acquisition: C.C.C. and I.C.A. All authors have read and agreed to the published version of the manuscript.

**Acknowledgments:** We would like to acknowledge the assistance of Margaret Nagai-Singer, Alissa Hendricks-Wenger, Jackie Sereno, and Audrey Rowe with conducting methods. Additionally, we are grateful to Jerod Skyberg for providing detailed information about LPS isolation from *Brucella* strains.

## References

1. Corbel, M.J. Brucellosis: An overview. *Emerg. Infect. Dis.* **1997**, *3*, 213. [CrossRef] [PubMed]
2. Lapaque, N.; Moriyon, I.; Moreno, E.; Gorvel, J.P. *Brucella* lipopolysaccharide acts as a virulence factor. *Curr. Opin. Microbiol.* **2005**, *8*, 60–66. [CrossRef] [PubMed]
3. De Figueiredo, P.; Ficht, T.A.; Rice-Ficht, A.; Rossetti, C.A.; Adams, L.G. Pathogenesis and immunobiology of brucellosis: Review of *Brucella*–Host Interactions. *Am. J. Pathol.* **2015**, *185*, 1505–1517. [CrossRef] [PubMed]
4. Janeway, C.A. Approaching the asymptote? Evolution and revolution in immunology. *Cold Spring Harb. Symp. Quant. Biol.* **1989**, *54*, 1–13. [CrossRef] [PubMed]
5. Franco, M.M.S.C.; Marim, F.M.; Alves-Silva, J.; Cerqueira, D.; Rungue, M.; Tavares, I.P.; Oliveira, S.C. AIM2 senses *Brucella abortus* DNA in dendritic cells to induce IL-1β secretion, pyroptosis and resistance to bacterial infection in mice. *Microbes Infect.* **2019**, *21*, 85–93. [CrossRef]
6. Gomes, M.T.R.; Campos, P.C.; Oliveira, F.S.; Corsetti, P.P.; Bortoluci, K.R.; Cunha, L.D.; Zamboni, D.S.; Oliveira, S.C. Critical role of ASC inflammasomes and bacterial type IV secretion system in caspase-1 activation and host innate resistance to *Brucella abortus* infection. *J. Immunol.* **2013**, *190*, 3629–3638. [CrossRef]
7. Marim, F.M.; Franco, M.M.C.; Gomes, M.T.R.; Miraglia, M.C.; Giambartolomei, G.H.; Oliveira, S.C. The role of NLRP3 and AIM2 in inflammasome activation during *Brucella abortus* infection. *Semin. Immunopathol.* **2017**, *39*, 215–223. [CrossRef]
8. Coutermarsh-Ott, S.; Eden, K.; Allen, I.C. Beyond the inflammasome: Regulatory NOD-like receptor modulation of the host immune response following virus exposure. *J. Gen. Virol.* **2016**, *97*, 825–838. [CrossRef]
9. Tupik, J.D.; Nagai-Singer, M.A.; Allen, I.C. To protect or adversely affect? The dichotomous role of the NLRP1 inflammasome in human disease. *Mol. Asp. Med.* **2020**, 100858. [CrossRef]
10. Oliveira, S.C.; de Oliveira, F.S.; Macedo, G.C.; de Almeida, L.A.; Carvalho, N.B. The role of innate immune receptors in the control of *Brucella abortus* infection: Toll-like receptors and beyond. *Microbes Infect.* **2008**, *10*, 1005–1009. [CrossRef]
11. Campos, M.A.; Rosinha, G.M.; Almeida, I.C.; Salgueiro, X.S.; Jarvis, B.W.; Splitter, G.A.; Qureshi, N.; Bruna-Romero, O.; Gazzinelli, R.T.; Oliveira, S.C. Role of Toll-like receptor 4 in induction of cell-mediated immunity and resistance to *Brucella abortus* infection in mice. *Infect. Immun.* **2004**, *72*, 176–186. [CrossRef] [PubMed]
12. De Almeida, L.A.; Macedo, G.C.; Marinho, F.A.; Gomes, M.T.; Corsetti, P.P.; Silva, A.M.; Cassataro, J.; Giambartolomei, G.H.; Oliveira, S.C. Toll-like receptor 6 plays an important role in host innate resistance to *Brucella abortus* infection in mice. *Infect. Immun.* **2013**, *81*, 1654–1662. [CrossRef] [PubMed]
13. Ferrero, M.C.; Hielpos, M.S.; Carvalho, N.B.; Barrionuevo, P.; Corsetti, P.P.; Giambartolomei, G.H.; Oliveira, S.C.; Baldi, P.C. Key role of Toll-like receptor 2 in the inflammatory response and major histocompatibility complex class II downregulation in *Brucella abortus*-infected alveolar macrophages. *Infect. Immun.* **2014**, *82*, 626–639. [CrossRef] [PubMed]
14. Campos, P.C.; Gomes, M.T.R.; Guimarães, E.S.; Guimarães, G.; Oliveira, S.C. TLR7 and TLR3 sense *Brucella abortus* RNA to induce proinflammatory cytokine production but they are dispensable for host control of infection. *Front. Immunol.* **2017**, *8*, 28. [CrossRef] [PubMed]
15. Macedo, G.C.; Magnani, D.M.; Carvalho, N.B.; Bruna-Romero, O.; Gazzinelli, R.T.; Oliveira, S.C. Central role of MyD88-dependent dendritic cell maturation and proinflammatory cytokine production to control *Brucella abortus* infection. *J. Immunol.* **2008**, *180*, 1080–1087. [CrossRef] [PubMed]
16. De Almeida, L.A.; Carvalho, N.B.; Oliveira, F.S.; Lacerda, T.L.S.; Vasconcelos, A.C.; Nogueira, L.; Bafica, A.; Silva, A.M.; Oliveira, S.C. MyD88 and STING signaling pathways are required for IRF3-mediated IFN-β induction in response to *Brucella abortus* infection. *PLoS ONE* **2011**, *6*, e23135. [CrossRef]
17. Hornung, V.; Latz, E. Critical functions of priming and lysosomal damage for NLRP3 activation. *Eur. J. Immunol.* **2010**, *40*, 620–623. [CrossRef]

18. Miao, E.A.; Andersen-Nissen, E.; Warren, S.E.; Aderem, A. TLR5 and Ipaf: Dual sensors of bacterial flagellin in the innate immune system. *Semin. Immunopathol.* **2007**, *29*, 275–288. [CrossRef]
19. Martinon, F.; Burns, K.; Tschopp, J. The inflammasome: A molecular platform triggering activation of inflammatory caspases and processing of proIL-beta. *Mol. Cell* **2002**, *10*, 417–426. [CrossRef]
20. Agostini, L.; Martinon, F.; Burns, K.; McDermott, M.F.; Hawkins, P.N.; Tschopp, J. NALP3 forms an IL-1beta-processing inflammasome with increased activity in Muckle-Wells autoinflammatory disorder. *Immunity* **2004**, *20*, 319–325. [CrossRef]
21. Davis, B.K.; Roberts, R.A.; Huang, M.T.; Willingham, S.B.; Conti, B.J.; Brickey, W.J.; Barker, B.R.; Kwan, M.; Taxman, D.J.; Accavitti-Loper, M.A.; et al. Cutting edge: NLRC5-dependent activation of the inflammasome. *J. Immunol.* **2011**, *186*, 1333–1337. [CrossRef] [PubMed]
22. Wlodarska, M.; Thaiss, C.A.; Nowarski, R.; Henao-Mejia, J.; Zhang, J.P.; Brown, E.M.; Frankel, G.; Levy, M.; Katz, M.N.; Philbrick, W.M.; et al. NLRP6 inflammasome orchestrates the colonic host-microbial interface by regulating goblet cell mucus secretion. *Cell* **2014**, *156*, 1045–1059. [CrossRef] [PubMed]
23. Vanaja, S.K.; Rathinam, V.A.K.; Fitzgerald, K.A. Mechanisms of inflammasome activation: Recent advances and novel insights. *Trends Cell Biol.* **2015**, *25*, 308–315. [CrossRef] [PubMed]
24. Shi, J.; Gao, W.; Shao, F. Pyroptosis: Gasdermin-mediated programmed necrotic cell death. *Trends Biochem. Sci.* **2017**, *42*, 245–254. [CrossRef]
25. Franco, M.M.C.; Marim, F.; Guimarães, E.S.; Assis, N.R.; Cerqueira, D.M.; Alves-Silva, J.; Harms, J.; Splitter, G.; Smith, J.; Kanneganti, T.-D. *Brucella abortus* triggers a cGAS-independent STING pathway to induce host protection that involves guanylate-binding proteins and inflammasome activation. *J. Immunol.* **2018**, *200*, 607–622. [CrossRef] [PubMed]
26. Yi, Y.-S. Regulatory Roles of the Caspase-11 Non-Canonical Inflammasome in Inflammatory Diseases. *Immune Netw.* **2018**, *18*, e41. [CrossRef]
27. Hielpos, M.S.; Fernández, A.G.; Falivene, J.; Alonso Paiva, I.M.; Muñoz González, F.; Ferrero, M.C.; Campos, P.C.; Vieira, A.T.; Oliveira, S.C.; Baldi, P.C. IL-1R and inflammasomes mediate early pulmonary protective mechanisms in respiratory *Brucella abortus* infection. *Front. Cell. Infect. Microbiol.* **2018**, *8*, 391. [CrossRef]
28. Cerqueira, D.M.; Gomes, M.T.R.; Silva, A.L.; Rungue, M.; Assis, N.R.; Guimarães, E.S.; Morais, S.B.; Broz, P.; Zamboni, D.S.; Oliveira, S.C. Guanylate-binding protein 5 licenses caspase-11 for Gasdermin-D mediated host resistance to *Brucella abortus* infection. *PLoS Pathog.* **2018**, *14*, e1007519. [CrossRef]
29. Lu, A.; Magupalli, V.G.; Ruan, J.; Yin, Q.; Atianand, M.K.; Vos, M.R.; Schröder, G.F.; Fitzgerald, K.A.; Wu, H.; Egelman, E.H. Unified polymerization mechanism for the assembly of ASC-dependent inflammasomes. *Cell* **2014**, *156*, 1193–1206. [CrossRef]
30. Broz, P. Caspase target drives pyroptosis. *Nature* **2015**, *526*, 642–643. [CrossRef]
31. Lacey, C.A.; Mitchell, W.J.; Dadelahi, A.S.; Skyberg, J.A. Caspase-1 and caspase-11 mediate pyroptosis, inflammation, and control of *Brucella* joint infection. *Infect. Immun.* **2018**, *86*. [CrossRef]
32. Liu, Z.; Zaki, M.H.; Vogel, P.; Gurung, P.; Finlay, B.B.; Deng, W.; Lamkanfi, M.; Kanneganti, T.D. Role of inflammasomes in host defense against *Citrobacter rodentium* infection. *J. Biol. Chem.* **2012**, *287*, 16955–16964. [CrossRef] [PubMed]
33. Armstrong, H.; Bording-Jorgensen, M.; Chan, R.; Wine, E. Nigericin Promotes NLRP3-Independent Bacterial Killing in Macrophages. *Front. Immunol.* **2019**, *10*, 2296. [CrossRef] [PubMed]
34. Khan, M.; Harms, J.S.; Liu, Y.; Eickhoff, J.; Tan, J.W.; Hu, T.; Cai, F.; Guimaraes, E.; Oliveira, S.C.; Dahl, R.; et al. *Brucella* suppress STING expression via miR-24 to enhance infection. *PLoS Pathog.* **2020**, *16*, e1009020. [CrossRef] [PubMed]
35. Englen, M.D.; Valdez, Y.E.; Lehnert, N.M.; Lehnert, B.E. Granulocyte/macrophage colony-stimulating factor is expressed and secreted in cultures of murine L929 cells. *J. Immunol. Methods* **1995**, *184*, 281–283. [CrossRef]
36. Sharma, A.; Puhar, A. Gentamicin Protection Assay to Determine the Number of Intracellular Bacteria during Infection of Human TC7 Intestinal Epithelial Cells by Shigella flexneri. *Bio-Protocol* **2019**, *9*. [CrossRef]

# Human *mecC*-Carrying MRSA: Clinical Implications and Risk Factors

**Carmen Lozano *, Rosa Fernández-Fernández, Laura Ruiz-Ripa, Paula Gómez, Myriam Zarazaga and Carmen Torres**

Area of Biochemistry and Molecular Biology, University of La Rioja, 26006 Logroño, Spain; rosa.fernandez.1995@gmail.com (R.F.-F.); laura_ruiz_10@hotmail.com (L.R.-R.); paula_gv83@hotmail.com (P.G.); myriam.zarazaga@unirioja.es (M.Z.); carmen.torres@unirioja.es (C.T.)

* Correspondence: carmen.lozano@unirioja.es

**Abstract:** A new methicillin resistance gene, named *mecC*, was first described in 2011 in both humans and animals. Since then, this gene has been detected in different production and free-living animals and as an agent causing infections in some humans. The possible impact that these isolates can have in clinical settings remains unknown. The current available information about *mecC*-carrying methicillin resistant *S. aureus* (MRSA) isolates obtained from human samples was analyzed in order to establish its possible clinical implications as well as to determine the infection types associated with this resistance mechanism, the characteristics of these *mecC*-carrying isolates, their possible relation with animals and the presence of other risk factors. Until now, most human *mecC*-MRSA infections have been reported in Europe and *mecC*-MRSA isolates have been identified belonging to a small number of clonal complexes. Although the prevalence of *mecC*-MRSA human infections is very low and isolates usually contain few resistance (except for beta-lactams) and virulence genes, first isolates harboring important virulence genes or that are resistant to non-beta lactams have already been described. Moreover, severe and even fatal human infection cases have been detected. *mecC*-carrying MRSA should be taken into consideration in hospital, veterinary and food safety laboratories and in prevention strategies in order to avoid possible emerging health problems.

**Keywords:** *Staphylococcus aureus*; methicillin resistance; human infection; CC130

## 1. Introduction

*Staphylococcus aureus* is an opportunistic pathogen that causes high morbidity and mortality. This microorganism is able to cause diverse diseases that range from having a relatively minor impact, such as a skin infection, to serious and life-threatening episodes, such as endocarditis, pneumonia or sepsis. The impact of *S. aureus* is enhanced by its great capacity to develop and acquire resistance to various antimicrobial agents. Among the antibiotic resistance of *S. aureus*, methicillin resistance mediated by the *mecA* gene is highly relevant as this mechanism provides this bacterium resistance to almost all beta-lactam antibiotics, seriously limiting therapeutic options [1,2]. Recently, the World Health Organization (WHO) outlined the greatest threats in terms of antimicrobial resistance and methicillin-resistant *S. aureus* (MRSA) was classified as a high-priority microorganism. For many years, MRSA infections were only reported in hospitals, with it being considered to be a nosocomial pathogen (hospital-associated MRSA or HA-MRSA). In the 1990s, community-associated MRSA (CA-MRSA) cases in healthy humans without any connection to healthcare settings started to be described and, nowadays, the distinction between CA-MRSA and HA-MRSA seems to be disappearing [3,4].

For the last two decades, a third epidemiological group known as livestock-associated MRSA (LA-MRSA) has been described. *S. aureus* has been considered to be an important zoonotic agent with a great capacity to cause infections in different animal species and in humans. Various studies have suggested that there is a high specificity of the different genetic lineages of *S. aureus* for the host [5]. However, many cases of clones related to animals have been detected and have caused infections in humans [6,7]. Presently, different clonal lineages associated with LA-MRSA have been described and, among these, the clonal complex (CC) CC398 stands out (Table S1). CC398 is related to production animals, mainly pigs, and has been detected worldwide [8]. Infection cases have been identified in humans, both in contact and without contact with animals [9–11]. In addition to CC398, there are other clonal complexes associated with animals such as CC5 in birds, CC9 in pigs, CC97 in cattle or CC133 in small ruminants [12–15].

Remarkably, a new methicillin resistance gene ($mecA_{LGA251}$, which shares only 70% similarity to *mecA* (Figure S1)) was first described in 2011 in both humans and animals [16,17]. Initially these strains were associated with dairy cows and these animals were considered to be a possible reservoir [16]. Since then, this gene has been detected in different production and free-living animals and as an agent causing infection in some humans [8,18]. This new gene was named *mecC* since *mecB* had previously been described in macrococci, but not in staphylococcal species [19]. Worryingly, *mecB* has been recently identified in *S. aureus* and future studies should determine the potential risk that this entails [20]. In the case of MRSA isolates carrying the *mecC* gene (*mecC*-MRSA isolates), these isolates have already been identified as belonging to diverse clonal lineages such as CC130, CC49, sequence type (ST) 151, ST425, CC599 or CC1943 and in very different hosts, including its detection in environmental samples [8,21–23]. There are different theories about the origin of the *mecC* gene and the possible impact that these isolates can have in clinical settings. In this review, the objective was to describe current knowledge about *mecC* detection in humans and its possible clinical implications, as well as to determine the infection types associated with this resistance mechanism, the characteristics of these *mecC*-carrying isolates, their possible relationship with animals and the presence of other risk factors.

## 2. Detection of *mecC*-MRSA Isolates in Humans

### 2.1. Human Studies Related to mecC-MRSA

Although the *mecC* gene was initially discovered in an isolate from bulk milk in England, the first human *mecC*-MRSA isolates were also identified in that same study [16]. These human isolates were obtained from patients from the United Kingdom and Denmark. Moreover, in a publication from the same year, two human MRSA isolates carrying this new resistance gene were independently identified in Ireland [17].

Since then, several retrospective and prospective studies using human *S. aureus* isolates/samples were carried out in order to search for *mecC*-MRSA isolates (Tables 1 and 2) [16,18,24–74]. Most of these studies were performed in European countries (Tables 1 and 2), and the UK and Denmark were the countries in which the highest levels of *mecC*-MRSA isolates were detected [16,24,25,39,41].

**Table 1.** Human studies related to *mecC*-MRSA isolates in which prevalence can be estimated [1]

| Reference | Country | Sampling Date | Prevalence: *mecC* Positive Isolates/*S. aureus* or Methicillin Resistant *S. aureus* (MRSA) (%) | Type of Sample/Infection (Number of Isolates) | Clonal Complex: Sequence Type [2] (Number of Isolates) | IEC [3] (Number of Isolates) | Non-beta lactam Resistance (Number of Isolates) [4] | Possible Relationship with Animals |
|---|---|---|---|---|---|---|---|---|
| [18,24] | Denmark | 1960–2011 | 112 (0.21%)/53746 MRSA | Wound (37), skin (26), blood (8), post-operative wound (5), urine (4), eye/ear (2), impetigo (2), unknown (28) | CC130 (98)/CC2361 (14): ST2173, ST2174 | Negative (2) | Q (NOR) (1), S (111) | Most were from rural areas (106): 4 with contact with animals |
| [16] | United Kingdom (UK) and Denmark | 1975–2011 | 51 (0.04%)/approximately 120500 *S. aureus* | Screen swab (10), Skin and soft tissue infections (7), nose (5), wound (5), blood (4), skin (4), nose/mouth (2), ear (1), eye/ear (1), finger (1), fluid (1), hand (1), PEG site(1), sputum (1), toe (1), unknown (6) | CC130: ST130 (18), ST1245 (3), ST1764 (3), ST1945 (3), ST1526 (1), ST1944 (1), n/d (17)/CC1943: ST1943 (1), ST1946 (1)/CC425: ST425 (3) | - | S (51) | - |
| [25] | Denmark | 1975–011 | 127/-: in routine testing 12 (5.91%)/203 MRSA | - | CC130 (107): ST130, ST1245, ST1526, ST1945/CC1943 (14): ST1943, ST1946, ST2173, ST2174/CC425 (6): ST425 | - | - | - |
| [26] | Ireland | 2000–2012 | 1 (1.14%)/88 MRSA | - | CC130 (1) | Negative (1) | Q (NOR) (1) | Patient lived on a Farm |
| [27] | Germany | 2000–2016 | 2 (0.16%)/1277 MRSA | - | CC130 (2) | - | - | - |
| [28] | Austria | 2002–2012 | 1 (0.31%)/327 MRSA | - | CC130 (1) | - | - | - |
| [29] | Belgium | 2003–2012 | 9 (0.18%)/4869 *S. aureus* | Screen swab (4), urine (2), wound (2), sputum (1) | CC130 (4)/CC49 (3)/CC1943 (2) | - | S (9) | Most were from a rural area with a high density of cattle farms |
| [30,31] | Germany and The Netherlands | 2004–2011 | 16/– | nasal swab (11), wound (2), joint aspirate (1), mouth swab (1), sputum (1) | CC130 (14)/CC1943: ST2361 (1)/CC599: ST599 (1) | Negative (16) | S (1) | - |
| [32] | Germany | 2004–2005 2010-2011 | 2 (0.06%)/3207 MRSA | Screen swab (1), sputum (1) | - | - | - | - |
| [33] | Germany | 2006–2011 | 11 (0.09%)/12691 MRSA | Wound (8), dermatitis (1), nasal swab (1), nosocomial pneumonia (1) | CC130 (11) | Negative (11) | Q [CIP (1), MFL (1)] (2), S (9) | Veterinarian (1) |
| [34,35] | UK | 2006–2012 | 2/– | Screen swab (2) | CC130 (2) | - | - | - |
| [36] | Slovenia | 2006–2013 | 6 (1.52%)/395 MRSA | Wound (4), Screen swab (2) | CC130: ST130 (6) | Negative (6) | S (6) | Most were from rural areas |
| [37] | Spain | 2008–2013 | 2 (0.04%)/5505 *S. aureus* | Joint fluid (1), wound (1) | CC130: ST1945 (2) | - | S (2) | -. |
| [38] | Austria | 2009-2013 | 6 (2%)/301 *S. aureus* | blood (2), screen swab (2), wound (2) | CC130: ST130 (3), new SLV (1)/ CC599: ST599 (2) | - | S (6) | Contact with pet rabbit (1), unknown (5) |

**Table 1.** *Cont.*

| Reference | Country | Sampling Date | Prevalence: *mecC* Positive Isolates/*S. aureus* or Methicillin Resistant *S. aureus* (MRSA) (%) | Type of Sample/Infection (Number of Isolates) | Clonal Complex: Sequence Type [2] (Number of Isolates) | IEC [3] (Number of Isolates) | Non-beta lactam Resistance (Number of Isolates) [4] | Possible Relationship with Animals |
|---|---|---|---|---|---|---|---|---|
| [39] | Denmark | 2010-2011 | 6 (6.32%)/95 MRSA | - | - | - | - | - |
| [40] | England | 2011-2012 | 9 (0.45%)/2010 MRSA | Screen swab (6), wound (2), leg ulcer (1) | CC130: ST130 (2), ST1245(4), ST2573 (1), ST2574 (1)/CC425: ST425 (1) | - | L (CLI) (1) -M (ERY) (1), S (8) | - |
| [41] | UK | 2012-2013 | 12 (0.53%)/2282 MRSA | Screen swab (9), SSTI (3) | CC130: ST1245 (6), ST130 (2), ST1945 (1), ST2574 (1)/CC425: ST425 (1)/CC1943: ST1943 (1) | Negative (11)/type E (1) | M (ERY) (1), S (11) | - |
| [42] | England | 2015 | 1 (0.08%)/1242 MRSA | Screen swab (1) | CC130: ST130 (1) | Negative (1) | S (1) | - |
| [43] | England | 2018-2019 | 1 (0.7%)/142 *S. aureus* | - | - | - | - | - |
| [44] | Germany, UK, Belgium | - | 80/- | - | - | - | - | - |

[1] Case reports were also analyzed in some other studies but, in this table, only results from prevalence studies are included. [2] CC, clonal complex; ST, sequence type; [3] IEC, immune evasion cluster; [4] L, resistant to lincosamides (CLI, clindamycin); M, resistant to macrolides (ERY, erythromycin); Q, resistant to fluoroquinolones (CIP, ciprofloxacin, MFL, moxifloxacin, NOR, norfloxacin); S, susceptible to all non-beta lactam agents tested. UK, United Kingdom.

**Table 2.** Studies performed on humans, in which *mec*C-MRSA isolates were sought but not detected.

| Reference | Country | Sampling Date | Type of Samples [1] | Number of Samples or (*S. aureus* or MRSA) Isolates |
|---|---|---|---|---|
| [45] | Switzerland | 2005–2012 | Clinical/screening | 1695 *S. aureus* isolates |
| [46] | Ghana | 2007–2012 | Clinical | 9834 blood samples |
| [47] | Turkey | 2007–2014 | Clinical | 1700 *S. aureus* isolates |
| [48] | Belgium | 2009–2011 | Screening | 149 farmers and family members (41 MRSA isolates) |
| [49] | Hungary | 2009–2011 | Screening | 878 children |
| [50] | United States | 2009–2011 | Clinical/screening | 364 *S. aureus* isolates (102 MRSA isolates) |
| [51] | Ireland | 2011 | Screening | 64 residents |
| [52] | UK | 2011 | Screening | 307 cattle veterinarians |
| [53] | Jordan | 2011–2012 | Screening | 716 humans (56 MRSA isolates) |
| [54] | Germany | 2011–2013 | Screening | 1878 non-hospitalized adults |
| [55] | Belgium | 2012–2013 | Clinical | 510 cystic fibrosis patients |
| [56] | Greece | 2012–2013 | Screening | 18 veterinary personnel |
| [57] | The Netherlands | 2012–2013 | Clinical/screening | 13,387 samples |
| [58] | UK | 2012–2013 | Clinical | 500 *S. aureus* isolates |
| [59] | Egypt | 2013 | Clinical/screening | 1300 dental patients |
| [60] | Taiwan | 2013–2014 | Clinical/screening | 3717 *S. aureus* isolates |
| [61] | Turkey | 2013–2014 | Screening | 7 MRSA isolates |
| [62] | Turkey | 2013–2016 | Clinical/screening | 494 MRSA isolates |
| [63] | Spain | 2014 | Screening | 15 humans in contact with animals |
| [64] | Poland | 2014–2016 | Screening | 955 students (only one MRSA isolate) |
| [65] | Germany | 2015 | Clinical | 140 Gram-positive isolates |
| [66] | UK | 2015 | Clinical | 520 *S. aureus* isolates |
| [67] | United States | 2015 | Screening | 479 patients |
| [68] | India | 2015–2017 | Screening | 32 animal handlers |
| [69] | Spain | 2016 | Clinical/screening | 45 non-beta-lactam susceptible MRSA isolates |
| [70] | Greece | 2016–2017 | Screening | 68 farmers |
| [71] | Denmark | 2017 | Screening | 16 workers at wildlife rehabilitation centres |
| [72] | Italy | 2017–2018 | Clinical/screening | 102 MRSA isolates |
| [73] | Egypt | - | Screening | 223 health care personnel |
| [74] | Madagascar | - | Screening | 1548 students and healthcare workers |

[1] Screening: isolates obtained in epidemiological studies for colonization detection.

Unfortunately, the design of these studies was very different, which complicates any comparison of the data obtained. Importantly, the criteria chosen for the selection of the initial isolates/samples varied significantly. While all *S. aureus* isolates were collected in some studies [29,38,45], only MRSA isolates were included in others [24,26–28,32,33,36,41,42]. Moreover, several studies were more restrictive and only used isolates that showed characteristics suspected of carrying the *mecC* gene such as *spa* types associated with *mecC*-positive clonal lineages previously described, *mecA*-negative MRSA isolates, isolates with antimicrobial susceptibility suspected to be *mecC*-positive or *pvl*-negative MRSA isolates [25,26,37,69] (Table S1). In any case, the *mecC*-MRSA human prevalence detected in most of the studies was very low. Several studies did not identify any *mecC*-positive *S. aureus* among included human isolates/samples (Table 2) [45–74]. In studies in which this gene was detected (Table 1), the prevalence identified, considering the total number of isolates/samples included, was < 1% in most of the cases [24,27–29,32,33,37,40–43], similar to that identified in the first study in which *mecC* was discovered (approximately 0.04%) [16]. In a few studies, the prevalence was > 1% but, in all of these, only a small number of initial isolates (<400 isolates) was used; this may be the reason for the high prevalence value obtained (up to 6.3%) [25,26,36,38,39]. Recently, a meta-analysis of the prevalence of *mecC*-MRSA, based on previously published results, estimated the prevalence of *mecC*-MRSA in the human subgroup at 0.004% (95% CI = 0.002–0.007), and the prevalence in the animal subgroup to be 0.098% (95% CI = 0.033–0.174) [75].

### 2.2. *mecC-MRSA Human Case Reports*

A total of 61 human case reports associated with *mecC*-MRSA isolates has been described (Table 3) [17,36,37,45,76–81]. Although *mecC*-positive isolates have been identified in Asia, Europe, and Oceania [21,82,83] in different hosts, all human case reports were described in European countries (Table 3). This was to be expected considering that the majority of the papers in which *mecC*-MRSA has been detected in both animals and humans, as well as in environmental samples, have been focused on countries on this continent [8,21–23].

In 4 of the 61 human case reports, *mecC*-MRSA was only identified in screen swabs (for colonization detection), with it not being related to the cause of the patient's admission [36,37,45], and the clinical information was not indicated in another two case reports [17]. In the remaining 56 studies, *mecC*-MRSA isolates were related to (number of cases): skin and wound infections (47 cases) [37,76,79,81], joint and bone infections (3 cases) [37,77,78], respiratory infections (2 cases) [76] and bacteremia (2 cases) [37,80]. Taking into consideration the type of samples in which *mecC*-positive isolates have been detected in humans (Tables 1–4), most *mecC* human cases were implicated in skin or wound infections. However, the detection of *mecC*-MRSA isolates in other types of samples such as blood, sputum or urine is remarkable (Table 4). Pertinently, some serious infections have been described, such as severe bone infections [78], nosocomial pneumonia [33] and bacteremia [16,24,80], in some cases ending with the death of the patient [37].

**Table 3.** Human *mecC* MRSA case reports.

| Reference [1] | Country | Sampling Date | Number of Described Case Reports | Year-Old Patient | Type of Sample/Infection [2] | Clonal Complex: Sequence Type [3] | IEC [4] | Non-Beta Lactam Resistance [5] | Possible Relationship with Animals |
|---|---|---|---|---|---|---|---|---|---|
| [76] | Sweden | 2005–2014 | 45 | Median age (range) 60 (2–86) | Wound, sputum, nasopharynx | CC130/CC2361 | - | L-M (1 isolate), S (44 isolates) | Most were from a rural area: farmer (1), patients lived on farms (4) |
| [77] | France | 2007 | 1 | 67 | Fluid of lesion heel | CC130: ST1945 | - | - | - |
| [37] | Spain | 2008–2013 | 7 | 3, 50, 63, 64, 76, 80, 85 | Blood, joint fluid, nasal screen swab, urine, wound | CC130: ST130, ST1945 | - | S | No epidemiological data were available except for one patient who did not have any contact with animals |
| [78] | France | 2010 | 1 | 48 | Blood, ear fluid, retrosternal abscess | CC130 | Negative | S | Contact with cows |
| [17] | Ireland | 2010 | 2 | 64, 85 | - | CC130: ST130, ST1764 | Negative | S (but detection of *tet* efflux) | - |
| [45] | Switzerland | 2011 | 1 | 59 | Groin, nose, and throat screen swab | CC130: ST130 | - | - | Contact with a cat |
| [79] | Spain | 2012 | 1 | 46 | Skin lesion swab | CC130 | - | S | Patient lived in rural area with high density of livestock animals |
| [36] | Slovenia | 2013 | 1 | 86 | Nose and skin screen swabs | CC130: ST130 | Negative | S | Patient lived on a farm and had contact with pigs, cats and dogs |
| [80] | Spain | 2013 | 1 | 76 | Blood | CC130: ST1945 | - | S | No contact with livestock |
| [81] | Spain | 2013–2014 | 1 | 34 | Superficial skin lesion swab | CC130: ST130 | Negative | S | Contact with livestock animals |

[1] Prevalence studies were also included in some papers but, in this table, only case report results are present. [2] Screen swab: sample for colonization detection. [3] CC, clonal complex; ST, sequence type; [4] IEC, immune evasion cluster. [5] L, resistant to lincosamides; M, resistant to macrolides; S, susceptible to all non-beta lactam agents tested.

**Table 4.** Type of sample/infection in which *mecC*-MRSA isolates have been identified among human patients.

| Type of Sample/Infection | Number of Isolates [2] | References |
|---|---|---|
| Screen swab [1] | 54 | [16,29–34,36,38,40–42,45,76] |
| Skin lesion/dermatitis/impetigo wound/post-operative wound/skin and soft tissue infections | 158 | 16,24,29,30,33,36-38,40,41,76,79,81 |
| Blood | 16 | [16,24,37,38,80] |
| Urine | 7 | [24,29,37] |
| Nosocomial pneumonia/sputum/ Tracheal aspirate | 7 | [29,30,32,33,76] |
| Nose | 5 | [16] |
| Eye/ear | 3 | [16,24] |
| Fluid of heel/joint fluid | 3 | [30,37,77] |
| Mouth/Nose | 2 | [16] |
| Ear | 1 | [16] |
| Finger | 1 | [16] |
| Fluid | 1 | [16] |
| Hand | 1 | [16] |
| Percutaneous endoscopic gastrostomy site | 1 | [16] |
| Retrosternal abscess | 1 | [78] |
| Toe | 1 | [16] |
| Unknown | 34 | [16,24] |

[1] In screen swab: all samples in which was clearly indicated that they did not cause infection were included. However, in several studies it was not indicated whether samples were screen samples or if these samples were taken in infection sites. [2] In human case reports, only one isolate from the most representative infection sample was considered.

## 3. Risk Factors for *mecC*-MRSA Infection

### 3.1. Contact with Animals

Since the first description of the *mecC* gene, contact with animals has been considered to be a risk factor for *mecC*-MRSA infection or carriage for several reasons [16,17]. This gene was identified in isolates belonging to CC130, and this clonal complex was predominantly detected among methicillin-susceptible *S. aureus* (MSSA) isolates from bovine sources [17]. Moreover, the discovery of this gene in isolates obtained from dairy cows suggested that these animals might provide a reservoir of this resistance mechanism [16]. Thereby, in some of the studies carried out since then, information about the possible contact of patients with animals was indicated (Tables 1 and 3). Many studies found out that most of the patients lived in rural areas or areas with a high density of farms [18,24,26,29,36,76,79]. In this sense, four studies indicated patient contact with livestock or farm animals [18,24,76,78,81], two referred to only contact with pets [38,45], two patients had no contact with animals and the authors did not have a plausible explanation for the detection of these isolates [37,80], one patient was a veterinarian [33], and in several studies this information was not indicated [16,17,27,30,31,41,77]. Interestingly, *mecC*-MRSA transmission between animals and humans was demonstrated in two human infection cases by whole genome sequencing. Specific clusters including isolates from each human infection case and their own livestock were detected. Thus, human and animal isolates from the same farm only differed by a small number of SNPs [18]. These findings highlight the role of livestock as a potential reservoir for *mecC*-MRSA.

### 3.2. *mecC*-MRSA Carriage in Humans

*S. aureus* shows a great capacity to colonize the skin and nares of hosts, being able to last over time and cause opportunistic infections [84,85]. *mecC*-MRSA isolates were identified as commensals in several prevalence and case report studies (see screen swab in Tables 1–4). At least 54 *mecC*-MRSA positive isolates were obtained from screen swabs, mainly from the nose, but also from throat and groin sites. Moreover, isolates obtained from other types of samples could also be considered as commensals, as in one human case report in Spain in which the isolate recovered from the urine of one patient was considered as a colonizer since the patient did not present urinary symptoms [37].

*mecC*-MRSA isolates implicated in both colonization and infection were obtained from the same patient in some studies [18,37]. Indeed, one patient with bacteremia due to an *mecC*-MRSA isolate also presented nasal colonization by the same *mecC*-MRSA isolate (with the same genetic characteristics) [18]. These results corroborated the importance of colonization being the previous step, which enables isolates causing severe disease. Interestingly, in another bacteremia case in which the patient died, a household transmission between grandfather and grandson was detected, with the grandson being colonized by the same isolate [37]. Nevertheless, in other studies, *mecC*-MRSA isolates were not identified as colonizers from patients with *mecC* infections [81], and it has been suggested that *mecC*-MRSA isolates might be worse colonizers and less contagious in humans than *mecA*-MRSA isolates [76]. In the study carried out in Sweden, only two out of the patient's 27 family members were positive for *mecC*-MRSA isolates and the median time for *mecC* carriage was 21 days [76].

### 3.3. *Patient Age*

Most of the patients described in case reports (Table 3) were middle-aged or elderly [17,36,37,45,77–80], except two patients: one of them was a 34 year-old farm worker with high contact with animals who presented a superficial skin lesion [81], and the other was a healthy 3 year-old child [37]. The average age of patients with *mecC*-MRSA detected in Denmark during 2007–2011 was 51 [24] and the average detected in Sweden in 2005–2014 was 60 [76]. In the Danish study, CA-MRSA *mecC* patients were significantly older than other CA-MRSA cases, indicating that *mecC*-MRSA seems to have a different origin and epidemiology to typical CA-MRSA [24].

### 3.4. *Underlying Chronic Disease*

Remarkably, in the 45 human cases detected in Sweden, most patients had some kind of underlying chronic disease (diabetes mellitus, cancer, autoimmune diseases or atherosclerotic diseases), or an existing skin lesion [76]. Infection by *mecC*-MRSA of wounds has also been suggested by others [79]. Moreover, *mecC*-MRSA infections were identified in patients with primary pathologies (diabetes, myelodysplastic syndrome, peripheral arterial occlusion disease, etc.) in one study in Austria [38], and in a patient with an urothelial carcinoma in Spain [80]. Unfortunately, information about other underlying diseases of *mecC*-MRSA positive patients is missing in most of the papers.

## 4. Characterization of *mecC*-MRSA Human Isolates

### 4.1. *Clonal Lineages of mecC-MRSA of Human Origin*

As in other hosts, most of the *mecC*-MRSA isolates obtained from human samples belonged to CC130 (Tables 1 and 3) (Figure 1). Other clonal complexes identified were CC49, CC425, CC599, CC1943 and CC2361 [16,24,25,29–31,38,40,41,76] (Table 1) (Figure 1). Worryingly, it has been hypothesized that SCC*mec* XI (the SCC element that contains the *mecC* gene) might have the potential to be transferred to other *S. aureus* clonal lineages due to the fact that it is bounded by integration site sequence repeats and that it has intact site specific recombination components [16] (Table S1 and S2). Until now, *mecC*-MRSA CC130 isolates have been identified in all countries in which clonal lineages were determined and it was the unique CC detected in Spain, France, Ireland, Slovenia and Switzerland [17,37,45,77–79,81] (Figure 1). Remarkably, in France and Spain there were several human infection reports, but in all of them the *mecC*-MRSA isolates belonged to CC130 (Table 3). After CC130, the clonal complexes

CC1943 and CC599 were the most widely detected in humans, being identified in four and three countries respectively [16,25,29,31,38,41] (Figure 1). Conversely, CC49 was only described in one study in Belgium [29]. While CC49, CC130, CC425, CC599 and CC1943 were also identified in *mecC*-MRSA isolates from a non-human origin, CC2361 has been only described in humans so far [24,76]. Thus, CC130 was described in farm, domestic and wild animals and in food samples; CC49 in horses and small mammals, CC425 in wild animals and food, CC599 in pets and farm animals and CC1943 in pets [8].

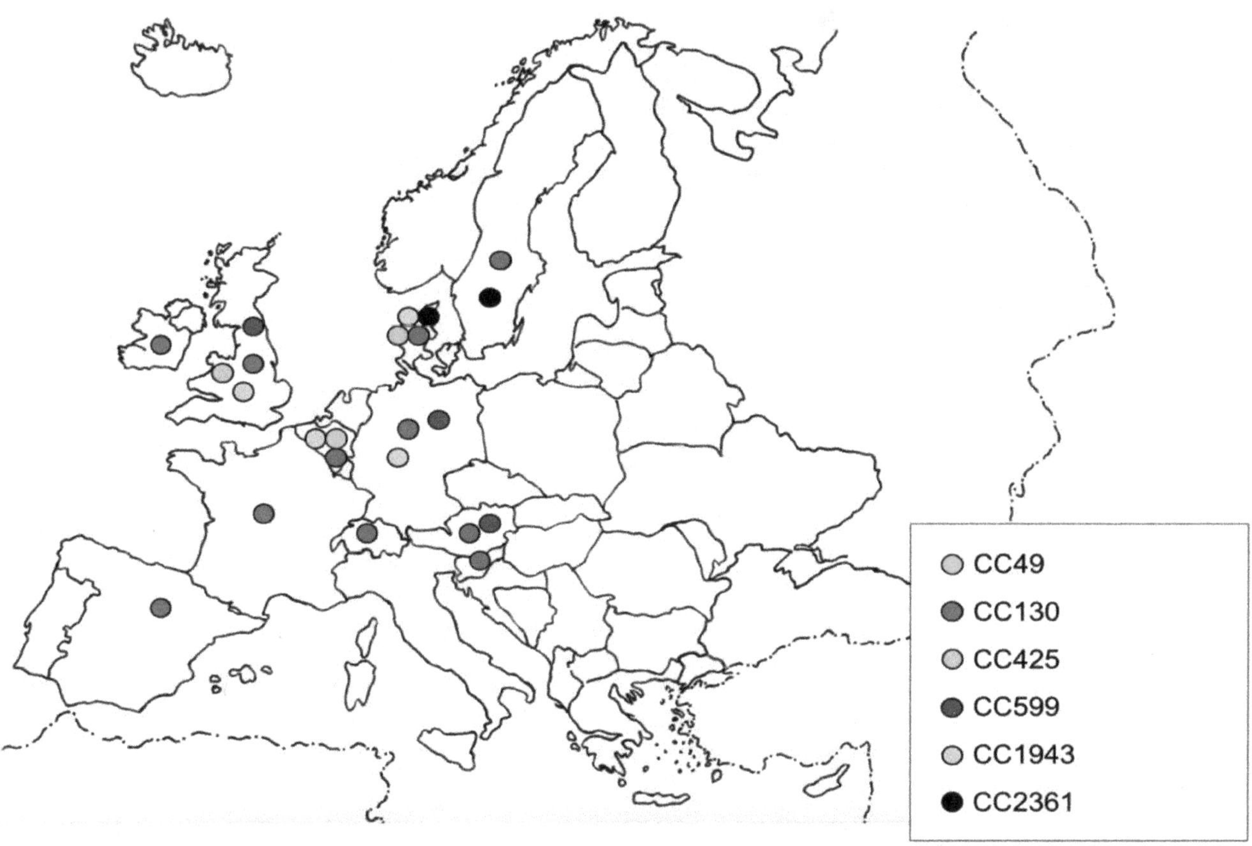


**Figure 1.** Clonal complexes (CCs) detected in *mecC*-MRSA human isolates.

A large variety of *spa* types was detected among the human *mecC*-MRSA isolates (Figure 2). The most predominant *spa* type was t843, which is associated with CC130 and was identified in a total of 260 human isolates. This *spa* type was detected in all countries in which human *mecC*-MRSA isolates were detected, except in Switzerland [45]. Other *spa* types were also described in several countries. Some of them were identified only in two countries, this is the case of t792, t1773, t5930, t6293, t6386, t7485, t7734, t7945, t7946, t7947 and t9397, but others were more widely spread as t978, t1535, t1736, t3391 or t6220 (Figure 2). Although there is a strong association among *spa* types and MLST clonal complexes [86], some *spa* types were associated with different clonal complexes. Two isolates obtained in screen swabs from two patients in two different hospitals from England presented the *spa* type t11706 [40]; one of these isolates belonged to ST1245 (CC130) and the other one to ST425 (CC425). Moreover, the *spa* types t978, t2345, t3391 and t8835 were associated in some studies with CC1943 [16,25,29], and in others with CC2361 [24,76]. Nevertheless, the founders of both clonal complexes, ST1943 and ST2361, are Single Locus Variant (SLV) of each other (and only differ at the *aroE* allele), which could explain these results (Table S1).

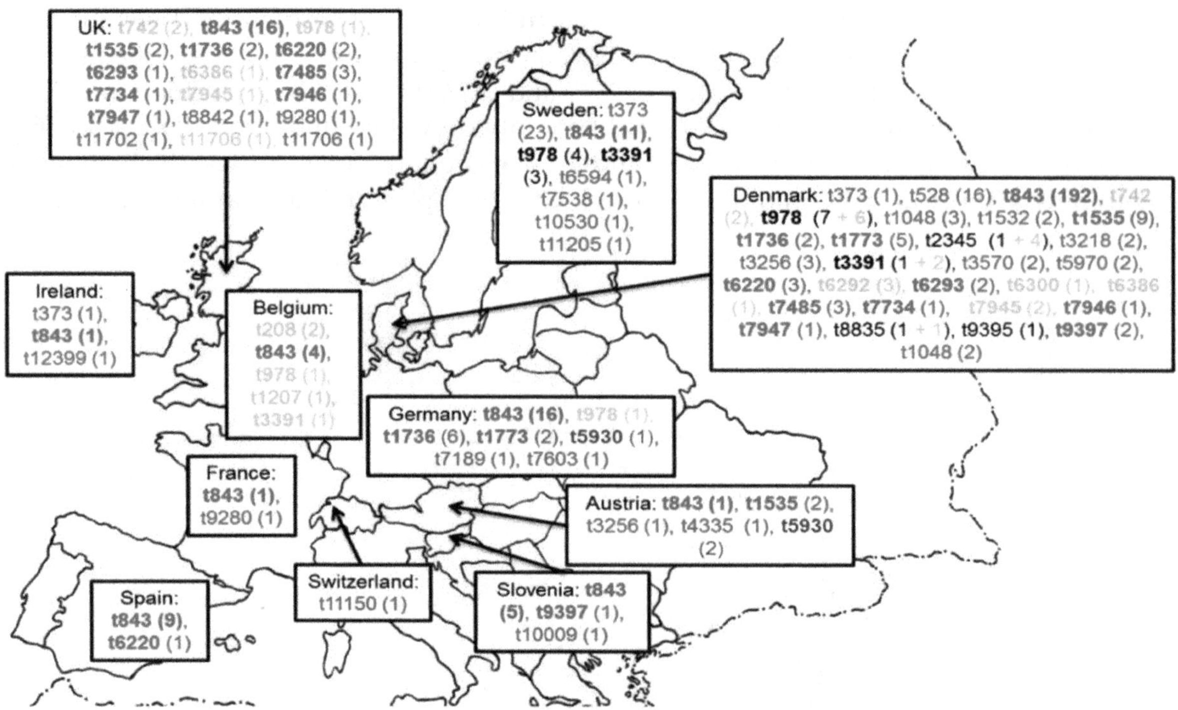


**Figure 2.** *spa* types detected in *mecC*-MRSA isolates of humans. Colors indicate the clonal complexes associated with each *spa* type: green CC49, red CC130, blue CC425, purple CC599, orange CC1943, black CC2361. The number of isolates of each *spa* type is indicated in parentheses (to calculate the number of isolates in human case reports, only one isolate from each *spa* type and each patient was considered)

### *4.2. Treatment and Antimicrobial Resistance Profile of mecC-MRSA Human Isolates*

Most of the human *mecC*-MRSA isolates detected were susceptible to all non-beta-lactam antimicrobials tested (Tables 1 and 3). This is in accordance with results obtained in *mecC*-MRSA isolates from other origins [21]. In one study performed in Spain, isolates using this criterion were selected in order to identify *mecC*-MRSA or CA-MRSA isolates [69]. Although *mecC*-MRSA was not detected, this resistance phenotype was a valuable marker for Panton-Valentine Leukocidin (PVL)-producer isolates (Table S1). Nevertheless, the low prevalence of *mecC*-MRSA isolates in this region could be responsible for this result and the use of this phenotype to suspect the presence of the *mecC* mechanism should not be discarded.

The most important problem for treating *mecC*-MRSA infections is that these isolates must be correctly identified. Although *mecC* isolates are considered to be MRSA, these isolates sometimes show borderline susceptibility results for oxacillin or cefoxitin, being identified as MSSA if the *mecA* gene is only tested [44]. This could lead to the implementation of inappropriate therapies. Resistance development to other antimicrobial agents could be added to this, considering the capacity of *S. aureus* to acquire resistance to various antimicrobial agents. Some *mecC*-MRSA isolates detected in humans have shown resistance to non-beta-lactam antimicrobials [24,33,40,41,76] (Table 1). Fluoroquinolone resistance was identified in two isolates in Germany [33] and in one isolate in Denmark [24]. Macrolide and lincosamide resistance was also detected in some studies: one erythromycin resistant isolated in the UK [41] and one erythromycin and clindamycin resistant isolate in Sweden [76] and England [40]. Regarding resistance mechanisms, in two studies carried out in Ireland, the gene *sdrM*, which encodes a multidrug efflux pump related to norfloxacin resistance and *tet* efflux related to tetracycline resistance, were identified in one and two *mecC*-MRSA CC130 isolates, respectively [17,26]. Although there has only been one study, whose objective was to compare different diagnostic tests, human *mecC*-MRSA isolates resistant to several antimicrobial families have been detected [87]. The presence of resistance to non beta-lactam agents in *mecC*-MRSA isolates significantly limits our therapeutic options.

### 4.3. Virulence of mecC-MRSA Human Isolates

The search for virulence genes in human *mecC*-MRSA isolates has been highly variable from one study to another. In any case, for the moment, the most common virulence genes detected have been *hla, hld, hlb, edinB, lukED, cap8* or *ica,* with these genes being highly associated with CC130 [17,26,31,33,36,38,41]. Fortunately, no *mecC*-MRSA isolates carrying the PVL genes were detected. However, other clonal lineages associated with animals have been able to acquire this virulence factor [88]. For this, their detection in *mecC*-MRSA isolates cannot be ruled out in the future. Significantly, some pyrogenic toxin superantigen (PTSAg) genes have been detected in *mecC*-MRSA isolates [29,31,38,41]. These genes might be related to specific clonal lineages such as CC599, CC1943 or CC2361. Thus, the gene *tst* encoding the toxic shock syndrome toxin has been found in three CC1943 isolates (two harboring *sec* gene and one containing *seg* and *sei* genes) [29,41], in three CC599 isolates (two of them positive for *sel* gene and one for *sec* and *sel*) [31,38] and in one *sec, seg, sei, sel, sen, seo* and *seu* positive CC2361 isolate [31] (Table S1).

## 5. *mecC*-MRSA Problem: What is its Origin? Is It an Emerging Problem?

The oldest known *mecC*-MRSA isolate, dated in 1975, was detected in the retrospective study performed by García-Álvarez et al. [16] This isolate was identified in a human blood sample from Denmark and its detection suggested a possible human origin for the *mecC* gene [16]. Later, in two other retrospective studies also carried out in Denmark, two *mecC*-positive isolates were identified in samples dated in 1975 [24,25], both were also identified in human blood samples [24,25]. However, in two of these studies, the oldest sample studied was obtained in 1975, so the presence of older isolates cannot be ruled out [16,25]. With respect to the remaining retrospective studies in which the presence of the *mecC* gene was sought, the dates of the samples were much later than these three studies, with them being isolates obtained from the year 2000 and later (Table 1). Regarding the earliest reported *mecC*-MRSA isolate in other hosts, 1975 also seems to be the key date [89,90]. Therefore, this resistance mechanism might have been present for over 45 years.

Moreover, this resistance mechanism is highly associated with CC130 since most of the *mecC*-MRSA isolates belong to this clonal lineage. A human-to-bovine host-jump of CC130, which occurred ~5429 years ago, has been suggested [91]. The time and host in which CC130 MSSA isolates acquired the *mecC* resistance gene remain unknown today. In order to establish a possible human or animal origin for the detected *mecC*-MRSA isolates in human samples, several studies analyzed the presence of IEC (immune evasion cluster) genes [17,24,26,31,33,36,41,42,78,81] (Tables 1 and 3). In all cases, human *mecC*-MRSA isolates were negative for *sak, chp* and *scn* except for one ST1945 (CC130) isolate obtained from a screen swab from a patient in the UK that was positive for *sak* and *scn* (IEC type E), suggesting a possible human origin [41]. Nevertheless, it has been suggested that IEC type E might be a conserved part of ST1945 since *mecC* MRSA ST1945 isolates from wild animals also showed IEC type E in several studies in Spain [41,63,92].

The newest human *mecC*-MRSA isolates detected so far in Europe were obtained in 2015, one of them in Germany [27], and the other in England [42]. Both strains showed similarities to those identified in the first studies [16,17] and both belonged to CC130. Nevertheless, after phylogenetic analysis, the strain identified in England seemed not to be highly related to any of the published sequenced *mecC*-MRSA CC130 isolates [42]. Despite the non-description of *mecC*-MRSA isolates in humans in the last 5 years in Europe, data provided by previous studies have detected an increasing tendency in the prevalence of *mecC*-positive isolates [24], indicating that surveillance in detecting this resistance mechanism must be maintained. The lack of detection could be due to the low prevalence of this resistance mechanism and/or problems in *mecC* diagnostic methods. Important difficulties in the detection of *mecC*-MRSA isolates have been indicated [44,93]. It has been shown that various clinical tests used in hospital labs might have failed to identify 0 to 41% of *mecC*-MRSA isolates [93]. It is important to optimize and develop new testing protocols and redefine currently available phenotypic testing methods [44]. In this regard, several commercial PCR-based tests that detect *mecC* and *mecA* genes have been developed. Moreover, recently, *mecA/mecC* MRSA isolates from cattle have been described [83]. The possible clinical impact of isolates carrying both genes is currently unknown.

## 6. Implications in Veterinary and Food Safety

Although this review is focused on the human health implications of *mecC*-MRSA isolates, the effects that these isolates can have on veterinary medicine should not be forgotten. *mecC*-MRSA isolates causing infections in domestic animals have been identified in several studies [8,94,95]. However, this resistance gene seems to be most frequently detected in livestock animals including cattle, sheep and rabbits [8,21]. Although *mecC*-MRSA rarely causes clinical disease in these food-producing animals, there are reports of bovine mastitis in several countries [96,97]. As observed in humans, most of the *mecC*-MRSA isolates detected in other hosts also belong to CC130, with the characteristics of these animal *mecC*-MRSA isolates being very similar to those detected in humans [8,21].

On the other hand, the presence of *mecC*-MRSA in dairy animals is highly relevant since it could be a route of entry of these isolates into the food chain. Indeed, milk samples have been identified carrying *mecC*-MRSA [8] with the consequent risk of colonization for food handlers. In this case, it is worth highlighting one of the clinical cases included in this review in which the patient was a cheese producer [81]. *mecC*-MRSA zoonotic transmission has been demonstrated in some studies [18], with the correct prevention, detection and control measures in veterinary and food safety laboratories being necessary.

## 7. Conclusions and Future Problems Associated with *mecC*

Currently, the prevalence of human *mecC*-MRSA infections is very low. However, *mecC*-MRSA isolate transmission between different hosts indicates the great capacity of these isolates for spreading. There is a wide range of reservoirs in wild, livestock and companion animals and zoonotic transmission of these isolates could increase the number of *mecC*-MRSA human clinical cases. Moreover, SCC*mec* XI might have the potential to be transferred to other clonal lineages in the future. Their transfer to more virulent and better-adapted human clones would be deeply troubling. Worryingly, the *mecC* gene has already been detected in clonal lineages in which important virulence genes were identified (CC599, CC1943 or CC2361) or in which IEC was described (ST1945-CC130). Moreover, *mecC*-MRSA isolates resistant to non-beta lactams have been detected. Acquisition of non-beta lactam resistance by *mecC*-MRSA isolates significantly limits our therapeutic options. *mecC*-MRSA should be taken into consideration in hospital and veterinary laboratories and in food safety institutions, and prevention strategies must be implemented in order to avoid possible emerging health problems.

**Author Contributions:** C.L. contributed to the search of articles and to their tabulation and classification into different categories. She also contributed to the design and the analysis of the review and to the writing of the paper. R.F.-F., L.R.-R., P.G. and M.Z. helped in the general review of the manuscript. C.T. contributed with project funding, and with the original idea and the design of the manuscript and reviewed the initial version of the manuscript. All authors have read and agreed to the published version of the manuscript.

## References

1. Cheng, M.; Antignac, A.; Kim, C.; Tomasz, A. Comparative study of the susceptibilities of major epidemic clones of methicillin-resistant *Staphylococcus aureus* to oxacillin and to the new broad-spectrum cephalosporin ceftobiprole. *Antimicrob. Agents Chemother.* **2008**, *52*, 2709–2717. [CrossRef] [PubMed]

2. Jacqueline, C.; Caillon, J.; Mabecque, V.L.; Miègeville, A.F.; Ge, Y.; Biek, D.; Batard, E.; Potel, G. In vivo activity of a novel anti-methicillin-resistant *Staphylococcus aureus* cephalosporin, ceftaroline, against vancomycin-susceptible and -resistant *Enterococcus faecalis* strains in a rabbit endocarditis model: A comparative study with linezolid and vancomycin. *Antimicrob. Agents Chemother.* **2009**, *53*, 5300–5302. [CrossRef] [PubMed]
3. Mediavilla, J.R.; Chen, L.; Mathema, B.; Kreiswirth, B.N. Global epidemiology of communityassociated methicillin resistant *Staphylococcus aureus* (CA-MRSA). *Curr. Opin. Microbiol.* **2012**, *15*, 588–595. [CrossRef] [PubMed]
4. Sung, J.Y.; Lee, J.; Choi, E.H.; Lee, H.J. Changes in molecular epidemiology of communityassociated and health care-associated methicillin-resistant *Staphylococcus aureus* in Korean children. *Diagn. Microbiol. Infect. Dis.* **2012**, *74*, 28–33. [CrossRef] [PubMed]
5. Herron-Olson, L.; Fitzgerald, J.R.; Musser, J.M.; Kapur, V. Molecular correlates of host specialization in *Staphylococcus aureus*. *PLoS ONE* **2007**, *2*, e1120. [CrossRef] [PubMed]
6. Aspiroz, C.; Lozano, C.; Vindel, A.; Lasarte, J.J.; Zarazaga, M.; Torres, C. Skin lesion caused by ST398 and ST1 MRSA, Spain. *Emerg. Infect. Dis.* **2010**, *16*, 157–159. [CrossRef] [PubMed]
7. Lozano, C.; Aspiroz, C.; Ezpeleta, A.I.; Gómez-Sanz, E.; Zarazaga, M.; Torres, C. Empyema caused by MRSA ST398 with atypical resistance profile, Spain. *Emerg Infect. Dis.* **2011**, *17*, 138–140. [CrossRef]
8. Zarazaga, M.; Gómez, P.; Ceballos, S.; Torres, C. Molecular epidemiology of *Staphylococcus aureus* lineages in the animal-human interface. In *Staphylococcus Aureus*; Fetsch, A., Ed.; Academic Press: Cambridge, MA, USA, 2018; pp. 189–214.
9. Benito, D.; Lozano, C.; Rezusta, A.; Ferrer, I.; Vasquez, M.A.; Ceballos, S.; Zarazaga, M.; Revillo, M.J.; Torres, C. Characterization of tetracycline and methicillin resistant *Staphylococcus aureus* Strains in a Spanish hospital: Is livestock-contact a risk factor in infections caused by MRSA CC398? *Int. J. Med. Microbiol.* **2014**, *304*, 1226–1232. [CrossRef]
10. Larsen, J.; Petersen, A.; Sørum, M.; Stegger, M.; van Alphen, L.; Valentiner-Branth, P.; Knudsen, L.K.; Larsen, L.S.; Feingold, B.; Price, L.B.; et al. Meticillin-resistant *Staphylococcus aureus* CC398 is an increasing cause of disease in people with no livestock contact in Denmark, 1999 to 2011. *Euro Surveill.* **2015**, *20*. [CrossRef]
11. Lekkerkerk, W.; van Wamel, J.B.; Snijders, S.B.; Willems, R.J.; van Duijkeren, E.; Broens, E.M.; Wagenaar, J.A.; Lindsay, J.A.; Vos, M.C. What is the origin of livestock-associated methicillin-resistant *Staphylococcus aureus* Clonal Complex 398 isolates from humans without livestock contact? an epidemiological and genetic analysis. *J. Clin. Microbiol.* **2015**, *53*, 1836–1841. [CrossRef]
12. Feltrin, F.; Alba, P.; Kraushaar, B.; Ianzano, A.; Argudín, M.A.; Di Matteo, P.; Porrero, M.C.; Aarestrup, F.M.; Butaye, P.; Franco, A.; et al. A Livestock-associated, multidrug-resistant, methicillin-resistant *Staphylococcus aureus* Clonal Complex 97 lineage spreading in dairy cattle and pigs in Italy. *Appl. Environ. Microbiol.* **2015**, *82*, 816–821. [CrossRef] [PubMed]
13. Lowder, B.V.; Guinane, C.M.; Ben Zakour, N.L.; Weinert, L.A.; Conway-Morris, A.; Cartwright, R.A.; Simpson, A.J.; Rambaut, A.; Nübel, U.; Fitzgerald, J.R. Recent human-to-poultry host jump, adaptation, and pandemic spread of *Staphylococcus aureus*. *Proc. Natl. Acad. Sci. USA* **2009**, *106*, 19545–19550. [CrossRef] [PubMed]
14. Monecke, S.; Gavier-Widén, D.; Hotzel, H.; Peters, M.; Guenther, S.; Lazaris, A.; Loncaric, I.; Müller, E.; Reissig, A.; Ruppelt-Lorz, A.; et al. Diversity of *Staphylococcus aureus* isolates in European wildlife. *PLoS ONE.* **2016**, *11*, e0168433. [CrossRef] [PubMed]
15. Ye, X.; Wang, X.; Fan, Y.; Peng, Y.; Li, L.; Li, S.; Huang, J.; Yao, Z.; Chen, S. Genotypic and phenotypic markers of livestock-associated methicillin-resistant *Staphylococcus aureus* CC9 in humans. *Appl. Environ. Microbiol.* **2016**, *82*, 3892–3899. [CrossRef]
16. García-Álvarez, L.; Holden, M.T.G.; Lindsay, H.; Webb, C.R.; Brown, D.F.J.; Curran, M.D.; Walpole, E.; Brooks, K.; Pickard, D.J.; Teale, C.; et al. Meticillin-resistant *Staphylococcus aureus* with a novel *mecA* homologue in human and bovine populations in the UK and Denmark: A descriptive study. *Lancet Infect. Dis.* **2011**, *11*, 595–603. [CrossRef]

17. Shore, A.C.; Deasy, E.C.; Slickers, P.; Brennan, G.; O'Connell, B.; Monecke, S.; Ehricht, R.; Coleman, D.C. Detection of staphylococcal cassette chromosome *mec* type XI carrying highly divergent *mecA, mecI, mecR1, blaZ,* and *ccr* genes in human clinical isolates of clonal complex 130 methicillin-resistant *Staphylococcus aureus. Antimicrob. Agents Chemother.* **2011**, *55*, 3765–3773. [CrossRef]
18. Harrison, E.M.; Paterson, G.K.; Holden, M.T.G.; Larsen, J.; Stegger, M.; Larsen, A.R.; Petersen, A.; Skov, R.L.; Christensen, J.M.; Zeuthen, A.B.; et al. Whole genome sequencing identifies zoonotic transmission of MRSA isolates with the novel *mecA* homologue *mecC. EMBO Mol. Med.* **2013**, *5*, 509–515. [CrossRef]
19. Becker, K.; Ballhausen, B.; Köck, R.; Kriegeskorte, A. Methicillin resistance in *Staphylococcus* isolates: The "*mec* alphabet" with specific consideration of *mecC*, a *mec* homolog associated with zoonotic *S. aureus* lineages. *Int. J. Med. Microbiol.* **2014**, *304*, 794–804. [CrossRef]
20. Becker, K.; van Alen, S.; Idelevich, E.A.; Schleimer, N.; Seggewiß, J.; Mellmann, A.; Kaspar, U.; Peters, G. Plasmid-Encoded Transferable *mecB*-Mediated Methicillin Resistance in *Staphylococcus aureus. Emerg. Infect. Dis.* **2018**, *24*, 242–248. [CrossRef]
21. Aires-de-Sousa, M. Methicillin-resistant *Staphylococcus aureus* among animals: Current overview. *Clin. Microbiol. Infect.* **2017**, *23*, 373–380. [CrossRef] [PubMed]
22. Gómez, P.; González-Barrio, D.; Benito, D.; García, J.T.; Viñuela, J.; Zarazaga, M.; Ruiz-Fons, F.; Torres, C. Detection of methicillin-resistant *Staphylococcus aureus* (MRSA) carrying the *mecC* gene in wild small mammals in Spain. *J. Antimicrobial Chemother.* **2014**, *69*, 2061–2064. [CrossRef] [PubMed]
23. Ruiz-Ripa, L.; Alcalá, L.; Simón, C.; Gómez, P.; Mama, O.M.; Rezusta, A.; Zarazaga, M.; Torres, C. Diversity of *Staphylococcus aureus* clones in wild mammals in Aragon, Spain, with detection of MRSA ST130-*mecC* in wild rabbits. *J. Appl. Microbiol.* **2019**, *127*, 284–291. [CrossRef] [PubMed]
24. Petersen, A.; Stegger, M.; Heltberg, O.; Christensen, J.; Zeuthen, A.; Knudsen, L.K.; Urth, T.; Sorum, M.; Schouls, L.; Larsen, J.; et al. Epidemiology of methicillin-resistant *Staphylococcus aureus* carrying the novel *mecC* gene in Denmark corroborates a zoonotic reservoir with transmission to humans. *Clin. Microbiol Infect.* **2013**, *19*, E16–E22. [CrossRef] [PubMed]
25. Stegger, M.; Andersen, P.S.; Kearns, A.; Pichon, B.; Holmes, M.A.; Edwards, G.; Laurent, F.; Teale, C.; Skov, R.; Larsen, A.R. Rapid detection, differentiation and typing of methicillin-resistant *Staphylococcus aureus* harbouring either *mecA* or the new *mecA* homologue *mecA*(LGA251). *Clin. Microbiol. Infect.* **2012**, *18*, 395–400. [CrossRef] [PubMed]
26. Kinnevey, P.M.; Shore, A.C.; Brennan, G.I.; Sullivan, D.J.; Ehricht, R.; Monecke, S.; Coleman, D.C. Extensive genetic diversity identified among sporadic methicillin-resistant *Staphylococcus aureus* isolates recovered in Irish hospitals between 2000 and 2012. *Antimicrob. Agents Chemother.* **2014**, *58*, 1907–1917. [CrossRef]
27. Monecke, S.; Jatzwauk, L.; Müller, E.; Nitschke, H.; Pfohl, K.; Slickers, P.; Reissig, A.; Ruppelt-Lorz, A.; Ehricht, R. Diversity of SCC*mec* elements in *Staphylococcus aureus* as observed in South-Eastern Germany. *PLoS ONE.* **2016**, *11*, e0162654. [CrossRef]
28. Zarfel, G.; Luxner, J.; Folli, B.; Leitner, E.; Feierl, G.; Kittinger, C.; Grisold, A. Increase of genetic diversity and clonal replacement of epidemic methicillin-resistant *Staphylococcus aureus* strains in South-East Austria. *FEMS Microbiol Lett.* **2016**, *363*, fnw137. [CrossRef]
29. Deplano, A.; Vandendriessche, S.; Nonhoff, C.; Denis, O. Genetic diversity among methicillin-resistant *Staphylococcus aureus* isolates carrying the *mecC* gene in Belgium. *J. Antimicrob. Chemother.* **2014**, *69*, 1457–1460. [CrossRef]
30. Kriegeskorte, A.; Ballhausen, B.; Idelevich, E.A.; Köck, R.; Friedrich, A.W.; Karch, H.; Peters, G.; Becker, K. Human MRSA isolates with novel genetic homolog, Germany. *Emerg. Infect. Dis.* **2012**, *18*, 1016–1018. [CrossRef]
31. Sabat, A.J.; Koksal, M.; Akkerboom, V.; Monecke, S.; Kriegeskorte, A.; Hendrix, R.; Ehricht, R.; Köck, R.; Becker, K.; Friedrich, A.W. Detection of new methicillin-resistant *Staphylococcus aureus* strains that carry a novel genetic homologue and important virulence determinants. *J. Clin. Microbiol.* **2012**, *50*, 3374–3377. [CrossRef]
32. Schaumburg, F.; Köck, R.; Mellmann, A.; Richter, L.; Hasenberg, F.; Kriegeskorte, A.; Friedrich, A.W.; Gatermann, S.; Peters, G.; von Eiff, C.; et al. Population dynamics among methicillin-resistant *Staphylococcus aureus* isolates in Germany during a 6-year period. *J. Clin. Microbiol.* **2012**, *50*, 3186–3192. [CrossRef] [PubMed]

33. Cuny, C.; Layer, F.; Strommenger, B.; Witte, W. Rare occurrence of methicillin-resistant *Staphylococcus aureus* CC130 with a novel *mecA* homologue in humans in Germany. *PLoS ONE.* **2011**, *6*, e24360. [CrossRef] [PubMed]
34. Cartwright, E.J.P.; Paterson, G.K.; Raven, K.E.; Harrison, E.M.; Gouliouris, T.; Kearns, A.; Pichon, B.; Edwards, G.; Skov, R.L.; Larsen, A.; et al. Use of Vitek 2 antimicrobial susceptibility profile to identify *mecC* in methicillin-resistant *Staphylococcus aureus. J. Clin. Microbiol.* **2013**, *51*, 2732–2734. [CrossRef] [PubMed]
35. Petersdorf, S.; Herma, M.; Rosenblatt, M.; Layer, F.; Henrich, B. A Novel Staphylococcal Cassette Chromosome *mec* Type XI primer for detection of *mecC*-harboring methicillin-resistant *Staphylococcus aureus* directly from screening specimens. *J. Clin. Microbiol.* **2015**, *53*, 3938–3941. [CrossRef]
36. Dermota, U.; Zdovc, I.; Strumbelj, I.; Grmek-Kosnik, I.; Ribic, H.; Rupnik, M.; Golob, M.; Zajc, U.; Bes, M.; Laurent, F.; et al. Detection of methicillin-resistant *Staphylococcus aureus* carrying the *mecC* gene in, human samples in Slovenia. *Epidemiol. Infect.* **2015**, *143*, 1105–1108. [CrossRef]
37. García-Garrote, F.; Cercenado, E.; Marín, M.; Bal, M.; Trincado, P.; Corredoira, J.; Ballesteros, C.; Pita, J.; Alonso, P.; Vindel, A. Methicillin-resistant *Staphylococcus aureus* carrying the *mecC* gene: Emergence in Spain and report of a fatal case of bacteraemia. *J. Antimicrob. Chemother.* **2014**, *69*, 45–50. [CrossRef]
38. Kerschner, H.; Harrison, E.M.; Hartl, R.; Holmes, M.A.; Apfalter, P. First report of *mecC* MRSA in human samples from Austria: Molecular characteristics and clinical data. *New Microbes New Infect.* **2014**, *3*, 4–9. [CrossRef]
39. Petersen, A.; Medina, A.; Rhod Larsen, A. Ability of the GENSPEED(®) MRSA test kit to detect the novel *mecA* homologue *mecC* in *Staphylococcus aureus. APMIS.* **2015**, *123*, 478–481. [CrossRef]
40. Paterson, G.K.; Morgan, F.J.E.; Harrison, E.M.; Cartwright, E.J.P.; Török, M.E.; Zadoks, R.N.; Parkhill, J.; Peacock, S.J.; Holmes, M.A. Prevalence and characterization of human *mecC* methicillin-resistant *Staphylococcus aureus* isolates in England. *J. Antimicrob. Chemother.* **2014**, *69*, 907–910. [CrossRef]
41. Harrison, E.M.; Coll, F.; Toleman, M.S.; Blane, B.; Brown, N.B.; Török, M.E.; Parkhill, J.; Peacock, S.J. Genomic surveillance reveals low prevalence of livestock-associated methicillin-resistant *Staphylococcus aureus* in the East of England. *Sci. Rep.* **2017**, *7*, 7406. [CrossRef]
42. Paterson, G.K. Low prevalence of livestock-associated methicillin-resistant *Staphylococcus aureus* clonal complex 398 and *mecC* MRSA among human isolates in North-West England. *J. Appl. Microbiol.* **2020**, *128*, 1785–1792. [CrossRef] [PubMed]
43. Ciesielczuk, H.; Xenophontos, M.; Lambourne, J. Methicillin-resistant *Staphylococcus aureus* harboring *mecC* still eludes us in East London, United Kingdom. *J. Clin. Microbiol.* **2019**, *57*, e00020-19. [CrossRef] [PubMed]
44. Kriegeskorte, A.; Idelevich, E.A.; Schlattmann, A.; Layer, F.; Strommenger, B.; Denis, O.; Paterson, G.K.; Holmes, M.A.; Werner, G.; Becker, K. Comparison of different phenotypic approaches to screen and detect *mecC*-harboring methicillin-resistant *Staphylococcus aureus. J. Clin. Microbiol.* **2017**, *56*, e00826-17. [CrossRef] [PubMed]
45. Basset, P.; Prod'hom, G.; Senn, L.; Greub, G.; Blanc, D.S. Very low prevalence of meticillin-resistant *Staphylococcus aureus* carrying the *mecC* Gene in Western Switzerland. *J. Hosp. Infect.* **2013**, *83*, 257–259. [CrossRef]
46. Dekker, D.; Wolters, M.; Mertens, E.; Boahen, K.G.; Krumkamp, R.; Eibach, D.; Schwarz, N.G.; Adu-Sarkodie, Y.; Rohde, H.; Christner, M.; et al. Antibiotic resistance and clonal diversity of invasive *Staphylococcus aureus* in the rural Ashanti Region, Ghana. *BMC Infect. Dis.* **2016**, *16*, 720. [CrossRef]
47. Kılıç, A.; Doğan, E.; Kaya, S.; Baysallar, M. Investigation of the presence of *mecC* and Panton-Valentine leukocidin genes in *Staphylococcus aureus* strains isolated from clinical specimens during seven years period. *Mikrobiyol. Bul.* **2015**, *49*, 594–599. [CrossRef]
48. Vandendriessche, S.; Vanderhaeghen, W.; Valente Soares, F.; Hallin, M.; Catry, B.; Hermans, K.; Butaye, P.; Haesebrouck, F.; Struelens, M.J.; Denis, O. Prevalence, risk factors and genetic diversity of methicillin-resistant *Staphylococcus aureus* carried by humans and animals across livestock production sectors. *J. Antimicrob. Chemother.* **2013**, *68*, 1510–1516. [CrossRef]
49. Laub, K.; Tóthpál, A.; Kardos, S.; Dobay, O. Epidemiology and antibiotic sensitivity of *Staphylococcus aureus* nasal carriage in children in Hungary. *Acta Microbiol. Immunol. Hung.* **2017**, *64*, 51–62. [CrossRef]
50. Ganesan, A.; Crawford, K.; Mende, K.; Murray, C.K.; Lloyd, B.; Ellis, M.; Tribble, D.R.; Weintrob, A.C. Evaluation for a novel methicillin resistance (*mecC*) homologue in methicillin-resistant *Staphylococcus aureus* isolates obtained from injured military personnel. *J. Clin. Microbiol.* **2013**, *51*, 3073–3075. [CrossRef]

51. Ludden, C.; Brennan, G.; Morris, D.; Austin, B.; O'Connell, B.; Cormican, M. Characterization of methicillin-resistant *Staphylococcus aureus* from residents and the environment in a long-term care facility. *Epidemiol. Infect.* **2015**, *143*, 2985–2988. [CrossRef]
52. Paterson, G.K.; Harrison, E.M.; Craven, E.F.; Petersen, A.; Larsen, A.R.; Ellington, M.J.; Török, M.E.; Peacock, S.J.; Parkhill, J.; Zadoks, R.N.; et al. Incidence and characterisation of methicillin-resistant *Staphylococcus aureus* (MRSA) from nasal colonisation in participants attending a cattle veterinary conference in the UK. *PLoS ONE.* **2013**, *8*, e68463. [CrossRef] [PubMed]
53. Aqel, A.A.; Alzoubi, H.M.; Vickers, A.; Pichon, B.; Kearns, A.M. Molecular epidemiology of nasal isolates of methicillin-resistant *Staphylococcus aureus* from Jordan. *J. Infect. Public Health.* **2015**, *8*, 90–97. [CrossRef] [PubMed]
54. Becker, K.; Schaumburg, F.; Fegeler, C.; Friedrich, A.W.; Köck, R. Prevalence of Multiresistant Microorganisms PMM Study. *Staphylococcus aureus from the German general population is highly diverse. Int. J. Med. Microbiol.* **2017**, *307*, 21–27. [CrossRef] [PubMed]
55. Dodémont, M.; Argudín, M.A.; Willekens, J.; Vanderhelst, E.; Pierard, D.; Miendje Deyi, V.Y.; Hanssens, L.; Franckx, H.; Schelstraete, P.; Leroux-Roels, I.; et al. Emergence of livestock-associated MRSA isolated from cystic fibrosis patients: Result of a Belgian national survey. *J. Cyst. Fibros.* **2019**, *18*, 86–93. [CrossRef] [PubMed]
56. Drougka, E.; Foka, A.; Koutinas, C.K.; Jelastopulu, E.; Giormezis, N.; Farmaki, O.; Sarrou, S.; Anastassiou, E.D.; Petinaki, E.; Spiliopoulou, I. Interspecies spread of *Staphylococcus aureus* clones among companion animals and human close contacts in a veterinary teaching hospital. A cross-sectional study in Greece. *Prev. Vet. Med.* **2016**, *126*, 190–198. [CrossRef]
57. Nijhuis, R.H.; van Maarseveen, N.M.; van Hannen, E.J.; van Zwet, A.A.; Mascini, E.M. A rapid and high-throughput screening approach for methicillin-resistant *Staphylococcus aureus* based on the combination of two different real-time PCR assays. *J. Clin. Microbiol.* **2014**, *52*, 2861–2867. [CrossRef] [PubMed]
58. Saeed, K.; Ahmad, N.; Dryden, M.; Cortes, N.; Marsh, P.; Sitjar, A.; Wyllie, S.; Bourne, S.; Hemming, J.; Jeppesen, C.; et al. Oxacillin-susceptible methicillin-resistant *Staphylococcus aureus* (OS-MRSA), a hidden resistant mechanism among clinically significant isolates in the Wessex region/UK. *Infection* **2014**, *42*, 843–847. [CrossRef]
59. Khairalla, A.S.; Wasfi, R.; Ashour, H.M. Carriage frequency, phenotypic, and genotypic characteristics of methicillin-resistant *Staphylococcus aureus* isolated from dental health-care personnel, patients, and environment. *Sci. Rep.* **2017**, *7*, 7390. [CrossRef]
60. Ho, C.M.; Lin, C.Y.; Ho, M.W.; Lin, H.C.; Chen, C.J.; Lin, L.C.; Lu, J.J. Methicillin-resistant *Staphylococcus aureus* isolates with SCC*mec* type V and spa types t437 or t1081 associated to discordant susceptibility results between oxacillin and cefoxitin, Central Taiwan. *Diagn. Microbiol. Infect. Dis.* **2016**, *86*, 405–411. [CrossRef]
61. van Duijkeren, E.; Hengeveld, P.; Zomer, T.P.; Landman, F.; Bosch, T.; Haenen, A.; van de Giessen, A. Transmission of MRSA between humans and animals on duck and turkey farms. *J. Antimicrob. Chemother.* **2016**, *71*, 58–62. [CrossRef]
62. Cikman, A.; Aydin, M.; Gulhan, B.; Karakecili, F.; Kurtoglu, M.G.; Yuksekkaya, S.; Parlak, M.; Gultepe, B.S.; Cicek, A.C.; Bilman, F.B.; et al. Absence of the *mecC* gene in methicillin-resistant *Staphylococcus aureus* isolated from various clinical samples: The first multi-centered study in Turkey. Multicenter Study. *J. Infect. Public Health.* **2019**, *12*, 528–533. [CrossRef] [PubMed]
63. Gomez, P.; Lozano, C.; González-Barrio, D.; Zarazaga, M.; Ruiz-Fons, F.; Torres, C. High prevalence of methicillin-resistant *Staphylococcus aureus* (MRSA) carrying the *mecC* gene in a semi-extensive red deer (*Cervus elaphus hispanicus*) farm in Southern Spain. *Vet. Microbiology.* **2015**, *177*, 326–331. [CrossRef]
64. Szymanek-Majchrzak, K.; Kosiński, J.; Żak, K.; Sułek, K.; Młynarczyk, A.; Młynarczyk, G. Prevalence of methicillin resistant and mupirocin-resistant *Staphylococcus aureus* strains among medical students of Medical University in Warsaw. *Przegl Epidemiol.* **2019**, *73*, 39–48. [CrossRef] [PubMed]
65. Rödel, J.; Bohnert, J.A.; Stoll, S.; Wassill, L.; Edel, B.; Karrasch, M.; Löffler, B.; Pfister, W. Evaluation of loop-mediated isothermal amplification for the rapid identification of bacteria and resistance determinants in positive blood cultures. *Eur. J. Clin. Microbiol. Infect. Dis.* **2017**, *36*, 1033–1040. [CrossRef] [PubMed]
66. Horner, C.; Utsi, L.; Coole, L.; Denton, M. Epidemiology and microbiological characterization of clinical isolates of *Staphylococcus aureus* in a single healthcare region of the UK, 2015. *Epidemiol. Infect.* **2017**, *145*, 386–396. [CrossRef]

67. Mehta, S.R.; Estrada, J.; Ybarra, J.; Fierer, J. Comparison of the BD MAX MRSA XT to the Cepheid™ Xpert®MRSA assay for the molecular detection of methicillin-resistant *Staphylococcus aureus* from nasal swabs. *Diagn. Microbiol. Infect. Dis.* **2017**, *87*, 308–310. [CrossRef]
68. Venugopal, N.; Mitra, S.; Tewari, R.; Ganaie, F.; Shome, R.; Rahman, H.; Shome, B.R. Molecular detection and typing of methicillin-resistant *Staphylococcus aureus* and methicillin-resistant coagulase-negative staphylococci isolated from cattle, animal handlers, and their environment from Karnataka, Southern Province of India. *Vet. World.* **2019**, *12*, 1760–1768. [CrossRef]
69. Ceballos, S.; Aspiroz, C.; Ruiz-Ripa, l.; Azcona-Gutierrez, J.M.; López-Cerero, L.; López-Calleja, A.I.; Álvarez, L.; Gomáriz, M.; Fernández, M.; Torres, C.; et al. Multicenter study of clinical non-β-lactam-antibiotic susceptible MRSA strains: Genetic lineages and Panton-Valentine leukocidin (PVL) production. *Enferm. Infecc. Microbiol. Clin.* **2019**, *37*, 509–513. [CrossRef]
70. Papadopoulos, P.; Angelidis, A.S.; Papadopoulos, T.; Kotzamanidis, C.; Zdragas, A.; Papa, A.; Filioussis, G.; Sergelidis, D. *Staphylococcus aureus* and methicillin-resistant *S. aureus* (MRSA) in bulk tank milk, livestock and dairy-farm personnel in north-central and north-eastern Greece: Prevalence, characterization and genetic relatedness. *Food Microbiol.* **2019**, *84*, 103249. [CrossRef]
71. Rasmussen, S.L.; Larsen, J.; van Wijk, R.E.; Jones, O.R.; Bjørneboe Berg, T.; Angen, O.; Rhod Larsen, A. European hedgehogs (*Erinaceus europaeus*) as a natural reservoir of methicillin-resistant *Staphylococcus aureus* carrying *mecC* in Denmark. *PLoS ONE.* **2019**, *14*, e0222031. [CrossRef]
72. Morroni, G.; Brenciani, A.; Brescini, L.; Fioriti, S.; Simoni, S.; Pocognoli, A.; Mingoia, M.; Giovanetti, E.; Barchiesi, F.; Giacometti, A.; et al. High rate of ceftobiprole resistance among clinical methicillin-resistant *Staphylococcus aureus* isolates from a hospital in central Italy. *Antimicrob. Agents Chemother.* **2018**, *62*, e01663-18. [CrossRef]
73. Hefzy, E.M.; Hassan, G.M.; El Reheem, F.A. Detection of Panton-Valentine Leukocidin-positive methicillin-resistant *Staphylococcus aureus* nasal carriage among Egyptian health care workers. *Surg. Infect. (Larchmt).* **2016**, *17*, 369–375. [CrossRef] [PubMed]
74. Hogan, B.; Rakotozandrindrainy, R.; Al-Emran, H.; Dekker, D.; Hahn, A.; Jaeger, A.; Poppert, S.; Frickmann, H.; Hagen, R.M.; Micheel, V.; et al. Prevalence of nasal colonisation by methicillin-sensitive and methicillin-resistant *Staphylococcus aureus* among healthcare workers and students in Madagascar. *B.M.C. Infect. Dis.* **2016**, *16*, 420. [CrossRef] [PubMed]
75. Diaz, R.; Ramalheira, E.; Afreixo, V.; Gago, B. Methicillin-resistant *Staphylococcus aureus* carrying the new *mecC* gene–a meta-analysis. *Diagn. Microbiol. Infect. Dis.* **2016**, *84*, 135–140. [CrossRef] [PubMed]
76. Lindgren, A.K.; Gustafsson, E.; Petersson, A.C.; Melander, E. Methicillin-resistant *Staphylococcus aureus* with *mecC*: A description of 45 human cases in southern Sweden. *Eur. J. Clin. Microbiol. Infect. Dis.* **2016**, *35*, 971–975. [CrossRef] [PubMed]
77. Laurent, F.; Chardon, H.; Haenni, M.; Bes, M.; Reverdy, M.E.; Madec, J.V.; Lagier, E.; Vandenesch, F.; Tristan, A. MRSA harboring *mecA* variant gene *mecC*, France. *Emerg. Infect. Dis.* **2012**, *18*, 1465–1467. [CrossRef] [PubMed]
78. Barraud, O.; Laurent, F.; François, B.; Bes, M.; Vignon, P.; Ploy, M.C. Severe human bone infection due to methicillin-resistant *Staphylococcus aureus* carrying the novel *mecC* variant. *J. Antimicrob. Chemother.* **2013**, *68*, 2949–2950. [CrossRef]
79. Cano García, M.E.; Monteagudo Cimiano, I.; Mellado Encinas, P.; Ortega Álvarez, C. Methicillin-resistant *Staphylococcus aureus* carrying the *mecC* gene in a patient with a wound infection. *Enferm. Infecc. Microbiol. Clin.* **2015**, *33*, 287–288. [CrossRef]
80. Romero-Gómez, M.P.; Mora-Rillo, M.; Lázaro-Perona, F.; Gómez-Gil, M.R.; Mingorance, J. Bacteraemia due to meticillin-resistant *Staphylococcus aureus* carrying the *mecC* gene in a patient with urothelial carcinoma. *J. Med. Microbiol.* **2013**, *62*, 1914–1916. [CrossRef]
81. Benito, D.; Gómez, P.; Aspiroz, C.; Zarazaga, M.; Lozano, C.; Torres, C. Molecular characterization of *Staphylococcus aureus* isolated from humans related to a livestock farm in Spain, with detection of MRSA-CC130 carrying *mecC* gene: A zoonotic case? *Enferm. Infecc. Microbiol. Clin.* **2016**, *34*, 280–285. [CrossRef]
82. Worthing, K.A.; Coombs, G.W.; Pang, S.; Abraham, S.; Saputra, S.; Trott, D.J.; Jordan, D.; Wong, H.S.; Abraham, R.J.; Norris, J.M. Isolation of *mecC* MRSA in Australia. *J. Antimicrob. Chemother.* **2016**, *71*, 2348–2349. [CrossRef]

83. Aklilu, E.; Ying, C.H. First *mecC* and *mecA* positive livestock-associated Methicillin resistant *Staphylococcus aureus* (*mecC* MRSA/LA-MRSA) from dairy cattle in Malaysia. *Microorganisms.* **2020**, *8*, 147. [CrossRef [PubMed]
84. Sakr, A.; Brégeon, F.; Mège, J.L.; Rolain, J.M.; Blin, O. *Staphylococcus aureus* nasal colonization: An update on mechanisms, epidemiology, risk factors, and subsequent infections. *Front. Microbiol.* **2018**, *9*, 2419. [CrossRef
85. Brown, A.F.; Leech, J.M.; Rogers, T.R.; McLoughlin, R.M. *Staphylococcus aureus* colonization: Modulation of host immune response and impact on human vaccine design. *Front. Immunol.* **2014**, *4*, 507. [CrossRef [PubMed]
86. Asadollahi, P.; Farahani, N.N.; Mirzaii, N.; Sajjad Khoramrooz, S.S.; van Belkum, A.; Asadollahi, K.; Dadashi, M.; Darban-Sarokhalil, D. Distribution of the most prevalent *spa* types among clinical isolates of Methicillin-Resistant and -Susceptible *Staphylococcus aureus* around the World: A Review. *Front. Microbiol.* **2018**, *9*, 163. [CrossRef] [PubMed] ]
87. Belmekki, M.; Mammeri, H.; Hamdad, F.; Rousseau, F.; Canarelli, B.; Biendo, M. Comparison of Xpert MRSA/SA Nasal and MRSA/SA ELITe MGB assays for detection of the *mecA* gene with susceptibility testing methods for determination of methicillin resistance in *Staphylococcus aureus* isolates. *J. Clin. Microbiol.* **2013**, *51*, 3183–3191. [CrossRef]
88. Stegger, M.; Lindsay, J.A.; Sørum, M.; Gould, K.A.; Skov, R. Genetic diversity in CC398 methicillin-resistant *Staphylococcus aureus* isolates of different geographical origin. *Clin. Microbiol. Infect.* **2010**, *16*, 1017–1019. [CrossRef]
89. Eriksson, J.; Espinosa-Gongora, C.; Stamphøj, I.; Rhod Larsen, A.; Guardabassi, L. Carriage frequency, diversity and methicillin resistance of *Staphylococcus aureus* in Danish small ruminants. *Vet. Microbiol.* **2013**, *163*, 110–115. [CrossRef]
90. Paterson, G.K.; Larsen, A.R.; Robb, A.; Edwards, G.E.; Pennycott, T.W.; Foster, G.; Mot, D.; Hermans, K.; Baert, K.; Peacock, S.J.; et al. The newly described *mecA* homologue, *mecA*(LGA251), is present in methicillin-resistant *Staphylococcus aureus* isolates from a diverse range of host species. *J. Antimicrob. Chemother.* **2012**, *67*, 2809–2813. [CrossRef]
91. Weinert, L.A.; Welch, J.J.; Suchard, M.A.; Lemey, P.; Rambaut, A.; Fitzgerald, J.R. Molecular dating of human-to-bovid host jumps by *Staphylococcus aureus* reveals an association with the spread of domestication. *Biol. Lett.* **2012**, *8*, 829–832. [CrossRef]
92. Ruiz-Ripa, L.; Gómez, P.; Alonso, C.A.; Camacho, M.C.; de la Puente, J.; Fernández-Fernández, R.; Ramiro, Y.; Quevedo, M.A.; Blanco, J.M.; Zarazaga, M.; et al. Detection of MRSA of Lineages CC130-*mecC* and CC398-*mecA* and *Staphylococcus delphini-lnu*(A) in Magpies and Cinereous Vultures in Spain. *Microb. Ecol.* **2019**, *78*, 409–415. [CrossRef] [PubMed]
93. Ford, A. *mecC*-harboring methicillin-resistant *Staphylococcus aureus*: Hiding in plain sight. *J. Clin. Microbiol.* **2017**, *56*, e01549-17. [CrossRef] [PubMed]
94. Haenni, M.; Châtre, P.; Dupieux, C.; Métayer, V.; Maillard, K.; Bes, M.; Madec, J.Y.; Laurent, F. *mecC*-positive MRSA in horses. *J. Antimicrob. Chemother.* **2015**, *70*, 3401–3402. [CrossRef]
95. Medhus, A.; Slettemeås, J.S.; Marstein, L.; Larssen, K.W.; Sunde, M. Methicillin-resistant *Staphylococcus aureus* with the novel *mecC* gene variant isolated from a cat suffering from chronic conjunctivitis. *J. Antimicrob. Chemother.* **2013**, *68*, 968–969. [CrossRef] [PubMed]
96. Haenni, M.; Châtre, P.; Tasse, J.; Nowak, N.; Bes, M.; Madec, J.Y.; Laurent, F. Geographical clustering of *mecC*-positive *Staphylococcus aureus* from bovine mastitis in France. *J. Antimicrob. Chemother.* **2014**, *69*, 2292–2293. [CrossRef]
97. Gindonis, V.; Taponen, S.; Myllyniemi, A.L.; Pyörälä, S.; Nykäsenoja, S.; Salmenlinna, S.; Lindholm, L.; Rantala, M. Occurrence and characterization of methicillin-resistant staphylococci from bovine mastitis milk samples in Finland. *Acta Vet. Scand.* **2013**, *55*, 61. [CrossRef]

# Combined Effect of Naturally-Derived Biofilm Inhibitors and Differentiated HL-60 Cells in the Prevention of *Staphylococcus aureus* Biofilm Formation

**Inés Reigada [1,*], Clara Guarch-Pérez [2], Jayendra Z. Patel [3], Martijn Riool [2], Kirsi Savijoki [1], Jari Yli-Kauhaluoma [3], Sebastian A. J. Zaat [2] and Adyary Fallarero [1]**

1 Drug Research Program, Division of Pharmaceutical Biosciences, Faculty of Pharmacy, University of Helsinki, FI-00014 Helsinki, Finland; kirsi.savijoki@helsinki.fi (K.S.); adyary.fallarero@helsinki.fi (A.F.)

2 Department of Medical Microbiology and Infection Prevention, Amsterdam institute for Infection and Immunity, Amsterdam UMC, University of Amsterdam, 1105 AZ Amsterdam, The Netherlands; c.m.guarchperez@amsterdamumc.nl (C.G.-P.); m.riool@amsterdamumc.nl (M.R.); s.a.zaat@amsterdamumc.nl (S.A.J.Z.)

3 Drug Research Program, Division of Pharmaceutical Chemistry and Technology, Faculty of Pharmacy, University of Helsinki, FI-00014 Helsinki, Finland; jayendra.patel@helsinki.fi (J.Z.P.); jari.yli-kauhaluoma@helsinki.fi (J.Y.-K.)

* Correspondence: ines.reigada@helsinki.fi

**Abstract:** Nosocomial diseases represent a huge health and economic burden. A significant portion is associated with the use of medical devices, with 80% of these infections being caused by a bacterial biofilm. The insertion of a foreign material usually elicits inflammation, which can result in hampered antimicrobial capacity of the host immunity due to the effort of immune cells being directed to degrade the material. The ineffective clearance by immune cells is a perfect opportunity for bacteria to attach and form a biofilm. In this study, we analyzed the antibiofilm capacity of three naturally derived biofilm inhibitors when combined with immune cells in order to assess their applicability in implantable titanium devices and low-density polyethylene (LDPE) endotracheal tubes. To this end, we used a system based on the coculture of HL-60 cells differentiated into polymorphonuclear leukocytes (PMNs) and *Staphylococcus aureus* (laboratory and clinical strains) on titanium, as well as LDPE surfaces. Out of the three inhibitors, the one coded **DHA1** showed the highest potential to be incorporated into implantable devices, as it displayed a combined activity with the immune cells, preventing bacterial attachment on the titanium and LDPE. The other two inhibitors seemed to also be good candidates for incorporation into LDPE endotracheal tubes.

**Keywords:** *Staphylococcus aureus*; biomaterials; medical devices; HL-60 cells; PMNs; biofilm; endotracheal tube; titanium; implantable devices; nosocomial diseases

---

## 1. Introduction

Over 2.6 million new cases of healthcare-associated infections are annually reported just in the European Union [1], and over 33,000 result in death [2] due to the increasing number of antimicrobial-resistance cases [3]. At least 25% of these infections are associated with the use of medical devices, and 80% of them are estimated to be caused by bacterial biofilms [4,5]. Biofilms are defined as a community of microorganisms encased within a self-produced matrix that adheres to biological or nonbiological surfaces [6,7]. They are currently regarded as the most important nonspecific mechanism of antimicrobial resistance [8,9].

During the worldwide crisis of SARS-CoV-2, as well as in many other pathological conditions, mechanical ventilation is used to assist or replace spontaneous breathing as a life-saving procedure in intensive care. However, the use of endotracheal intubation also poses major risks in prolonged ventilation. The endotracheal tube provides an ideal opportunity for bacteria to form biofilms on both the outer and luminal surface of the tube, increasing the risk of pulmonary infection by 6 to 10 times [10–12], with *Staphylococcus* spp. and *Pseudomonas aeruginosa* among the most frequent colonizing agents [13]. Colonization by microorganisms and the subsequent formation of a biofilm can happen within hours [13], but these kinds of devices are relatively easy to replace. On the other hand, infection of orthopedic implants is particularly problematic, as these devices remain in the body, often causing chronic and/or recurring infections mediated by biofilms. These infections also frequently require removal of the infected implant, thereby causing implant failure [14–16]. Given the rising number of implantations, the absolute number of complications is inevitably increasing at the same pace, causing not only distress for the patients but also an increasing economic burden [15,16].

The most common causative agents of infection in orthopedic implants are Gram-positive cocci of the genus *Staphylococcus*, e.g., *Staphylococcus aureus* and *Staphylococcus epidermidis* [17]. In the absence of a foreign body, contaminations caused by these opportunistic pathogens are usually cleared by the immune system. In contrast, the placement of an implant per se represents a risk factor for the development of a chronic infection. This is due to the fact that the surgical procedure causes tissue damage resulting in the local generation of damage-associated molecular patterns (DAMPs), endogenous danger molecules that are released from damaged or dying cells and activate the innate immune system by interacting with pattern recognition receptors (PRRs) [18]. This is sensed by host neutrophils, which migrate to the injured tissue sites, activating defense mechanisms, such as the generation of oxygen-derived and nitrogen-derived reactive species as well as phagocytosis, to unsuccessfully attempt to clear the foreign material. These events lead to immune cell exhaustion and death, and tissue damage caused by the triggered inflammation eventually leads to a niche of immune suppression around the implant [19]. Under these specific conditions, the clearance of planktonic bacteria by immune cells becomes impaired [14], which predisposes the implant to microbial colonization and biofilm-mediated infection.

However, it is possible that not only host immune cells and bacterial cells can be present at the moment of implantation, but also the cells of the tissue where the material is being implanted. The "race for the surface" concept describes the competition between bacteria and the host cells of the tissue where the device is implanted to colonize the surface of the material [20]. The rapid integration of the material into the host tissue is a key component in the success of an implant, as the colonization of the surface by the cells of the host not only ensures correct integration, but also prevents bacterial colonization [21]. However, if bacterial adhesion occurs first, the host defense system is unable to prevent the colonization and subsequent biofilm formation [22].

One of the current challenges to prevent and resolve biofilm-mediated infections is the limited repertoire of compounds that are able to act on them at sufficiently low concentrations [23]. Because of this, there is intense ongoing research focused on the search for new biofilm inhibitors by means of chemical screenings. However, for compounds to be truly effective when used for the protection of biomaterials in translational applications, they must be tested in meaningful experimental models. Based on this, we previously optimized an in vitro system based on the coculture of SaOS-2, osteogenic cells, and *S. aureus* (laboratory and clinical strains) on a titanium surface (Reigada; et al. [24]), and studied the effect of three naturally derived biofilm inhibitors (Figure 1). Two of them were dehydroabietic acid (DHA) derivatives, namely, *N*-(abiet-8,11,13-trien-18-oyl)cyclohexyl-L-alanine and *N*-(abiet-8,11,13-trien-18-oyl)-D-tryptophan, coded **DHA1** (Figure 1a) and **DHA2** (Figure 1b), respectively. The third one was a flavan derivative, 6-chloro-4-(6-chloro-7-hydroxy-2,4,4-trimethylchroman-2-yl)benzene-1,3-diol, coded **FLA1** (Figure 1c). All of them were previously reported by our group and demonstrated to display activity in preventing biofilm formation, as well as disrupting preformed *S. aureus* biofilms on 96-well plates [25,26], but the testing in the coculture model with osteogenic cells provided with new insights into their applicability as part of anti-infective implantable devices.

The coculture model developed previously [24] involving SaOS-2 cells and *S. aureus* strains offered an in vitro environment that was closer in terms of host–bacteria interactions and substrate materials to the in vivo scenario of the implanted device. Moreover, in terms of antimicrobial evaluation, it provided information not only regarding the antibiofilm activity but also about the effects on tissue integration. However, this model did not assess the effect of the antimicrobials on immune cells. As mentioned earlier, chronic inflammation not only lowers the antimicrobial efficacy, but also complicates the integration of the implant material as a consequence of maintained inflammation and tissue damage. Therefore, it is essential to assess the effect that antimicrobials might have in the presence of immune cells, particularly for those intended to be used in medical devices. Endotracheal tubes significantly differ from orthopedic implants in material, function, and implantation procedure, but they both cause impairment of the host immune antimicrobial capacity.

Because of this, in this contribution, we move one step forward toward emulating in vivo conditions encountered by medical devices in an in vitro setting. In this case, we introduce immune cells, specifically neutrophils, in a coculture model with bacterial cells. Neutrophils were selected as immune cells as they are the first cell types to migrate toward damaged tissue cells [27]. Our aim was to develop cocultures of *S. aureus* and differentiated HL-60 cells, grown on titanium coupons or low-density polyethylene (LDPE) tubes, to simulate biofilm formation on orthopedic implants or endotracheal tubes, respectively. As a proof-of-concept, we also study the possible antimicrobial effects of naturally derived antibiofilm inhibitors **DHA1**, **DHA2**, and **FLA1**, and their possible applicability as part of medical devices.

## 2. Materials and Methods

### 2.1. Compounds

The dehydroabietic acid derivatives coded **DHA1** and **DHA2** (previously coded 11 and 9b, respectively) were synthesized according to [25]. Their spectral data were identical to those reported in [28]. The flavan derivative coded **FLA1** (previously coded 291 in [26]), was purchased from TimTec (product code: ST075672, www.timtec.net). These compounds were selected given their low minimum inhibitory concentrations (MICs) and the low concentrations needed in order to prevent *S. aureus* biofilm formation (Table S1). The control antibiotic rifampicin was purchased from Sigma-Aldrich (CAS number 13292-46-1) and coded **RIF**. Molecular structures are shown in Figure 1.

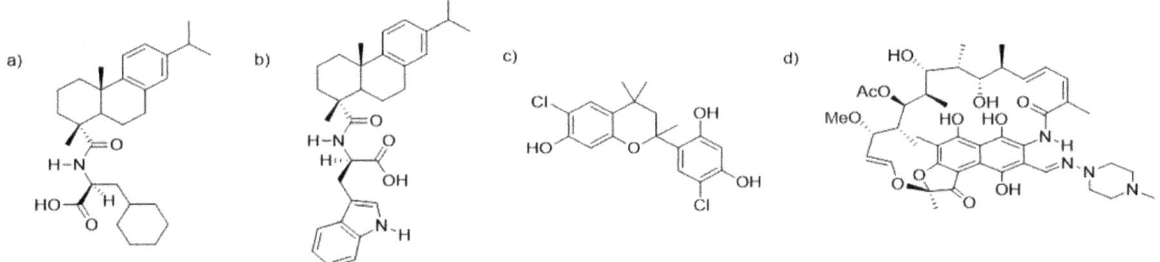


**Figure 1.** Chemical structures of the two dehydroabietic acid (DHA) derivatives, (**a**) *N*-(abiet-8,11,13-trien-18-oyl)cyclohexyl-L-alanine, (**b**) *N*-(abiet-8,11,13-trien-18-oyl)-D-tryptophan, coded **DHA1** and **DHA2**, the flavan derivative, (**c**) 6-chloro-4-(6-chloro-7-hydroxy-2,4,4-trimethylchroman-2-yl)benzene-1,3-diol, coded **FLA1**, and (**d**) rifampicin, coded **RIF**.

### 2.2. Bacterial Strains

Bacterial studies were performed using the laboratory strain *S. aureus* ATCC 25923 (American Type Culture Collection, Manassas, VA, USA) and one clinical strain isolated from hip prostheses and osteosynthesis implants at the Hospital Fundación Jiménez Díaz (Madrid, Spain) [29] (coded *S. aureus* P2).

### 2.3. *HL-60 Cell Culture and Differentiation*

HL-60 (ATCC CCL-240) cells were grown and maintained in Roswell Park Memorial Institute (RPMI) 1640 Medium (R8758, Sigma Aldrich, St. Louis, MO, USA), supplemented with 20% (*v*/*v*) heat-inactivated fetal bovine serum (FBS) (Sigma Aldrich, St. Louis, MO, USA) and 1% (*v*/*v*) penicillin/streptomycin (Sigma Aldrich, St. Louis, MO, USA). Cells were maintained in suspension at a concentration between $10^5$–$10^6$ cells/mL in 72 $cm^2$ culture flasks (VWR, Radnor, PA, USA) at 37 °C in 5% $CO_2$ in a humidified incubator. *N,N*–Dimethylformamide (DMF)(Sigma Aldrich, St. Louis, MO, USA) was utilized in order to differentiate the cells into polymorphonuclear-like cells [30]. In order to carry out the differentiation, cells were cultured for 6 days in the maintenance medium at a concentration of 100 mM DMF. The success of the differentiation was assessed visually after Giemsa staining using a Leica DMLS microscope (Leica, Wetzlar, Germany). Differentiated cells were neutrophil-like, with a multilobar nucleus and a fairly clear cytoplasm.

### 2.4. *Biofilm Prevention Efficacy of Differentiated HL-60 Cells against Different Bacterial Concentrations of S. aureus ATCC 25923*

#### 2.4.1. Bacterial Inoculum Preparation

Pure colonies (2–3) of *S. aureus* ATCC 25923 were added to 5 mL of tryptic soy broth (TSB, Neogen®, Lansing, MI, USA) and incubated overnight at 37 °C, 120 rpm. After incubation, 10 µL of the preculture was added to 5 mL fresh TSB and incubated for 3–4 h (37 °C, 120 rpm), in order to obtain a midlogarithmic growth phase of bacteria. Bacterial cultures were washed twice with sterile phosphate-buffered saline (PBS; 140 mM NaCl, pH 7.4) and the concentration was adjusted to $2 \times 10^8$ in RPMI 1640 based on the optical density of the suspension at 600 nm. From this stock, serial dilutions were made in RPMI 1640 between $2 \times 10^8$ and $2 \times 10^2$ CFU/mL.

#### 2.4.2. Immune Cell (HL-60) Preparation

Twenty-four hours prior to starting the experiments, the media of the differentiated HL-60 cells were refreshed with nonsupplemented RPMI 1640 in order to remove possible traces of antibiotics from the maintenance media. Differentiated HL-60 cells (after 6 days exposure to a concentration of 100 mM DMF) were counted with a Countess™ II automated cell counter (Thermo Scientific, Waltham, MA, USA) and adjusted to a concentration of $2 \times 10^5$ cells/mL. In order to test the influence of activating the cells with phorbol 12-myristate 13-acetate (PMA) (Sigma Aldrich, St. Louis, MO, USA) on the antimicrobial capacity, half of the suspension of differentiated HL-60 cells was incubated for 20 min in RPMI 1640 supplemented with 25 nM PMA. PMA-activated and nonactivated HL-60 cells were separately added (100 µL) to flat-bottomed, 96-well microplates (Nunclon Δ surface, Nunc, Roskilde, Denmark).

#### 2.4.3. Coculture of *S. aureus* ATCC 25923 and HL-60 Cells

*S. aureus* ATCC 25923 suspensions at the different concentrations (100 µL) were added to the wells of the 96-well microplate already containing 100 µL of one of the two different HL-60 cell suspensions (activated or nonactivated). In the bacterial control wells, 100 µL of the *S. aureus* ATCC 25923 suspension at different concentrations were added to wells containing 100 µL of RPMI 1640 alone. The wells corresponding to the HL-60 cell control consisting of a suspension of HL-60 cells at a concentration $10^5$ cells/mL in RPMI 1640 were used to observe the cell morphology after 18 h of incubation, but no quantitative viability test was carried out. The 96-well microplates were incubated for 18 h at 37 °C and 5% $CO_2$ in a humidified incubator.

#### 2.4.4. Biofilm Quantification in 96-Well Microplates

After *S. aureus* ATCC 25923 biofilms were grown, the media were carefully removed and each well was washed twice with 125 µL sterile PBS, followed by the addition of 150 µL of TSB. Each 96-well microplate was closed with a plastic seal and parafilm and placed in a plastic bag, which was sealed with heat in order to prevent leakage. The plate was sonicated for 10 min at 35 kHz in an Ultrasonic Cleaner 3800 water bath sonicator (Branson Ultrasonics, Danbury, CT, USA) at 25 °C. This procedure did not affect the viability of the staphylococci (Figure S1).

*S. aureus* ATCC 25923 was then resuspended in RPMI 1640 using 3 pipetting cycles (up/down). Samples (10 µL) from each tested condition were transfer to 90 µL of TSB and serial dilutions were made from $10^{-1}$ up to $10^{-7}$. Aliquots (10 µL) of all dilutions were transferred to sheep blood agar plates (Amsterdam UMC, Amsterdam, The Netherlands) and incubated at 37 °C overnight. Viable plate colonies were counted the next day.

### *2.5. Influence of Opsonizing S. aureus ATCC 25923 on the Efficacy of HL-60 Cells in Preventing Bacterial Attachment on Titanium Coupons*

*S. aureus* ATCC 25923 suspensions were prepared as described in Section 2.4.1. The inoculum was adjusted to $2 \times 10^7$ CFU/mL and opsonized using 50% (*v*/*v*) pooled human serum in PBS (pooled from 15 healthy blood donors) by incubating at 37 °C for 30 min with gentle agitation, washing twice with PBS, and resuspending in RPMI 1640 [31]. As a control, a nonopsonized suspension of *S. aureus* ATCC 25923 was used. The two different suspensions (opsonized and nonopsonized), were then added to sterilized titanium coupons (0.4 cm height, 1.27 cm diameter, BioSurface Technologies Corp., Bozeman, MT, USA).

On the other hand, HL-60 cells were prepared as described in Section 2.4.2 and added to the titanium coupons, to which the bacterial suspension was previously added. The final volume covering each coupon was 1 mL, made up of 500 µL of a suspension of $2 \times 10^7$ CFU/mL *S. aureus* ATCC 25923, and 500 µL of a suspension of $2 \times 10^5$ HL-60 cells/mL in RPMI 1640.

### *2.6. Effect of the Antimicrobial Compounds on the Prevention of S. aureus ATCC 25923 and S. aureus P2 Adhesion in Coculture with Differentiated HL-60 Cells on Titanium Coupons*

#### 2.6.1. Culture of Staphylococci and HL-60 Cells

Bacterial inocula of *S. aureus* ATCC 25923 and *S. aureus* P2 were prepared as described in Section 2.4.1. The concentration was adjusted to $2 \times 10^7$ CFU/mL in RPMI 1640. For the differentiated HL-60 cells, the media were refreshed with RPMI 1640 24 h before the experiments started in order to remove possible traces of antibiotics from the maintenance media. On the day of the experiment, cells were counted with a Countess™ II automated cell counter (Thermo Scientific, Waltham, MA, USA) and adjusted to a concentration of $2 \times 10^5$ cells/mL in RPMI 1640. Each suspension (bacterial and HL-60 cells suspensions, 500 µL of each) was added to sterilized titanium coupons that were inserted in the different wells of a 24-well plate (Nunclon Δ surface, Nunc, Roskilde, Denmark) and to which the different tested compounds or the control antibiotics were added at a concentration of 50 µM (0.25% DMSO).

The 24-well plates containing the titanium coupons were maintained in cocultures with a solution containing $10^7$ CFU of *S. aureus* ATCC 25923 or *S. aureus* P2 and $10^5$ HL-60 cells in a total volume of 1 mL of RPMI 1640 for 24 h. Titanium coupons with added **RIF** (50 µM, 0.25% DMSO) were used as positive antibiotic controls. As coculture controls, titanium coupons without the addition of the tested antimicrobial compounds or control antibiotics were exposed simultaneously to both cellular systems (*S. aureus* and HL-60 cells) at the concentrations previously described. In addition, bacterial controls (exposed or not to the antimicrobial compounds in the absence of differentiated HL-60 cells) were also included.

#### 2.6.2. Bacterial Adherence on Titanium Coupons

Titanium coupons were gently washed with TSB to remove remaining adhering planktonic cells and transferred to Falcon tubes containing 1 mL of 0.5% (*w*/*v*) Tween® 20-TSB solution. Next, the tubes were sonicated in an Ultrasonic Cleaner 3800 water bath sonicator (Branson Ultrasonics, Danbury, CT, USA) at 25 °C for 5 min at 35 kHz. The tubes were vortexed for 20 s prior to and after the sonication step. Serial dilutions of the resulting bacterial suspensions were made from $10^{-1}$ up to $10^{-7}$ and plated on tryptic soy agar (Neogen®, Lansing, MI, USA) plates.

### 2.7. *Effect of the Antimicrobial Compounds on the Prevention of S. aureus ATCC 25923 Adhesion in Coculture with Differentiated HL-60 Cells on LDPE Tubes*

The assay was carried out as described in Section 2.6 but instead of titanium coupons the materials added to the 24-well plates were 1 cm long sections of a sterilized fine bore LDPE tubing (Smiths Medical ASD, Minneapolis, MN, USA).

### 2.8. *Scanning Electron Microscopy (SEM)*

In order to visualize the effect of the antimicrobial compounds on the prevention of *S. aureus* ATCC 25923 adhesion in coculture with immune cells on titanium coupons, *S. aureus* ATCC 25923 was cocultured with differentiated HL-60 cells on titanium coupons, as described in Section 2.6. Prior to SEM, samples were fixed in 4% (*v/v*) paraformaldehyde and 1% (*v/v*) glutaraldehyde (Merck, Kenilworth, NJ, USA) overnight at room temperature. Samples were rinsed twice with distilled water, with the duration of each cycle being 10 min. The dehydration procedure consisted of 2 cycles of incubation for 15 min in 50% (*v/v*) ethanol, 2 cycles of incubation for 20 min in 70% (*v/v*) ethanol, 2 cycles of incubation for 30 min in 80% (*v/v*) ethanol, 2 cycles of incubation for 30 min in 90% (*v/v*) ethanol, and 2 cycles of incubation for 30 min in 100% (*v/v*) ethanol. In order to reduce the sample surface tension, samples were immersed in hexamethyldisilazane (Polysciences Inc., Warrington, FL, USA) overnight and air-dried. Before imaging, samples were mounted on aluminum SEM stubs and sputter-coated with a 4 nm platinum–palladium layer using a Leica EM ACE600 sputter coater (Leica Microsystems, Wetzlar, Germany). Images were acquired at 2 kV using a Zeiss Sigma 300 SEM (Zeiss, Oberkochen, Germany) at the Electron Microscopy Center Amsterdam (Amsterdam UMC).

Of each coupon, 8–10 fields were viewed and photographed at magnifications of 250×, 500×, 1000×, and 3000×. Representative images are shown in the results.

### 2.9. *Statistical Analysis*

The quantitative data are reported as the mean and standard deviation (SD) of at least three independent experiments. Data were analyzed using GraphPad Prism 8 for Windows. For statistical comparisons, Tukey's multiple comparison test and Welch's unpaired *t*-test were applied, and $p < 0.05$ was always considered as statistically significant.

## 3. Results

### 3.1. *Effect of PMA Activation of Differentiated HL-60 Cells on Prevention of S. aureus ATCC 25923 Biofilm Formation*

Before performing the experiments with the titanium coupons and the antimicrobial compounds, the initial concentration of *S. aureus* ATCC 25923 was determined where the bacteria were able to form a biofilm in absence of the HL-60 cells, but were prevented from forming a biofilm when cocultured with $10^5$ HL-60 cells. At the same time, it was assessed whether the activation of the HL-60 cells with PMA enhanced their bacterial clearance capability (Figure 2). At an initial *S. aureus* ATCC 25923 concentration of $10^8$ CFU/mL, HL-60 cells did not significantly affect the bacterial attachment in 96-well microplates. A reduction on the adhered *S. aureus* ATCC 25923 viable cell counts was observed at an *S. aureus* ATCC 25923 concentration of $10^7$ CFU/mL and below ($p < 0.001$ in all cases when comparing the bacterial control with bacteria cocultured with HL-60 cells, and $p = 0.001$ for cells activated with PMA and a starting inoculum of $10^3$ CFU/Ml), with the exception of the lowest bacterial concentration tested, i.e., $10^2$ CFU/mL, where no difference was found. Both PMA-activated and nonactivated

HL-60 cells (gray and green columns, respectively) showed similarly reduced numbers of adherent *S. aureus* ATCC 25923 viable cell counts at 24 h, so it was concluded that activation with PMA does not significantly enhance *S. aureus* ATCC 25923 clearance.

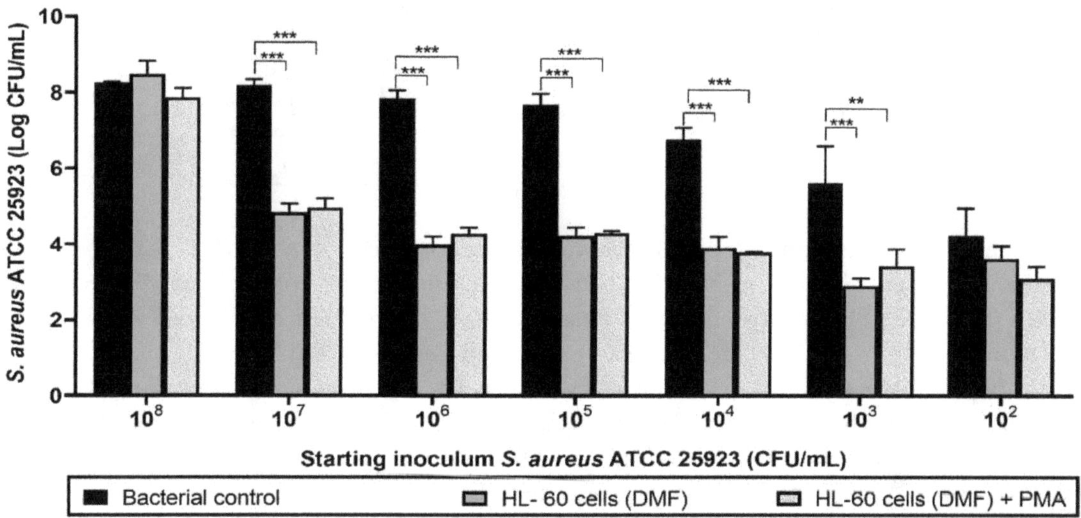


**Figure 2.** Viable counts of 24-hour-old biofilms formed by different concentrations of *S. aureus* ATCC 25923 cocultured with HL-60 cells on 96-well microplates. Black columns represent the bacterial control (viable attached cells in the absence of HL-60 cells). Green columns show the coculture of *S. aureus* ATCC 25923 with HL-60 cells differentiated with *N,N*-dimethylformamide (DMF). Gray columns show the coculture of *S. aureus* ATCC 25923 with HL-60 cells differentiated with DMF and activated with phorbol 12-myristate 13-acetate (PMA). Values are means and SD of three independent experiments (*** $p < 0.001$; ** $p < 0.01$).

Based upon these results, the optimal initial *S. aureus* ATCC 25923 concentration was found to be $10^7$ CFU/mL, which was then selected for the rest of the experiments. Since PMA activation of the HL-60 cells did not influence their phagocytic activity, no PMA stimulation was performed in subsequent experiments.

### *3.2. Influence of Opsonizing S. aureus ATCC 25923 on the Efficacy of HL-60 Cells in Preventing Bacterial Attachment on Titanium Coupons*

One of the most relevant mechanisms of the host defense against *Staphylococcus* spp. is phagocytosis. Given that opsonization of *S. aureus* is important for neutrophils to be able to clear planktonic *S. aureus* [32], the phagocytic efficacy of HL-60 cells, as well as the impact of opsonization of *S. aureus* ATCC 25923, in preventing *S. aureus* ATCC 25923 biofilm formation on titanium coupons was simultaneously explored (Figure 3). The left part of Figure 3 shows how the preventive capability of HL-60 cells was preserved when tested on titanium surfaces ($p < 0.001$, when comparing *S. aureus* + HL-60 (gray column) with the bacterial control (black column)). This bacterial clearance activity was also observed when *S. aureus* ATCC 25923 was opsonized ($p = 0.007$), but no significant differences were found when comparing the effects of the HL-60 cells on opsonized versus nonopsonized *S. aureus* ATCC 25923 ($p = 0.260$). The antibacterial effects of PMA-activated HL-60 cells against opsonized *S. aureus* ATCC 25923 was additionally explored, but no differences were found with the antimicrobial activity of nonactivated HL-60 against nonopsonized *S. aureus* ATCC 25923 (Figure S2). Because opsonization did not enhance the bacterial clearance capacity of HL-60, the use of opsonized *S. aureus* during the rest of the experiments was decided against.

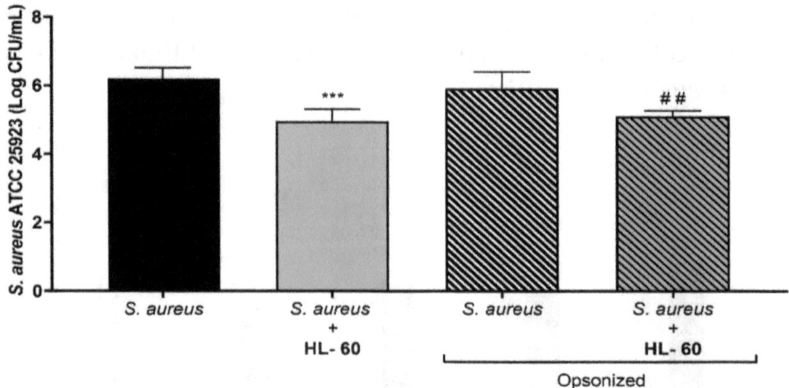


**Figure 3.** Viable counts of adhered opsonized and nonopsonized *S. aureus* ATCC 25923 on titanium coupons after coculture with HL-60 cells for 24 h. The left half of the graph includes the nonopsonized *S. aureus* ATCC 25923 biofilm formation, in absence (black column) or cocultured with HL-60 cells (green column). The right half includes the opsonized *S. aureus* ATCC 25923 biofilm formation, in absence (white/black column) or cocultured with HL-60 cells (green/black column). "*" indicates statistical differences with the nonopsonized *S. aureus* ATCC 25923 in monoculture, while "#" represents statistical differences with the opsonized *S. aureus* ATCC 25923 in monoculture with Welch's unpaired *t*-test (*** $p < 0.001$; ## $p < 0.01$). Values are means and SD of three independent experiments.

### 3.3. *Effect of the Antimicrobial Compounds on the Prevention of S. aureus Adhesion in Coculture with Differentiated HL-60 Cells on Titanium Coupons*

The effects of the three biofilm inhibitors **DHA1**, **DHA2**, and **FLA1** and one control antibiotic, **RIF**, on the prevention of *S. aureus* attachment on titanium coupons were investigated using two strains, the laboratory strain ATCC 25923 (Figure 4a) and the clinical isolate P2 (Figure 4b), either in the absence (gray bars) or presence of HL-60 cells (green bars). In the absence of HL-60 cells, all of the tested antimicrobial compounds, as well as the control antibiotic, significantly reduced the attachment of *S. aureus* ATCC 25923 ($p < 0.001$, $p = 0.040$, $p < 0.001$, and $p < 0.001$, for **DHA1**, **DHA2**, **FLA1**, and **RIF** versus the control, respectively). In the case of the clinical strain (Figure 4b), all the antimicrobial compounds, except **DHA2**, also showed antimicrobial activity in the absence of HL-60 cells ($p = 0.0067$, $p < 0.001$, and $p = 0.0087$ versus control for **DHA1**, **FLA1**, and **RIF**, respectively).

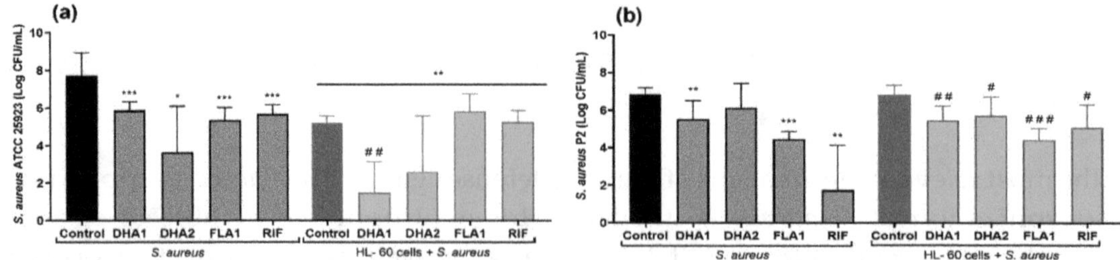


**Figure 4.** Viable counts of adhered *S. aureus* ATCC 25923 (**a**) and *S. aureus* P2 (**b**) on titanium coupons exposed to different antimicrobial compounds (tested at 50 μM) after 24 h of incubation. The gray bars show the attached viable bacteria when exposed just to the antimicrobial compounds, while the green bars show the results when *S. aureus* strains were cocultured with HL-60 cells. "*" indicates statistical differences with the control in monoculture while "#" represents statistical differences with the control in cocultured controls with Welch's unpaired *t*-test (* $p < 0.05$; ** $p < 0.01$; *** $p < 0.001$)/ # $p < 0.05$; ## $p < 0.01$; ### $p < 0.001$). Values are means and SD of three independent experiments.

The right part of Figure 4a (green bars) corresponds to the same experiments performed in the presence of HL-60 cells. Under such conditions, there was also a significant reduction of the attached viable *S. aureus* ATCC 25923 when compared with the material incubated without HL-60 cells (green control bar versus black control bar, $p < 0.001$). Moreover, this reduction was further increased by compound **DHA1** ($p = 0.025$ when comparing the coculture control column with the **DHA1** column in the HL-60 cells + *S. aureus* ATCC 25923 section of Figure 4, and $p < 0.001$ when comparing the **DHA1** gray column with the **DHA1** green column). The same tendency was observed for **DHA2**, but no statistical differences were found when comparing this to the control (dark green control: HL-60 cells + *S. aureus* ATCC 25923 section of Figure 4). Despite the fact that **FLA1** and the model antibiotic (**RIF**) successfully prevented the adhesion of *S. aureus* ATCC 25923, they did not cause a further increase in the bacterial clearance activity of the HL-60 cells. In contrast, the mere presence of HL-60 cells did not result in a significant reduction of *S. aureus* P2 attachment (Figure 4b), since no differences were found between the *S. aureus* P2 control in monoculture and in coculture with HL-60 cells. In this case, no differences were found between the antibacterial effects of the compounds in monoculture when compared with the same treatment in coculture with HL-60 cells (**DHA1**, **DHA2**, **FLA1**, and **RIF** gray columns versus their corresponding green columns).

Using SEM imaging, it was visually confirmed that **DHA1** reduced the number of *S. aureus* ATCC 25923 adherent to the titanium surface (Figure 5). In fact, almost no cocci were observed on the **DHA1**-treated titanium across the entire coupon. In addition, the presence of HL-60 cells also reduced the bacterial attachment, which was further enhanced by the treatment of **DHA1**. Adherent HL-60 cells were observed, as seen in the last row of images.

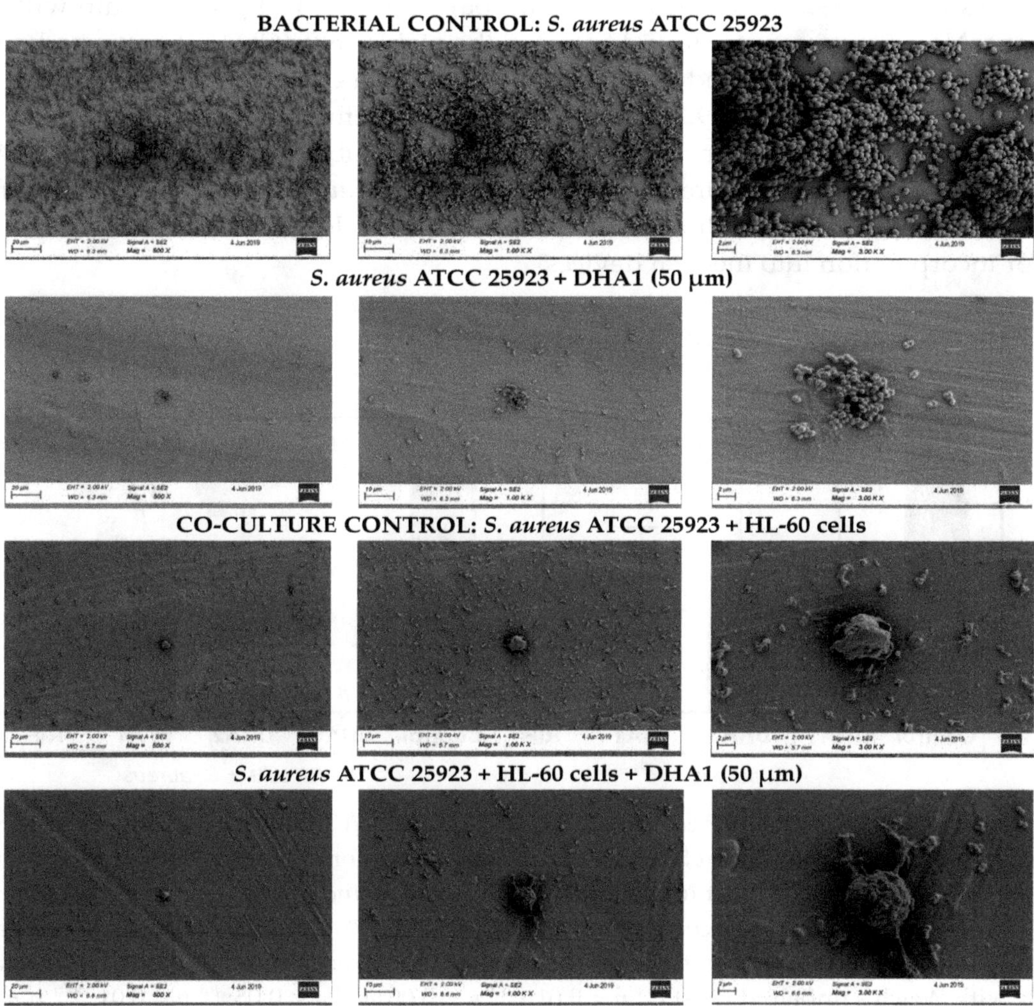


**Figure 5.** Representative images acquired by SEM. From left to right the same section of the titanium coupon is shown with different magnification, 500, 1000 and 3000×. Upper row of images, bacterial control, i.e., coupons incubated with *S. aureus* ATCC 25923 only. Second row, titanium coupons incubated with *S. aureus* ATCC 25923 and treated with **DHA1**. Third row, cocultured control, titanium coupons incubated with *S. aureus* ATCC 25923 and HL-60 cells simultaneously. Fourth row, titanium coupons cocultured with *S. aureus* ATCC 25923 and HL-60 and treated with **DHA1**.

### 3.4. *Effect of the Antimicrobial Compounds on the Prevention of S. aureus ATCC 25923 Adhesion in Coculture with Differentiated HL-60 Cells on LDPE tubes*

The adhesive capacity of *S. aureus*, as well as the antimicrobial effect of different compounds, is known to significantly differ depending on the material [33]. For this reason, the applicability of the compounds as part of endotracheal tubes was tested on a clinically relevant material, LDPE. Figure 6 shows the effects of the tested compounds and the control antibiotic (**RIF**) on the prevention of *S. aureus* ATCC 25923 attachment on LDPE tubes. The left part of the figure shows that all the antimicrobial compounds significantly reduced the numbers of attached viable *S. aureus* ATCC 25923 cells, in the absence of HL-60 ($p < 0.001$, $p = 0.0035$, $p < 0.001$, and $p < 0.001$, when comparing, respectively, **DHA1**, **DHA2**, **FLA1**, and **RIF** with the control). Adding HL-60 cells resulted in significant prevention of *S. aureus* ATCC 25923 attachment to LDPE tubes ($p < 0.001$ when comparing the coculture control (dark green column) with the bacterial control (black column)). In this case, all the antimicrobial compounds looked to be able to further potentiate this bacterial clearance capability ($p < 0.001$, $p = 0.023$, $p = 0.002$, and $p < 0.001$, when comparing, respectively, **DHA1, DHA2, FLA1**, and **RIF** light green columns with the control dark green column). However, similarly to what was observed on the titanium model, it was only the **DHA1** treatment that further potentiated the action of HL-60 cells against *S. aureus* ATCC 25923 ($p = 0.015$, when comparing the **DHA1** gray column with the **DHA1** green column). No differences were found between the viable cells (CFU/mL) attached on the LDPE exposed to *S. aureus* ATCC 25923 and treated with **DHA2, FLA1**, and **RIF** in monoculture and those exposed to both *S. aureus* ATCC 25923 and HL-60 with the same treatments (gray columns versus green columns). Full bacterial clearance was detected in LDPE tubes in the presence of **DHA1** and **RIF** (i.e., no viable bacterial counts measured), where cocultures of *S. aureus* ATCC 25923 with differentiated HL-60 cells were formed. These findings further highlight the relevance of **DHA1** as an antimicrobial candidate for incorporation into medical devices.

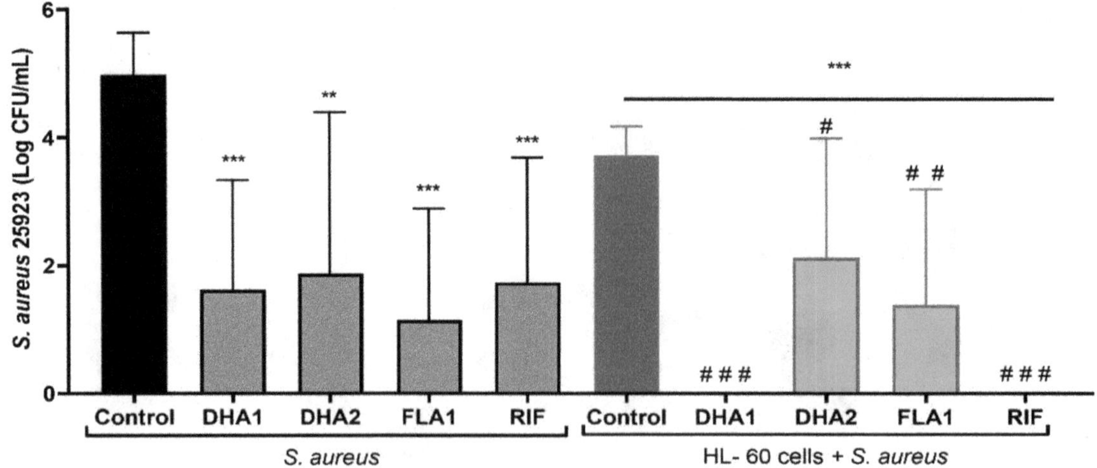


**Figure 6.** Viable counts of adhered *S. aureus* ATCC 25923 on LDPE tubes exposed to different antimicrobial compounds (tested at 50 μM) after 24 h of incubation. The gray bars show the attached viable bacteria when exposed just to the antimicrobial compounds, while the green bars show the results when *S. aureus* strains were cocultured with HL-60 cells. "*" indicates statistical differences with the control in monoculture, while "#" represents statistical differences with the control in cocultured controls with Welch's unpaired *t*-test (** $p < 0.01$; *** $p < 0.001$)/(# $p < 0.05$; ## $p < 0.01$; ### $p < 0.001$). Values are means and SD of three independent experiments.

## 4. Discussion

In this study, we explored the potential of incorporation of three previously identified naturally derived biofilm inhibitors into medical devices, particularly for titanium implantable devices and LDPE endotracheal tubes. From the three tested antimicrobial compounds, in line with previous findings [24], **DHA1** appeared to be the best candidate for incorporation as part of implantable medical devices. All of the compounds proved to be interesting candidates to include into anti-infective endotracheal tubes, but it was **DHA1** that again showed itself to be the most promising candidate, as it was the only one that significantly increased the bacterial clearance capacity of HL-60 cells against *S. aureus* ATCC 25923.

For compounds to be truly effective when used for protection of biomaterials in translational applications, they must be tested in meaningful experimental models. The insertion of any device provokes an acute inflammatory response that may cause ineffectiveness of innate immune cells such as neutrophils in cleaning planktonic bacteria, since these cells are directed to degrade the material. For this reason, it is vitally important to study the effects of antimicrobial compounds on neutrophils to assess their suitability as part of medical devices.

Before this investigation, no published reports existed on the effect of the two DHA derivatives (**DHA1** and **DHA2**) on neutrophils, however, the parent compound (dehydroabietic acid, DHA) was previously reported to have slight toxicity toward this cell type [34]. This prior knowledge further justified the need for an assessment of the effects of **DHA1** and **DHA2** on the bacterial clearance capacity of neutrophils. Similarly, no data existed on effects of **FLA1** on neutrophils, but other flavan derivatives were studied. Out of 10 different flavan-3-ol derivatives tested on human neutrophils, only two presented a slight toxic effect toward the neutrophils, but all of them reduced reactive oxygen species (ROS) and interleukin-8 production [35]. On the other hand, flavan-3-ol derivatives extracted from *Bistorta officinalis* (Delarbre) were reported to inhibit tumor necrosis factor-$\alpha$ (TNF-$\alpha$) release from neutrophils [36]. These earlier findings were promising in terms of applying the flavan derivative **FLA1** as part of medical devices, since its antimicrobial capacity in combination with its anti-inflammatory effects could result in prevention of infection while providing ideal cues toward material integration and resolution of inflammation.

In this study, we utilized HL-60 cells differentiated to polymorphonuclear-like cells in order to study the effects of the antimicrobial compounds in the presence of neutrophils. Alternatively, freshly extracted neutrophils could also have been used, but in such case differences would be encountered between individual donors in terms of reproducibility, the total number of cells that can be harvested, their short lifespan, or the disturbances in their physiology due to isolation procedures [37,38].

The three tested antimicrobial compounds and the control antibiotic (**RIF**) reduced *S. aureus* ATCC 25923 adhesion to titanium in the absence of HL-60 at the tested concentration of 50 μM. In the case of the clinical *S. aureus* strain, all compounds, except **DHA2**, significantly reduced *S. aureus* P2 biofilm formation on titanium in the absence of HL-60. The prevention of biofilm formation by **DHA1** and **FLA1** was as expected, as the compounds were used at concentrations higher than their MIC values [25,26]. Compound **DHA2** showed some prevention activity, despite being tested at a concentration slightly below its MIC (i.e., 60 μM) [26]. As with most antimicrobials, cytotoxicity is a concern, and **DHA2** was shown to reduce viability of HL cells (originating from the human respiratory tract) at a concentration of 100 μM [25].

The mere presence of HL-60 cells significantly reduced the bacterial attachment of *S. aureus* ATCC 25923. In contrast, this effect was not observed for the clinical *S. aureus* strain P2. This was also observed in our previous study, with the results obtained with the laboratory strain *S. aureus* ATCC 25923 significantly differing from the ones obtained with *S. aureus* P2 under the same experimental conditions. The latter has a key relevance in assessing the applicability of biofilm inhibitors as part of titanium implantable devices, as it was isolated from patients with orthopedic device-related infections. For this reason, the additional measurement of the preventive capacity of the biofilm inhibitors against the clinical *S. aureus* P2 strain is of great relevance.

Compound **DHA1** was shown to further potentiate *S. aureus* ATCC 25923 clearance caused by HL-60, and this effect was further confirmed by SEM. This compound also managed to effectively prevent the adherence of the clinical strain (*S. aureus* P2), but in this case it was difficult to establish if the reduction in bacterial attachment caused by **DHA1** (as well as **FLA1** and **RIF**) was due to a combined antimicrobial effect with the HL-60 cells or if it was due to their intrinsic antimicrobial capacity, as their effects in monoculture were equal to those observed in coculture with the HL-60 cells. These results emphasized the importance of not limiting the in vitro experimentation to laboratory strains, especially in cases aimed at finding compounds effective against medical device-associated infections, as the results obtained with laboratory strains may overestimate the efficacy of the compound.

These results demonstrate that **DHA1** seems to further increase *S. aureus* ATCC 25923 clearance by HL-60, while none of the other compounds negatively affect the antimicrobial effect of these immune cells. This is also of high relevance, as adverse effects on the immune response could be detrimental, and even increase the risk of infection. As an example, Croes et al. [39] biofunctionalized the surface of titanium implants with chitosan-based coatings that were incorporated with different concentrations of silver nanoparticles. Despite the good antimicrobial results obtained in the in vitro tests, these coatings did not demonstrate antibacterial effects *in vivo*. Due to the toxicity of the silver nanoparticles on the immune cells, these coatings aggravated infection-mediated bone remodeling, including increased osteoclast formation and inflammation-induced new bone formation.

Similar results were obtained in LDPE tubes, with the bacterial clearance capacity of the HL-60 cells against *S. aureus* ATCC 25923 also observed, and **DHA1** further potentiated this activity. The prevention of *S. aureus* adherence on the surface of endotracheal tubes may have potential to significantly reduce the rates of ventilator-associated pneumonia caused by these bacteria [40]. Additionally, by preventing the attachment of *S. aureus*, the attachment of *P. aeruginosa* may also be hampered, as several infection models demonstrated how early colonization by *S. aureus* facilitated subsequent *P. aeruginosa* colonization [41,42]. The development of a dual-species biofilm is expected to not only strongly worsen the pathology but significantly complicate the treatment [43].

In this study, and in concordance with our previous findings, **DHA1** was identified as the best candidate to be incorporated into implantable devices. This is because, in addition to its intrinsic antibiofilm capacity against both laboratory and clinical strains of *S. aureus*, it also seems to be able to enhance *S. aureus* ATCC 25923 clearance by HL-60 cells. Further mechanistic studies should be performed in order to elucidate if **DHA1** has a direct effect on the antimicrobial activity of PMNs. In the near future, we plan to assess the effects of **DHA1** on phagocytosis, ROS production, and formation of neutrophil extracellular traps (NETs).

Given the promising results that **DHA1** showed, both in our previous publications and in the current one, this biofilm inhibitor is involved in plans to be integrated as part of a titanium coating by means of 3D printing, with the coating formulation consisting of **DHA1**-loaded poly(lactic-co-glycolic acid) micro particles that suspended in a gelatine–methacrylate gel inkjet-printed onto titanium coupons [44]. The printing procedure is already validated and the prototype materials are currently being tested. In the case of endotracheal tubes, our current results suggest that all of the tested antimicrobial compounds would be beneficial for the prevention of *S. aureus* adhesion and subsequent biofilm formation, but **DHA1** appears to be the best candidate.

To the best of our knowledge, this is one of the first studies showing a positive effect of novel antimicrobials on the antibiofilm-clearing capacity of immune cells. This is of particular relevance because it does not only provide new alternatives to fight against the immense burden of bacterial biofilms, but it also sets the basis for a new in vitro system to accelerate the drug discovery process, thereby enabling better selection of antimicrobial incorporation into medical devices.

## 5. Conclusions

We showed the suitability of a coculture of *S. aureus* and differentiated HL-60 cells as an in vitro assay to assess the applicability of antimicrobial compounds for the protection of medical devices. As a proof-of-concept, we tested three antimicrobials, concluding that the DHA derivative **DHA1** is the best candidate for incorporation into implantable devices, as it does not only prevent biofilm formation on titanium but also seems to enhance the antibacterial capability of immune cells. On the other hand, according to our results, all of the antimicrobial compounds studied here, i.e., the two DHA derivatives, **DHA1** and **DHA2,** and the flavan derivative, **FLA1**, can tentatively be regarded as promising candidates to form part of anti-infective endotracheal tubes.

**Author Contributions:** Conceptualization, I.R., A.F. and S.A.J.Z.; methodology, J.Z.P., C.G.-P., M.R. and I.R.; software, I.R.; validation, I.R. and A.F.; formal analysis, I.R.; investigation, I.R.; resources, J.Y.-K., J.Z.P., K.S., A.F. and S.A.J.Z.; data curation, I.R.; writing—original draft preparation, I.R.; writing—review and editing, A.F., M.R., S.A.J.Z., C.G.-P. and J.Y.-K.; visualization, I.R., and A.F.; supervision, A.F., K.S. and S.A.J.Z.; project administration, A.F.; funding acquisition, A.F., S.A.J.Z. and J.Y.-K. All authors read and agreed to the published version of the manuscript.

**Acknowledgments:** We thank Teemu J. Kinnari and Ramón Pérez-Tanoira for their support during experimentation with the clinical strain. We thank Firas Hamdan for his help obtaining the human serum.

## References

1. Friedrich, A.W. Control of hospital acquired infections and antimicrobial resistance in Europe: The way to go. *Wien. Med. Wochenschr.* **2019**, *169*, 25–30. [CrossRef]
2. Cassini, A.; Högberg, L.D.; Plachouras, D.; Quattrocchi, A.; Hoxha, A.; Simonsen, G.S.; Colomb-Cotinat, M.; Kretzschmar, M.E.; Devleesschauwer, B.; Cecchini, M.; et al. Attributable deaths and disability-adjusted life-years caused by infections with antibiotic-resistant bacteria in the EU and the European Economic Area in 2015: A population-level modelling analysis. *Lancet Infect. Dis.* **2019**, *19*, 56–66. [CrossRef]
3. Ferri, M.; Ranucci, E.; Romagnoli, P.; Giaccone, V. Antimicrobial resistance: A global emerging threat to public health systems. *Crit. Rev. Food Sci. Nutr.* **2017**, *57*, 2857–2876. [CrossRef] [PubMed]
4. Percival, S.L.; Suleman, L.; Vuotto, C.; Donelli, G. Healthcare-associated infections, medical devices and biofilms: Risk, tolerance and control. *J. Med. Microbiol.* **2015**, *64*, 323–334. [CrossRef] [PubMed]
5. Khan, H.A.; Baig, F.K.; Mehboob, R. Nosocomial infections: Epidemiology, prevention, control and surveillance. *Asian Pac. J. Trop. Biomed.* **2017**, *7*, 478–482. [CrossRef]
6. Paharik, A.E.; Horswill, A.R. The Staphylococcal Biofilm: Adhesins, Regulation, and Host Response. *Microbiol. Spectr.* **2016**, *4*. [CrossRef]
7. Hall-Stoodley, L.; Costerton, J.W.; Stoodley, P. Bacterial biofilms: From the Natural environment to infectious diseases. *Nat. Rev. Microbiol* **2004**, *2*, 95. [CrossRef]
8. Kumar, A.; Alam, A.; Rani, M.; Ehtesham, N.Z.; Hasnain, S.E. Biofilms: Survival and defense strategy for pathogens. *Int. J. Med. Microbiol.* **2017**, *307*, 481–489. [CrossRef]
9. Singh, S.; Singh, S.K.; Chowdhury, I.; Singh, R. Understanding the Mechanism of Bacterial Biofilms Resistance to Antimicrobial Agents. *Open Microbiol. J.* **2017**, *11*, 53–62. [CrossRef]
10. Ferreira Tde, O.; Koto, R.Y.; Leite, G.F.; Klautau, G.B.; Nigro, S.; Silva, C.B.; Souza, A.P.; Mimica, M.J.; Cesar, R.G.; Salles, M.J. Microbial investigation of biofilms recovered from endotracheal tubes using sonication in intensive care unit pediatric patients. *Braz. J. Infect. Dis. Off. Publ. Braz. Soc. Infect. Dis.* **2016**, *20*, 468–475. [CrossRef] [PubMed]
11. Bardes, J.M.; Waters, C.; Motlagh, H.; Wilson, A. The Prevalence of Oral Flora in the Biofilm Microbiota of the Endotracheal Tube. *Am. Surg.* **2016**, *82*, 403–406. [CrossRef]
12. Li, H.; Song, C.; Liu, D.; Ai, Q.; Yu, J. Molecular analysis of biofilms on the surface of neonatal endotracheal tubes based on 16S rRNA PCR-DGGE and species-specific PCR. *Int. J. Clin. Exp. Med.* **2015**, *8*, 11075–11084. [PubMed]
13. Vandecandelaere, I.; Matthijs, N.; Van Nieuwerburgh, F.; Deforce, D.; Vosters, P.; De Bus, L.; Nelis, H.J.; Depuydt, P.; Coenye, T. Assessment of microbial diversity in biofilms recovered from endotracheal tubes using culture dependent and independent approaches. *PLoS ONE* **2012**, *7*, e38401. [CrossRef] [PubMed]
14. Arciola, C.R.; Campoccia, D.; Montanaro, L. Implant infections: Adhesion, biofilm formation and immune evasion. *Nat. Rev. Microbiol.* **2018**, *16*, 397–409. [CrossRef]
15. Kurtz, S.M.; Lau, E.; Watson, H.; Schmier, J.K.; Parvizi, J. Economic burden of periprosthetic joint infection in the United States. *J. Arthroplast.* **2012**, *27*, 61–65.e1. [CrossRef] [PubMed]

16. Kaufman, M.G.; Meaike, J.D.; Izaddoost, S.A. Orthopedic Prosthetic Infections: Diagnosis and Orthopedic Salvage. *Semin. Plast. Surg.* **2016**, *30*, 66–72. [CrossRef]
17. Arciola, C.R.; Campoccia, D.; Ehrlich, G.D.; Montanaro, L. Biofilm-based implant infections in orthopaedics. *Adv. Exp. Med. Biol.* **2015**, *830*, 29–46. [CrossRef]
18. Roh, J.S.; Sohn, D.H. Damage-Associated Molecular Patterns in Inflammatory Diseases. *Immune Netw.* **2018**, *18*, e27. [CrossRef]
19. Franz, S.; Rammelt, S.; Scharnweber, D.; Simon, J.C. Immune responses to implants—A review of the implications for the design of immunomodulatory biomaterials. *Biomaterials* **2011**, *32*, 6692–6709. [CrossRef]
20. Gristina, A.G.; Naylor, P.T.; Myrvik, Q. The Race for the Surface: Microbes, Tissue Cells, and Biomaterials. In *Molecular Mechanisms of Microbial Adhesion*; Springer: New York, NY, USA, 1989; pp. 177–211.
21. Perez-Tanoira, R.; Han, X.; Soininen, A.; Aarnisalo, A.A.; Tiainen, V.M.; Eklund, K.K.; Esteban, J.; Kinnari, T.J. Competitive colonization of prosthetic surfaces by *Staphylococcus aureus* and human cells. *J. Biomed. Mater. Res. Part A* **2017**, *105*, 62–72. [CrossRef] [PubMed]
22. Stones, D.H.; Krachler, A.M. Against the tide: The role of bacterial adhesion in host colonization. *Biochem. Soc. Trans.* **2016**, *44*, 1571–1580. [CrossRef] [PubMed]
23. Jaskiewicz, M.; Janczura, A.; Nowicka, J.; Kamysz, W. Methods Used for the Eradication of Staphylococcal Biofilms. *Antibiotics* **2019**, *8*, 174. [CrossRef]
24. Reigada, I.; Perez-Tanoira, R.; Patel, J.Z.; Savijoki, K.; Yli-Kauhaluoma, J.; Kinnari, T.J.; Fallarero, A. Strategies to Prevent Biofilm Infections on Biomaterials: Effect of Novel Naturally-Derived Biofilm Inhibitors on a Competitive Colonization Model of Titanium by *Staphylococcus aureus* and SaOS-2 Cells. *Microorganisms* **2020**, *8*, 345. [CrossRef] [PubMed]
25. Manner, S.; Vahermo, M.; Skogman, M.E.; Krogerus, S.; Vuorela, P.M.; Yli-Kauhaluoma, J.; Fallarero, A.; Moreira, V.M. New derivatives of dehydroabietic acid target planktonic and biofilm bacteria in *Staphylococcus aureus* and effectively disrupt bacterial membrane integrity. *Eur. J. Med. Chem.* **2015**, *102*, 68–79. [CrossRef] [PubMed]
26. Manner, S.; Skogman, M.; Goeres, D.; Vuorela, P.; Fallarero, A. Systematic Exploration of Natural and Synthetic Flavonoids for the Inhibition of *Staphylococcus aureus* Biofilms. *Int. J. Mol. Sci.* **2013**, *14*, 19434–19451. [CrossRef]
27. Jhunjhunwala, S. Neutrophils at the Biological–Material Interface. *ACS Biomater. Sci. Eng.* **2018**, *4*, 1128–1136. [CrossRef]
28. Hochbaum, A.I.; Kolodkin-Gal, I.; Foulston, L.; Kolter, R.; Aizenberg, J.; Losick, R. Inhibitory effects of D-amino acids on *Staphylococcus aureus* biofilm development. *J. Bacteriol.* **2011**, *193*, 5616–5622. [CrossRef]
29. Esteban, J.; Gomez-Barrena, E.; Cordero, J.; Martin-de-Hijas, N.Z.; Kinnari, T.J.; Fernandez-Roblas, R. Evaluation of quantitative analysis of cultures from sonicated retrieved orthopedic implants in diagnosis of orthopedic infection. *J. Clin. Microbiol.* **2008**, *46*, 488–492. [CrossRef]
30. Collins, S.J.; Ruscetti, F.W.; Gallagher, R.E.; Gallo, R.C. Terminal differentiation of human promyelocytic leukemia cells induced by dimethyl sulfoxide and other polar compounds. *Proc. Natl. Acad. Sci. USA* **1978**, *75*, 2458–2462. [CrossRef]
31. Lu, T.; Porter, A.R.; Kennedy, A.D.; Kobayashi, S.D.; DeLeo, F.R. Phagocytosis and killing of *Staphylococcus aureus* by human neutrophils. *J. Innate Immun.* **2014**, *6*, 639–649. [CrossRef]
32. van Kessel, K.P.M.; Bestebroer, J.; van Strijp, J.A.G. Neutrophil-Mediated Phagocytosis of *Staphylococcus aureus*. *Front. Immunol.* **2014**, *5*, 467. [CrossRef] [PubMed]
33. Hiltunen, A.K.; Savijoki, K.; Nyman, T.A.; Miettinen, I.; Ihalainen, P.; Peltonen, J.; Fallarero, A. Structural and Functional Dynamics of *Staphylococcus aureus* Biofilms and Biofilm Matrix Proteins on Different Clinical Materials. *Microorganisms* **2019**, *7*, 584. [CrossRef]
34. Sunzel, B.; Söderberg, T.A.; Reuterving, C.O.; Hallmans, G.; Holm, S.E.; Hänström, L. Neutralizing effect of zinc oxide on dehydroabietic acid-induced toxicity on human polymorphonuclear leukocytes. *Biol. Trace Elem.* **1991**, *30*, 257–266. [CrossRef]
35. Czerwińska, M.E.; Dudek, M.K.; Pawłowska, K.A.; Pruś, A.; Ziaja, M.; Granica, S. The influence of procyanidins isolated from small-leaved lime flowers (Tilia cordata Mill.) on human neutrophils. *Fitoterapia* **2018**, *127*, 115–122. [CrossRef] [PubMed]

36. Pawłowska, K.A.; Hałasa, R.; Dudek, M.K.; Majdan, M.; Jankowska, K.; Granica, S. Antibacterial and anti-inflammatory activity of bistort (*Bistorta officinalis*) aqueous extract and its major components. Justification of the usage of the medicinal plant material as a traditional topical agent. *J. Ethnopharmacol.* **2020**, *260*, 113077. [CrossRef]
37. Manda-Handzlik, A.; Bystrzycka, W.; Wachowska, M.; Sieczkowska, S.; Stelmaszczyk-Emmel, A.; Demkow, U.; Ciepiela, O. The influence of agents differentiating HL-60 cells toward granulocyte-like cells on their ability to release neutrophil extracellular traps. *Immunol. Cell Biol.* **2018**, *96*, 413–425. [CrossRef]
38. Yaseen, R.; Blodkamp, S.; Luthje, P.; Reuner, F.; Vollger, L.; Naim, H.Y.; von Kockritz-Blickwede, M. Antimicrobial activity of HL-60 cells compared to primary blood-derived neutrophils against *Staphylococcus aureus*. *J. Negat. Results Biomed.* **2017**, *16*, 7. [CrossRef] [PubMed]
39. Croes, M.; Bakhshandeh, S.; van Hengel, I.A.J.; Lietaert, K.; van Kessel, K.P.M.; Pouran, B.; van der Wal, B.C.H.; Vogely, H.C.; Van Hecke, W.; Fluit, A.C.; et al. Antibacterial and immunogenic behavior of silver coatings on additively manufactured porous titanium. *Acta Biomater.* **2018**, *81*, 315–327. [CrossRef]
40. Seitz, A.P.; Schumacher, F.; Baker, J.; Soddemann, M.; Wilker, B.; Caldwell, C.C.; Gobble, R.M.; Kamler, M.; Becker, K.A.; Beck, S.; et al. Sphingosine-coating of plastic surfaces prevents ventilator-associated pneumonia. *J. Mol. Med.* **2019**, *97*, 1195–1211. [CrossRef]
41. Lyczak, J.B.; Cannon, C.L.; Pier, G.B. Lung infections associated with cystic fibrosis. *Clin. Microbiol. Rev.* **2002**, *15*, 194–222. [CrossRef]
42. Alves, P.M.; Al-Badi, E.; Withycombe, C.; Jones, P.M.; Purdy, K.J.; Maddocks, S.E. Interaction between *Staphylococcus aureus* and *Pseudomonas aeruginosa* is beneficial for colonisation and pathogenicity in a mixed biofilm. *Pathog. Dis.* **2018**, *76*. [CrossRef]
43. Beaudoin, T.; Yau, Y.C.W.; Stapleton, P.J.; Gong, Y.; Wang, P.W.; Guttman, D.S.; Waters, V. *Staphylococcus aureus* interaction with *Pseudomonas aeruginosa* biofilm enhances tobramycin resistance. *NPJ Biofilms Microbiomes* **2017**, *3*, 25. [CrossRef] [PubMed]
44. Yang, Y.; Chu, L.; Yang, S.; Zhang, H.; Qin, L.; Guillaume, O.; Eglin, D.; Richards, R.G.; Tang, T. Dual-functional 3D-printed composite scaffold for inhibiting bacterial infection and promoting bone regeneration in infected bone defect models. *Acta Biomater.* **2018**, *79*, 265–275. [CrossRef]

# 10

# Adhesins of *Brucella*: Their Roles in the Interaction with the Host

**Magalí G. Bialer [1], Gabriela Sycz [1,†], Florencia Muñoz González [2,3], Mariana C. Ferrero [2,3], Pablo C. Baldi [2,3,*] and Angeles Zorreguieta [1,4,*]**

[1] Fundación Instituto Leloir (FIL), IIBBA (CONICET-FIL), Buenos Aires 1405, Argentina; mbialer@leloir.org.ar (M.G.B.); gabrielasycz@gmail.com (G.S.)

[2] Cátedra de Inmunología, Facultad de Farmacia y Bioquímica, Universidad de Buenos Aires, Buenos Aires 1113, Argentina; fmgonzalez@ffyb.uba.ar (F.M.G.); ferrerom@ffyb.uba.ar (M.C.F.)

[3] Instituto de Estudios de la Inmunidad Humoral (IDEHU), CONICET-Universidad de Buenos Aires, Buenos Aires 1113, Argentina

[4] Departamento de Química Biológica, Facultad de Ciencias Exactas y Naturales, Universidad de Buenos Aires, Buenos Aires 1428, Argentina

* Correspondence: pablobal@ffyb.uba.ar (P.C.B.); azorreguieta@leloir.org.ar (A.Z.)

† Current address: Department of Molecular Microbiology, Washington University School of Medicine in St. Louis, 660 S Euclid Ave (63110), St. Louis, MO 63110, USA.

**Abstract:** A central aspect of *Brucella* pathogenicity is its ability to invade, survive, and replicate in diverse phagocytic and non-phagocytic cell types, leading to chronic infections and chronic inflammatory phenomena. Adhesion to the target cell is a critical first step in the invasion process. Several *Brucella* adhesins have been shown to mediate adhesion to cells, extracellular matrix components (ECM), or both. These include the sialic acid-binding proteins SP29 and SP41 (binding to erythrocytes and epithelial cells, respectively), the BigA and BigB proteins that contain an Ig-like domain (binding to cell adhesion molecules in epithelial cells), the monomeric autotransporters BmaA, BmaB, and BmaC (binding to ECM components, epithelial cells, osteoblasts, synoviocytes, and trophoblasts), the trimeric autotransporters BtaE and BtaF (binding to ECM components and epithelial cells) and Bp26 (binding to ECM components). An in vivo role has also been shown for the trimeric autotransporters, as deletion mutants display decreased colonization after oral and/or respiratory infection in mice, and it has also been suggested for BigA and BigB. Several adhesins have shown unipolar localization, suggesting that *Brucella* would express an adhesive pole. Adhesin-based vaccines may be useful to prevent brucellosis, as intranasal immunization in mice with BtaF conferred high levels of protection against oral challenge with *B. suis*.

**Keywords:** *Brucella*; adhesins; Ig-like domain; monomeric autotransporters; trimeric autotransporters; extracellular matrix; polar localization; virulence factors; vaccine candidates; fibronectin

---

## 1. Introduction

*Brucella* spp. are Gram-negative bacteria that infect several animal species and can be transmitted to humans by several routes, producing one of the most common zoonotic diseases worldwide. A central aspect of *Brucella* pathogenicity is its ability to invade, survive, and replicate in several phagocytic and non-phagocytic cell types, leading not only to chronic infections but also to chronic inflammatory phenomena in different tissues. For both phagocytic and non-phagocytic cells, the first step of the invasion process involves interactions between surface molecular factors of *Brucella* and the host cell, leading to the cellular adhesion of the pathogen. Several *Brucella* proteins have been shown to be involved in the adhesion of this bacterium to different cell types and/or to extracellular matrix

(ECM) components. In this review, we describe the main characteristics of these *Brucella* adhesins in the broader context of bacterial adhesins, and how they contribute to the cellular infectious process of brucellae.

## 2. *Brucella* Infection and Clinical Manifestations

With more than 500,000 new cases annually, human brucellosis continues to be one of the commonest zoonotic diseases worldwide. Although this disease has been eradicated in some developed countries, it still constitutes a public health problem in Latin America, the Middle East, North and East Africa, and South and Central Asia [1]. Moreover, the disease is still present in some European countries.

*Brucella* spp. are Gram-negative non-capsulated and non-sporulated bacilli or cocobacilli that lack cilia or flagella. Despite the great number of *Brucella* species identified, which may infect domestic animals and wild animals, only *B. melitensis* (goats and sheep), *B. suis* (pigs), *B. abortus* (cattle), and, to a minor extent, *B. canis* (dogs) are linked to human brucellosis. Different *Brucella* species yield smooth (S) or rough (R) colonies when cultured in agar, which is a difference that is directly related to the structure of the lipopolysaccharide (LPS). The LPS is divided into three regions: the lipid A (the innermost portion), a polysaccharidic core, and the O polysaccharide (the outermost portion). Smooth species (*B. melitensis*, *B. suis*, *B. abortus*, and others) produce a "complete" LPS (S-LPS) containing the three portions, whereas rough species (*B. canis* and *B. ovis*) produce a rough LPS (R-LPS) that lacks the O polysaccharide.

*Brucella* infection in humans is mainly acquired through the consumption of raw animal products, inhalation of contaminated aerosols in slaughterhouses, rural settings or laboratories [2], and contact of the abraded skin with contaminated tissues or materials. Less frequently, accidental infection with attenuated vaccine strains [3–6] and vertical transmission have been reported [7].

Human brucellosis has a wide spectrum of clinical manifestations, which depend on the stage of the disease and the organs and systems involved. The disease usually presents as a febrile illness accompanied by myalgia, arthralgia, and hepatomegaly, and may evolve with an uncomplicated course or may present complications involving particular organs or systems [8]. Osteoarticular involvement is the most common focal complication [9]. The diversity of tissues that can be affected by *Brucella* is likely related to its ability to invade, survive, and replicate in several phagocytic and non-phagocytic cell types, as explained below. Despite its tendency to produce chronic illness and even disabling disease, human brucellosis is only rarely fatal. In animals, the most prevalent manifestations are abortions, reduced fertility, weight loss, and reduced milk production [10].

## 3. *Brucella* Entry into Host Cells

A central aspect of *Brucella* pathogenicity is its ability to invade, survive, and replicate in several cell types, leading not only to chronic infections but also to chronic inflammatory phenomena that explain most of the clinical manifestations of brucellosis [11]. Most initial research was performed using murine macrophagic cell lines [12], bovine macrophages [13], human monocytes [14], and widely used non-phagocytic cell lines such as HeLa (human cervical cells) or Vero (kidney, African green monkey) [15,16]. However, further in vitro studies revealed that *Brucella* is capable to infect and replicate in human osteoblasts [17], synoviocytes [18], trophoblasts [19], endothelial cells [20], lung epithelial cells [21], dendritic cells [22], and hepatocytes [23] as well as in murine alveolar macrophages [24], canine trophoblasts and phagocytes [25], and ovine testis cells lines [26].

The internalization of *Brucella* into the different cell types is a complex multi-stage process. Whatever the host cell (phagocytic or non-phagocytic) and the *Brucella* strain involved (smooth or rough), the first step in this process involves interactions between surface molecular factors of both the host cell and the pathogen leading to cellular binding of the bacterium. In fact, it was shown by scanning electron microscopy that *B. abortus* adheres (as early as 1 h after infection) and forms bacterial aggregates on the surface of host cells in a time-dependent manner [27]. Whereas several of the surface molecular factors involved have been identified, the full repertoire of molecular components and mechanisms acting on either active bacterial penetration or passive uptake of *Brucella* spp. are not fully characterized [28]. For non-opsonized bacteria, internalization into macrophages seems to depend on lipid rafts present in the plasma membrane of these cells [29,30]. It has been shown that lipid rafts-associated molecules, including cholesterol and the ganglioside GM1, are involved in the entry of *B. suis* into murine macrophages under non-opsonic conditions [31]. In addition, a class A scavenger receptor (SR-A) seems to be required for *B. abortus* internalization into macrophages through a lipid raft-mediated mechanism [32]. These three host molecules have been involved in the ability of naturally rough *Brucella* species (*B. ovis, B. canis*) to infect murine macrophages [33]. Although these lipid raft-associated molecules have a role in *Brucella* internalization in macrophages, it has not been determined whether they participate in bacterial adhesion or, alternatively, only contribute to bacterial penetration or uptake.

In addition to these lipid raft-associated molecules, other host components have been identified as being involved in the interaction between brucellae and the host cells. Sialic acid-containing molecules were proposed to be involved in the interaction of brucellae with macrophages and epithelial cells [27]. GM1 is a sialylated molecule, which may perhaps explain its role in lipid raft-mediated internalization in macrophages. This study also produced evidence suggesting that cell surface heparan sulphate molecules may be involved in *Brucella* binding to epithelial cells. Based on the hypothesis that the interactions of *Brucella* with the ECM contribute to the spread of the bacteria through tissue barriers, the ability of the pathogen to bind to ECM constituents was also explored. It was shown that *B. abortus* binds in a dose-dependent manner to immobilized fibronectin and vitronectin and, to a lesser extent, to chondroitin sulphate, collagen, and laminin [27].

As mentioned above, adhesion to host cells is the first step of the infectious cycle of many pathogens. Most bacterial pathogens express adhesins and other molecules that mediate the binding to a wide range of cell surface molecules and ECM components depending on the lifestyle of the microorganism. The fact that *Brucella* species can bind to the cell surface and ECM components strongly suggests the expression of bacterial molecules involved in such an interaction. Although not formally shown to be involved in adhesion, *Brucella* LPS has been linked to the internalization of the pathogen in macrophages. Smooth *B. abortus* strains expressing a complete LPS (including the O-polysaccharide) enter macrophages through lipid-rafts, whereas a rough mutant does not [30,34]. However, naturally rough *Brucella* species (*B. ovis, B. canis*) seem to use lipid rafts for entry [33], suggesting that lipid raft-mediated internalization of brucellae does not depend on O-polysaccharide expression. A role for some outer membrane proteins, namely Omp22 and Omp25, in *Brucella* binding or internalization has also been suggested. Targeted inactivation of their corresponding genes impaired internalization of rough *B. ovis* but not that of *B. abortus* [35,36]. Moreover, *B. abortus* mutants were more adherent than the wild-type strain. While the role of the LPS and outer membrane proteins in the ability of *Brucella* to adhere to cells or ECM requires further clarification, more recent studies have led to the identification of bacterial adhesins clearly involved in these adhesion processes (see below).

Upon entry into the host cells, *Brucella* organisms initiate an intracellular cycle that involves a sequential traffic through the endocytic, secretory, and autophagic compartments. Bacterial effectors delivered inside the infected cells through a type IV secretion system encoded by the *virB* operon are essential to accomplish these steps [15,37–39]. The O polysaccharide of the LPS is also involved in the ability of *Brucella* to establish intracellular infections. Phagosomes containing smooth strains of *B. suis* do not fuse with lysosomes, at least in murine macrophages, whereas those harboring rough mutants rapidly fuse [40]. This seems to be related to the fact that only the naturally smooth strains enter the cells through lipid-rafts and can inhibit phagosome-lysosome fusion [30,34].

## 4. Bacterial Adhesins

Most pathogenic bacteria interact with their hosts through adhesive molecules (adhesins) that are exposed on their cell surfaces. Since adhesion to host cells can also stimulate immune activation, several bacteria produce a surface layer (i.e., capsular polysaccharide) that prevents immune recognition or phagocytosis. For this reason, they often express adhesins on polymeric structures that extend out from the cell surface at a prudential distance. For some bacteria, attachment to the host cell surface is also crucial for effector injection through complex secretion systems. Lastly, adhesion to the host cell is the previous step to internalization for those bacteria whose strategy to achieve proliferation and survival is the intracellular life [41,42]. The adhesins can be grouped into two types: (1) filamentous (fimbrial) adhesins consisting of complex structures made up of multiple subunits and (2) non-fimbrial adhesins that can be monomeric or trimeric proteins.

Fimbrial adhesins are a varied group of polymeric fibers that are visible using electron microscopy. In Gram-negative bacteria, these adhesins can be classified into: (1) the chaperone-usher pili (CUP), (2) the alternative chaperone-usher pathway pili, (3) Type IV pili, and (4) pili assembled by the extracellular nucleation-precipitation pathway (curli) [41,43]. The subunit at the tip of the CUP pili is a lectin that can bind sugar-containing molecules on the host cell surfaces [44]. The Type IV pili are long filaments composed of pilin subunits assembled into bundles, which are involved in diverse functions including bacterial twitching motility, auto-aggregation, and attachment to host cells [45]. Curli are involved in many physiological and pathogenic processes such as biofilm formation and host cell adhesion and invasion. The curli are assembled via the nucleation-precipitation pathway and display structural similarities with functional amyloids [46,47]. As mentioned below, *Brucella* spp. do not seem to express fimbrial adhesins.

Non-fimbrial adhesins include adhesins that belong to the RTX (repeat in toxin) protein family and those that correspond to type V secretion systems (T5SS), which are also called autotransporter proteins.

RTX adhesins are secreted by a type 1 secretion system (T1SS) that has three components: an inner-membrane ABC (ATP binding cassette) transporter, a membrane fusion protein, and an outer-membrane pore from the TolC family. The substrates of T1SSs do not harbor an N-terminal cleavable signal peptide but share a structural C-terminal domain that is not cleaved off during the secretion process [48]. The RTX adhesins are usually loosely attached to the bacterial surface and have been implicated in bacteria-to-bacteria interactions during biofilm formation and adhesion to epithelial cells [49].

The T5SSs (subfamilies Va–Ve) play important roles in the interaction of several pathogens with their hosts [50]. Originally, the term "autotransporter" was proposed because it was thought that all the information for its translocation from the inner membrane to the extracellular medium was mostly contained in the protein itself. This concept has changed since other factors, such as chaperones and the BAM (β-barrel Assembly Machinery) system are required for secretion of these proteins. Furthermore, more recently, it was shown that another system, the TAM (Translocation and Assembly Module) complex, is also required for the correct translocation of autotransporters into the outer membrane [51–53]. It was proposed that this complex spanning the periplasmic space might solve the energy problem to translocate proteins through the outer membrane [52]. Therefore, the current model proposes that the TAM and BAM systems would act in a concerted manner [51,54].

The T5SS or autotransporter proteins share common structural and functional characteristics: (1) an N-terminal Sec-dependent signal peptide that mediates the transport from the cytoplasm to the periplasm, (2) a passenger (and functional) domain, and (3) a C-terminal β-barrel domain that forms a pore in the outer membrane through which the passenger domain is translocated to the cell surface [55]. In the subclass Va, the autotransporters are monomeric and the passenger and secretion domains are integrated into the same protein, the β-barrel domain forms a pore of 12 antiparallel β-strands, and the passenger regions consist of highly variable repetitive amino acid motifs. Some of these autotransporters are important virulence factors, playing diverse functions in the interaction with the host. The passenger domains have often enzymatic activity and usually adopt a repetitive β-helix fold extending away from the bacterial cell surface, as demonstrated by the crystal structure of the Pertactin passenger domain [56]. Passengers with enzymatic activity are cleaved off from the surface while adhesion passengers can be retained on the cell surface without cleavage (for a comprehensive review, see Reference [50]). Some adhesins of the monomeric autotransporter family have been described in *Brucella* spp., as explained below.

In the two-partner secretion systems (T5SS type Vb), the passenger and β-translocator domains are encoded by two different genes. Filamentous haemagglutinin adhesins are exported by this type of system. These adhesins are often involved in a tight interaction with a host cell receptor and also in biofilm formation [50].

All members of the trimeric autotransporter Type Vc group that have been characterized so far are implicated in adhesion functions. They usually bind to host receptors or to host ECM components. As detailed in the next section, *Brucella* spp. express adhesins that belong to this subclass of autotransporters. While the overall organization of these proteins is similar to that of the monomeric autotransporters, they contain a shorter C-terminal translocation domain of 50–100 amino acids and the 12 β-strand pore is achieved by protein trimerization. Usually, the passenger domain harbors conserved structural elements named as head, connector, and stalk domains. The combinations of these repeats result in either "lollipop structures" like YadA or as "beads-on-a-string" like BadA [52]. Although the head domains typically mediate adhesion to host targets, the stalk domains can also participate in adhesion functions. Internal regions may serve to extend the head domain away from the bacterial cell surface. Unlike several monomeric autotransporters, trimeric autotransporters are not released into the extracellular space [50].

The Type Ve of T5SS harbors a 12-stranded β-barrel domain and a secreted, monomeric passenger domain that remains attached after translocation. The main difference with type Va autotransporters is that the type Ve have an inverted domain order with the β-barrel at the N-terminal end and the passenger domain at the C-terminus, and, thus, are named as "inverse autotransporters" [57]. Well-known examples are the intimin and invasin from pathogenic *Escherichia coli* and *Yersinia* spp., respectively. The passenger domains of this type of T5SS contain domains with Immunoglobulin (Ig)-like or lectin-like structures. The intimin of enteropathogenic and enterohemorrhagic *E. coli* strains mediates an intimate contact with the Tir receptor, which is delivered by the bacterium to the surface of the host cell. The invasin of *Yersinia* spp. binds directly to β1-integrins on the apical side of gut epithelial cells, which promotes bacterial internalization via endocytosis [58,59].

## 5. Adhesins of *Brucella*

The genomes of *Brucella* spp. do not harbor loci associated with components of pili or curli that could function as fimbrial adhesins. Furthermore, by electron microscopy, no pilus-like structures have been observed. However, several non-fimbrial adhesins have been identified that were shown to have a role in the interaction with the host. A diagram of these adhesins, showing their domains, is depicted in Figure 1, and additional information is presented in Table 1.

**Table 1.** Adhesins described in *Brucella* spp.

| Adhesin | Organism | KEGG Entry | NCBI Protein ID | Protein Class | Host Ligands Detected | Cellular Adhesion Role | In Vivo Infection Role | Reference |
|---|---|---|---|---|---|---|---|---|
| **SP29** | *B. abortus* 9-941 | BruAb2_0373 | WP_002965789.1 | D-ribose ABC transporter substrate-binding protein | Sialic acid-containing proteins | Erythrocytes | ND | [60] |
| **SP41** | *B. abortus* 9-941 | BruAb2_0571 | WP_002965982.1 | ATP-binding cassette transporter | Sialic acid-containing proteins | Epithelial (HeLa) | No role detected in *B. ovis* infections | [61,62] |
| **BigA** | *B. abortus* 2308 | BAB1_2009 | EEP62646.1 | Ig-like domain- containing protein | Cell adhesion molecules | Epithelial (HeLa, Caco.2, MDCK) | Potential role in oral infections * | [63,64] |
| **BigB** | *B. abortus* 2308 | BAB1_2012 | WP_002967016.1 | Ig-like domain- containing protein | Cell adhesion molecules | Epithelial (HeLa) | Potential role in oral infections * | [63,65] |
| **Bp26** | *B. melitensis 16M* | BMEI0536 | WP_002964581.1 | Uncharacterized | Type I collagen, vitronectin, fibronectin | ND | ND | [66] |
| **BmaC** | *B. suis* 1330 | BRA1148 | WP_006191504.1 | Monomeric autotransporter | Fibronectin, type I collagen | Epithelial (HeLa, A549). Synoviocytes. Osteoblasts | ND | [67,68] |
| **BmaA** | *B. suis* 1330 | BR0173 | AAN33380.1 | Monomeric autotransporter | Fibronectin, type I collagen | Epithelial (HT 29, Caco.2). Synoviocytes. Osteoblasts. Trophoblasts | ND | [68] |
| **BmaB** | *B. suis* 1330 | BR2013 | AAN30903.1 | Monomeric autotransporter | Fibronectin | Synoviocytes. Osteoblasts. Trophoblasts | ND | [68] |
| **BtaE** | *B. suis* 1330 | BR0072 | WP_006191142.1 | Trimeric autotransporter | Fibronectin, hyaluronic acid | Epithelial (HeLa, A549) | Mutants display decreased colonization after oral infection | [69] |
| **BtaF** | *B. suis* 1330 | BR1846 | A0A0H3G4K1.1 | Trimeric autotransporter | Fibronectin, hyaluronic acid, fetuin, type I collagen | Epithelial (HeLa, A549) | Mutants display decreased colonization after oral or respiratory infection | [70,71] |

ND: not determined. (*) A mutant lacking the Bab1-2009-2012 genomic island is attenuated in oral infections in mice.

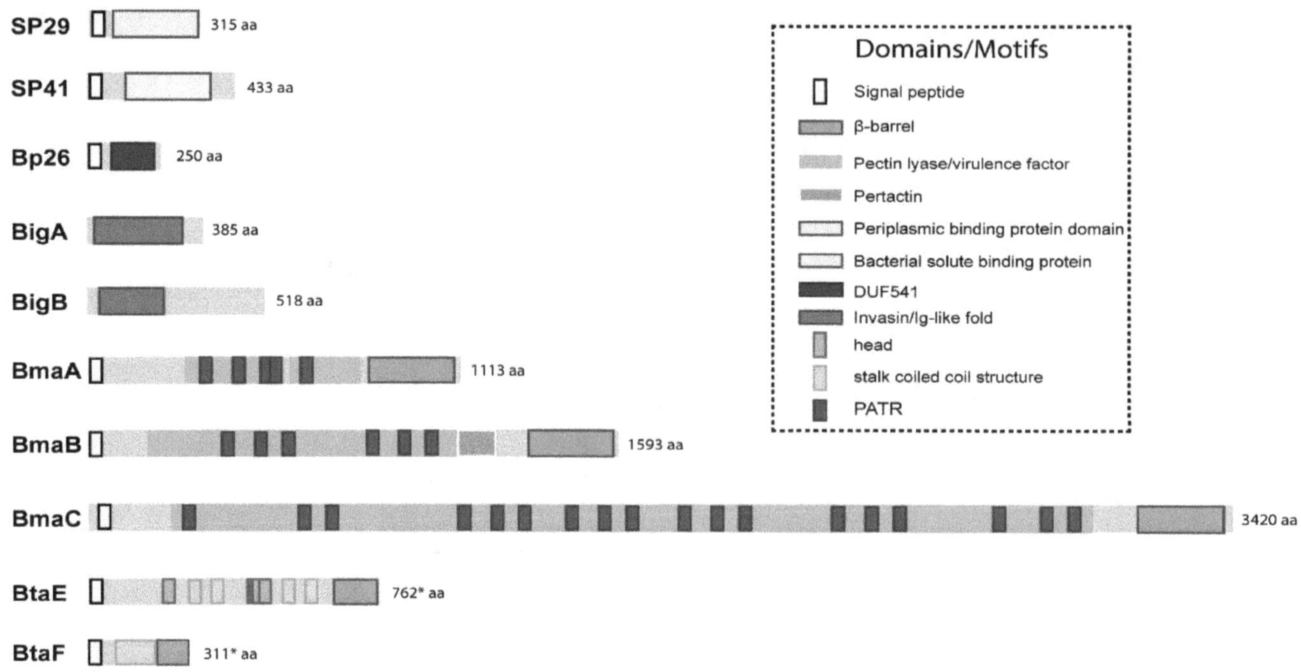


**Figure 1.** Domain organization of described *Brucella* adhesins. Schematic representation of the adhesins described in *Brucella* spp, showing functional and structural domains predicted by bioinformatics (SignalP 5.0, Pfam, BLAST, InterProScan). Asterisks indicate the cases for which start codons upstream of the annotated ORF were identified and the translation product of the new ORF contained an N-terminal signal peptide with a reliable score. aa: amino acids.

### 5.1. Unclassified Adhesins

Due to the abundance of carbohydrates in the surface of red cells, hemagglutination tests have been used for the detection and characterization of many lectin-like adhesins in bacterial pathogens. Using this approach, Rocha-Gracia et al. found that *B. abortus* and *B. melitensis* can agglutinate human (A+ and B+), hamster, and rabbit erythrocytes, and that this activity was associated with a bacterial 29-kDa surface protein (SP29) that binds to these cells [60]. Purified SP29 bound directly to rabbit erythrocytes, and this binding was abolished by neuraminidase treatment of red cells, indicating that SP29 binds to sialic acid-containing receptors. The analysis of an internal fragment obtained by peptic digestion suggested that SP29 is a D-ribose-binding periplasmic protein precursor found in *B. melitensis* (BruAb2_0373) (Figure 1). No further characterization of this protein or its importance for *Brucella* pathogenesis has been reported despite the demonstration that *B. melitensis* is able to invade erythrocytes in vivo at least in the mouse model [72]. This later study revealed that *B. melitensis* can adhere to murine erythrocytes as early as 3-h post-infection but is later found mainly in the cytoplasm of these cells. Moreover, erythrocytes represented the major fraction of infected cells in the bloodstream. Purified erythrocytes from infected mice were able to transmit *B. melitensis* infection to naïve mice.

To our best knowledge, the first *Brucella* adhesin for which a functional role was fully characterized in vitro was SP41 (Figure 1, Table 1) [61]. This protein is the predicted product of the *ugpB locus*, which encodes a protein of 433 amino acids with similarity to a periplasmic glycerol-3-phosphate-binding ATP-binding cassette (ABC) transporter protein found in several bacterial species, and harbors a bacterial solute-binding protein domain. Immunofluorescence studies indicated that SP41 is surface exposed, and antibodies directed to SP41 inhibited *B. suis* adherence to HeLa cells. Notably, a Δ*ugpB B. suis* mutant exhibited a significant reduction in the adherence to epithelial cells, supporting the contention that SP41 is an adhesin. Treatment of HeLa cells with neuraminidase abolished SP41 binding to these cells, suggesting the involvement of sialic acid residues in this interaction (Figure 2, Table 1). In contrast, a further study in *B. ovis* did not reveal an effect of *ugpB* deletion on early internalization or intracellular survival of this rough species in murine macrophages (J774.A1 cell line) or HeLa cells [62]. In addition, the deletion had no effect on the ability of *B. ovis* to colonize the spleen after intraperitoneal inoculation in mice. The *ugpB* gene seems to be functional in *B. ovis* as revealed by RT-PCR assays, and the encoded protein differs only by five amino acids from that of *B. suis*. It was argued that other adhesins would be more exposed on the bacterial surface of *B. ovis* due to the absence of O-polysaccharide chains, favoring their interaction with the host cell.

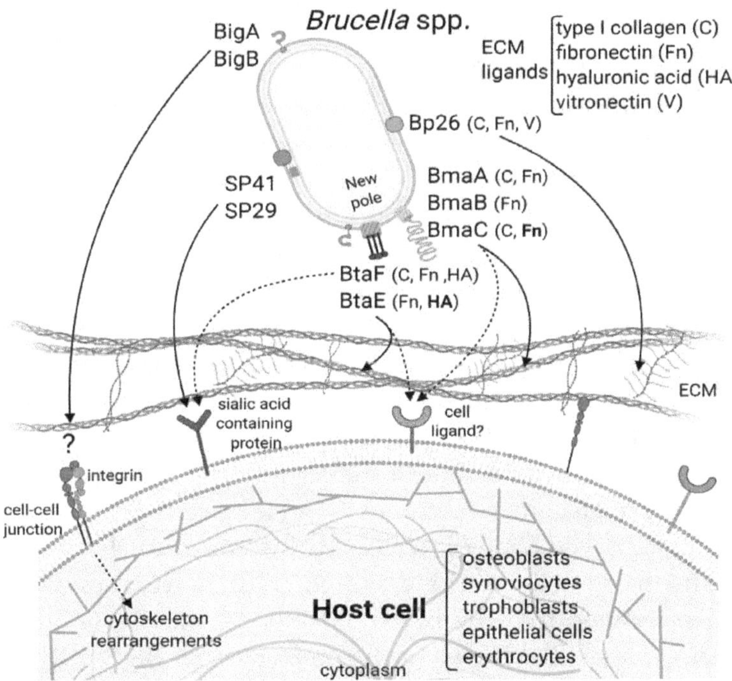


**Figure 2.** Model of the interaction of *Brucella* adhesins with host cells. The described *Brucella* spp. adhesins are depicted on the bacterial surface and their interactions with ECM components and ligands on the host cell are represented by black arrows. ECM components and cell types to which these adhesins bind are mentioned. ECM ligands in bold are supported by consistent evidence while those in a normal font are supported by indirect evidence. The Bma and Bta proteins are mostly localized at the new pole generated after asymmetric bacterial division. Bipolar localization was shown for BigA, but BigB polarity has not been determined. SP29 is predicted to be periplasmic while SP41 was shown to be exposed on the bacterial surface. It is not clear if Bp26 is localized to the outer membrane or in the periplasm. The cell ligands for Big and Bp26 adhesins have not yet been identified. In addition to ECM, Bma and Bta adhesins could interact with cell surface ligands. Dotted arrows represent putative interactions. Importantly, while all the *Brucella* adhesins characterized to date are shown, they may be not simultaneously expressed on bacteria.

The potential role of the *Brucella* Bp26 protein as an adhesin was recently tested in vitro [66]. The rationale exposed by the authors for testing this protein was not related to structural or homology criteria, but to the fact that Bp26 induces strong antibody responses in infected individuals [73]. Bp26 is a 250 amino acid-predicted protein with a domain of unknown function (DUF541) (Figure 1). The binding properties of Bp26 to ECM components such as type I collagen, fibronectin, vitronectin, and laminin were tested by ELISA and biolayer interferometry. According to the results of these assays, Bp26 binds to both immobilized and soluble type I collagen and vitronectin, and to soluble (but not immobilized) fibronectin, but does not bind to laminin (Figure 2, Table 1). The relevance of Bp26 for in vitro adhesion of *Brucella* to cells or for the outcome of in vivo infections has not been tested.

### *5.2. Adhesins Containing Ig-Like Domains*

A study by Czibener and Ugalde allowed the identification of a pathogenicity island in *B. abortus* (BAB1_2009-2012) whose deletion resulted in a reduced attachment of the bacterium to HeLa cells [63]. Furthermore, the deletion mutant also displayed a reduced capacity to colonize the spleen of mice after oral infection as compared to the wild-type strain. In particular, BAB1_2009 was found to encode a protein that harbors a bacterial Ig-like (BIg-like) domain present in adhesins from the invasin/intimin family [74] (Figure 1). In the following study, the role of this protein (named as BigA) in adhesion to epithelial cells was demonstrated [64]. This study also revealed that BigA is an exposed outer membrane protein and that incubation of the bacteria with antibodies against the Ig-like domain of BigA before infection of HeLa cells reduces the number of intracellular bacteria. While a deletion mutant strain displayed a significant defect in both adhesion and invasion to polarized epithelial cell lines such as Caco-2 (human colon) and Madin–Darby canine kidney (MDCK), overexpression of the *bigA* gene greatly increased them (Table 1). Confocal microscopy analyses showed that the BigA adhesin targets the bacteria to the cell-cell junction membrane in confluent epithelial cells and also induces cytoskeleton rearrangements (Figure 2). A recent publication by the same research group showed that other Ig-like (Blg-like) domain-containing protein (BAB1_2012, named BigB) (Figure 1), encoded by the same locus (BAB1_2009-2012), is also involved in adhesion to epithelial cells and targets proteins involved in cell-cell and cell-matrix interactions (Figure 2) [65]. The Δ*bigB* mutant showed a significant reduction in intracellular bacteria at the early stages of infection both in HeLa cells and in polarized MDCK cells (Table 1). It was further demonstrated by counting fluorescent bacteria that the phenotype on HeLa cells was due to a defect in adhesion. Similar to BigA, recombinant BigB induced profound cytoskeleton rearrangements in HeLa cells (Figure 2). HeLa cells transfected with focal adhesion markers showed changes in focal adhesion sites. It was proposed that, similar to BigA, BigB targets proteins in cell-cell junctions, which, in turn, triggers changes in the cytoskeleton (Figure 2). This work also showed that the BAB1_2011 gene encodes a periplasmic protein (PalA), which is necessary for the proper display of both the BigA and BigB adhesins, indicating that the genomic island is dedicated to the adhesion of *Brucella* to host cells. Although the phenotypes of the *big* mutants have not been tested in vivo, the previous result obtained with the BAB1_2009-2012 deletion mutant strongly suggests that the Big adhesins have a role in vivo.

### *5.3. Autotransporters*

Numerous virulence factors of bacterial pathogens contain domains or motifs related to adhesion to biotic or abiotic surfaces. A comprehensive search for conserved adhesion-associated domains/motifs in the *B. suis* 1330 genome and subsequent phylogenetic analyses revealed the presence of three clearly separated groups of adhesins: (1) monomeric autotransporters (BRA0173, BR2013, BRA1148), (2) trimeric autotransporters (BR0072 and BR1846), and (3) Ig-like domain containing-proteins (BR2009 and BR2012) [75]. Other proteins with no clear associated functions were also identified but not described in this work. Group 2 also included the protein BR0049. We have recently shown that BR0049 is certainly not an adhesin, but is required for the correct insertion of proteins from the autotransporter families (see below) [53]. Group 3 comprised orthologous proteins of the *B. abortus* BigA and BigB proteins described above.

#### 5.3.1. Monomeric Autotransporters

As mentioned above, *B. abortus* binds in a dose-dependent manner to components of the ECM, such as fibronectin and vitronectin [27]. In an attempt to identify *B. suis* genes encoding proteins that might be involved in the binding of brucellae to fibronectin, Posadas et al. [67] panned an M13 phage display library of the *B. suis* 1330 genome against immobilized fibronectin. Several recombinant phages showed affinity to immobilized fibronectin. However, only one expressed a portion of a protein that was predicted to be exposed on the cell surface. This protein corresponded to one of the monomeric autotransporters described above (BRA1148) and was named as BmaC for *Bucella* monomeric autotransporter. This protein exhibits all the characteristics of this protein family. BmaC is a large 340 kDa-protein with a long N-terminal cleavable 72 amino acid signal peptide and several adhesion-related motifs within the passenger domain, an extended pectin lyase virulence factor domain, and several passenger-associated-transport-repeats (PATR) (Figure 1). The portion of BmaC expressed on the phage also exhibits affinity (although much less) for type I collagen, suggesting that this very large protein might contribute to the interaction of *B. suis* with other ECM ligands.

Several lines of evidence have shown that BmaC of *B. suis* 1330 mediates the binding of *B. suis* to epithelial cells through cellular fibronectin. The attachment of *B. suis* to HeLa cells was inhibited by the ϕ-BRA1148 recombinant phage in a dose-dependent manner. A *bmaC* deletion mutant was impaired in the attachment to immobilized fibronectin and to the surface of HeLa and A549 epithelial cells (Table 1). Furthermore, the *bmaC* defective mutant was outcompeted by the wild-type strain in co-infection experiments, and anti-BmaC and anti-fibronectin antibodies significantly inhibited the binding of *B. suis* to HeLa cells [67]. Immunofluorescence microscopy showed that all bacteria with a detectable fluorescent signal displayed BmaC at only one pole, indicating that BmaC is polarly exposed on the cell surface (Figure 2). This is not surprising since several monomeric autotransporters are exposed on the bacterial surface at one pole [76,77]. Confocal microscopy analysis showed the presence of some small GFP-labelled bacterial aggregates on the surface of HeLa cells, mostly in cell boundaries. Single bacteria were found interacting through one of their poles with the cell surface on both the cell body and cell protrusion. Occasionally, polar BmaC was located at the pole interacting with the cell. These observations suggested that the polar localization of BmaC could be relevant in the interaction with host cells in vivo [67].

The monomeric autotransporter proteins encoded by BR0173 and BR2013 (BmaA and BmaB, respectively) of *B. suis* 1330, although much smaller, share significant sequence similarities with BmaC (Figure 1). It was reported that a mutant of *B. suis* 1330, deficient in BmaB (previously called OmaA), is cleared from spleens of BALB/c mice faster than the wild-type strain (between the third and the ninth week post infection), suggesting that BmaB is required during the chronic phase of infection [78]. A recent study [68] indicated that the *bmaB* locus from all *B. abortus* strains analyzed and both the *bmaA* and *bmaC* loci from all *B. melitensis* strains seem to correspond to pseudogenes, while, in *B. suis*, all the Bma proteins could be functional in several strains of this species. In line with these observations, gain or loss of function studies indicated that, at least in *B. suis* strain 1330, BmaA, BmaB, and BmaC proteins contribute, to a greater or lesser degree, to bacterial adhesion among different cell types, such as epithelial (HT-29 and Caco2), synoviocytes, osteoblasts, and trophoblasts (Table 1, Figure 2). These observations show that there are variations in the repertoire of functional adhesins in *Brucella* spp. and open the possibility that these adhesins are involved in host preferences. BmaB was also found at the new pole generated after cell division [68].

#### 5.3.2. Trimeric Autotransporters

As mentioned above, the search for conserved adhesion-associated domains/motifs in *B. suis* 1330 identified a group of trimeric autotransporters, including BR0072 and BR1846, which were named BtaE and BtaF, respectively. The *B. suis* BtaE trimeric autotransporter is a 740 amino acid protein that harbors several regions corresponding to the head and the neck subdomains in addition to a connector region and the β-barrel translocator domain (Figure 1). Different genetic approaches showed that BtaE of *B. suis* is involved in the adhesion to ECM components and host cells [69]. The BtaE-defective strain exhibited a decreased ability to adhere to HeLa and A549 epithelial cells and was outcompeted by the wild-type *B. suis* strain for the adhesion to HeLa cells (Table 1). Expression of BtaE in a "non-adherent" *E. coli* strain increased the binding of this heterologous bacterium to immobilized hyaluronic acid and fibronectin. On the other hand, *btaE* deletion impaired bacterial adhesion to hyaluronic acid but had no effect in the adhesion to fibronectin, suggesting that other fibronectin-binding adhesins (such as BmaC) could compensate somehow for the absence of BtaE. The adhesion of the wild-type strain to HeLa cells decreased in the presence of hyaluronic acid, while this compound had almost no effect in the attachment of the *btaE* mutant to these cells, supporting the hypothesis that BtaE mediates the binding of *Brucella* to hyaluronic acid. Therefore, BtaE could also participate in *Brucella* dissemination to different target tissues such as cartilage, heart, and bone, which may result in brucellosis complications. In vivo experiments using the mouse model indicate that the BtaE adhesin is necessary for a successful infection. In effect, a significantly lower number of bacteria were recovered from spleens of animals inoculated through the intragastric route with the *btaE* mutant compared to those inoculated with the wild-type strain [69].

In a subsequent work, it was shown that the BtaE orthologue of *B. abortus* 2308 is also involved in adhesion to epithelial cells. Compared to *B. suis* BtaE, the ortologue of *B. abortus* is much larger and contains a higher number of repetitive adhesion motifs. Furthermore, the *btaE* gene of *B. suis* and *B. abortus* are under the regulation of different mechanisms (see below) [79]. The *btaE* deletion mutant of *B. abortus* 2308 showed a significant reduction in the adhesion to HeLa cells when compared with the wild-type strain, demonstrating that the BtaE variant of *B. abortus* 2308 contributes to the interaction of *Brucella* with the host cell surface to a similar extent to that observed for the *B. suis* 1330 orthologue [69].

The regulation of *btaE* at a promoter level was analyzed by Sieira et al. [79]. Comparison of *btaE* promoter sequences among different *Brucella* species revealed that a novel HutC binding site in the promoter region of *btaE* from *B. abortus* 2308 was generated de novo recently in the evolution of the genus. HutC, which is a regulator of the histidine metabolism, also acts as a co-activator contributing to modulation of expression of the *Brucella virB* operon [80]. Moreover, additional transcriptional factors (MdrA and IHF) binding sites were identified in the *btaE* promoter of *B. abortus*. The target-DNA sequences were confirmed by EMSA and DNAseI footprinting assays.

The HutC binding site is not present in the *btaE* promoters of other *Brucella* strains since it is interrupted by a cytosine. In effect, an electrophoretic mobility shift assay showed that HutC is not able to interact with the *btaE* promoter of *B. suis* 1330, even though the IHF and MdrA showed a binding pattern similar to that observed for the *btaE* promoter of *B. abortus* 2308. Based on these findings, it was proposed that, as a result of the cis regulatory gain of function, the *btaE* promoter acquired the ability to fine-tune its transcriptional output in response to changes in environmental parameters such as nutrient availability. Thus, differential *btaE* expression might generate phenotypic diversity at the regulatory level of adhesins, which might contribute to reciprocal selection between *Brucella* species and their mammalian hosts.

The BtaF trimeric autotransporter of *B. suis* 1330, encoded by the BR1846 annotated locus, is a small protein that harbors an N-terminal peptide signal, a 170 amino acid passenger domain, and a YadA-like C-terminal translocator region (β-barrel translocator domain) (Figure 1). Unlike BtaE, the BtaF protein does not show the presence of conserved adhesion motifs. Analysis with the TA Domain Annotation (daTAA) server [81] showed that most of the passenger domains correspond to a coiled coil stalk but none associated with the "head" structural region [70]. In addition, a careful analysis of the annotated *btaF* upstream region indicated that the ORF starts earlier, adding 33 additional amino acids at the N-terminal sequence. Using a new version of the program to predict the structural features of trimeric autotransporters and the alternative ORF, it was possible to identify a region at the N-terminus that would correspond to the head, even though the structure would be different from those described so far [82].

BtaF of *B. suis* 1330 has shown to be promiscuous in its ability to bind to different substrates. The heterologous expression of this small trimeric protein markedly increased the adhesion of non-adherent *E. coli* to HeLa cells and various substrates such as fibronectin, fetuin (a sialic acid-rich protein), hyaluronic acid, and collagen I and also to an abiotic surface such as polystyrene [70] (Table 1, Figure 2). In agreement with these observations, the *btaF* deletion mutant of *B. suis* showed a significant reduction in the ability to bind to fetuin, hyaluronic acid, and collagen I, and in the adhesion to an abiotic surface, even though the adhesion to fibronectin was not affected. Again, as it was observed for the *btaE* mutant, overlapping functions with other adhesins (such as Bma proteins and BtaE) may account for the lack of a Δ*btaF* phenotype toward fibronectin. The BtaF-defective strain showed a significant reduction in the attachment to HeLa and A549 epithelial cells.

Notably, BtaF was required for complete virulence in mice infected through the oral route (intragastric administration) [70]. The strain lacking BtaF showed a reduction of about one log in the number of bacteria recovered from spleen at early stages of infection. The absence of both trimeric adhesins (BtaE and BtaF) resulted in a more severe phenotype in vivo compared with the attenuation observed for the single mutants. It is possible that some of the functions might be shared or complementary between BtaE and BtaF, while others could be exclusive to BtaF or BtaE. An indirect ELISA assay on sera from healthy and sick pigs infected with *B. suis* suggested that both adhesins are expressed in vivo in the natural host (swine), supporting the role of these adhesins in the infection process. Recently, it was shown that BtaF is also required for virulence in mice after inoculation via a respiratory (intratracheal administration) route [71]. In this case, the splenic load of the deletion mutant was significantly reduced at 7-days and 30-days post-infection as compared to the wild-type strain.

Smooth *Brucella* strains prevent detection by complement partly due to a distinctive structure of its LPS [83,84]. However, it was proposed that, in addition to the LPS, other surface factors mediate the varied sensitivity of *Brucella* species to the bactericidal action of serum [84]. In addition, it has been shown that, in contrast to human serum, components present in murine normal serum do not opsonize smooth *B. abortus* [85]. The *btaF* mutant showed a significantly reduced survival in the presence of 50% porcine serum compared with the wild-type strain [70]. Furthermore, an *E. coli* strain expressing BtaF showed a more than ten-fold increase in the survival percentage in 8% porcine serum as compared with the control strain. Both strains showed similar levels of survival in heat-inactivated porcine serum, suggesting that BtaF is involved in the resistance to complement-mediated serum killing.

Similar to BmaC and BmaB, BtaE and BtaF adhesins were found to be polarly localized on the bacterial surface [69,70] (Figure 2). Again, the trimeric adhesins were detected in a low proportion of bacteria but, in all cases, the signal showed unipolar localization and, in some cases, sub-polar localization. As it was observed for BmaC and BmaB, they were found at the same pole as AidB-YFP (a new pole marker) [86] and at the opposite of the PdhS-eGFP labeling (old pole marker) [87]. It was proposed that the new pole generated after the asymmetric division would be functionally differentiated for adhesion. An attractive hypothesis is that the initial adhesion of *Brucella* to the host cell would be mediated by adhesins located at the new pole and that adhesin expression only occurs in an infectious bacterial subpopulation. Various cellular mechanisms such as asymmetric division, polar growth, and polar functions generate two functionally differentiated cells [88,89]. Polar localization could be a way of increasing the adhesive power by concentrating the adhesins in a particular region. In fact, host invasion by a bacterial pole can facilitate entry because of the bacterial shape [90]. It is important to note that polar adherence to surfaces is a conserved mechanism shared by several *Alphaproteobacteria* [91,92].

#### 5.3.3. Autotransporters Insertion in the Outer Membrane

Autotransporter translocation into the outer membrane is assisted by the BAM machinery and associated chaperones. More recently, it was shown that the TAM system, made up of TamA and TamB, is also required for the correct insertion of autotransporters from the *Gammaproteobacteria* group (see above) [93].

As mentioned, during the construction of the phylogenetic tree, the BR0049 protein came out as a possible adhesin from the autotransporter family likely due to some structural similarities with this family of proteins. However, our in silico analysis and other reports [94] indicated that BR0049 and its orthologues from other *Alphaproteobacteria* are phylogenetically related to members of the TamB family from *Gammaproteobacteria*. TamB is a large protein mostly periplasmic but inserted in the inner membrane through a non-cleavable signal peptide. BR0049 of *Brucella* spp. shares a relatively low identity (around 22%) with TamB from *Gammaproteobacteria*, but, similarly to this protein, it contains a membrane anchor signal at the N-terminus, which is followed by a region with an abundant β-helix structure, and, at the C-terminus, a short β-barrel structure within the conserved DUF490 domain. As it was proposed for TamB from *Gammaproteobacteria*, it was demonstrated that BR0049 is required for the correct insertion in the OM of the *B. suis* BmaB monomeric autotransporter. In addition, BR0049 was required for complete virulence in mice infected through the intragastric route [53]. The BR0049 mutant showed an increased sensitivity to polymyxin B, lysozyme, and Triton X-100, and, thus, BR0049 was named as MapB (Membrane altering protein). Several results indicated that MapB of *Brucella* plays functions that go beyond that of assisting in autotransporter assembly, suggesting that the TAM machinery would be involved in cell envelope biogenesis.

### 5.4. *Brucella Adhesins as Vaccine Candidates*

Vaccination is a key health measure in the control and prevention of infectious diseases. In the case of brucellosis, it is necessary to control bacterial dissemination by vaccination of natural hosts as well as vaccination of people professionally exposed to *Brucella* spp. infection. Commercially available *Brucella* vaccines approved for use in animals are based on attenuated strains. These vaccines have serious disadvantages such as producing abortion in pregnant females, being virulent for humans [95–97], and inducing immune responses that interfere with animal serological diagnosis [98]. Thus, there is a need to develop safer and more efficient vaccines [99]. In this sense, acellular vaccines provide great advantages, mainly in terms of safety, not only in its production but also in its administration. Nevertheless, the selection of appropriate vaccine candidates requires the study of host-pathogen interaction.

An essential step in establishing a successful infection is the adhesion of microorganisms to eukaryotic cells, resulting in colonization of the tissue involved. Therefore, the molecules involved in this initial interaction have been widely studied as targets for the development of vaccines against various pathogens such as *E. coli* [100,101], *Haemophilus ducreyi* [102], and *Neisseria meningitidis* [103,104] among others. Despite the extensive knowledge of adhesins' role in the pathogenicity of various bacteria, this group of proteins was not studied well in *Brucella* spp. in terms of immunogenicity and its potential as vaccine candidates.

Al-Mariri et al. studied the immunogenicity and protective efficacy of a DNA vaccine encoding SP41 adhesin from *B. melitensis* in BALB/c mice [105]. Intramuscular (i.m.) administration of pCISP41, a plasmid construct for SP41 expression in mammalian cells, induced SP41-specific serum immunoglobulin G (IgG) antibodies. Moreover, spleen cells from pCISP41-vaccinated mice showed significant T cell proliferation after in vitro stimulation with recombinant SP41 (rSP41) and lysed *B. melitensis*. Splenocytes from pCISP41-immunized animals also responded to rSP41 and bacterial lysate stimulation secreting high levels of gamma interferon (IFN-$\gamma$), even though no interleukin-5 (IL-5) was detected. This suggests a predominant T-helper-1 (Th1) immune response.

After an intraperitoneal challenge with *B. melitensis* 16M, mice immunized with pCISP41 exhibited a reduction of 1.25 and 1.14 log in their spleen burden when compared with control mice at 4-weeks and 8-weeks post-challenge, respectively. Nevertheless, vaccination with attenuated *B. melitensis* Rev-1 strain induced better protection levels than pCISP41 vaccination at both time points, achieving a reduction in spleen burden of 1.79 and 3.17 log, respectively. Although SP41 has been shown to be involved in adhesion to epithelial cells, no studies have been made to evaluate the potential of mucosal or systemic vaccination with this adhesin to protect against *Brucella* infection acquired through mucosae.

*Brucella* infection is frequently acquired through the oral and respiratory routes, and adhesins are anticipated to have a relevant role in these infectious processes. As mentioned above, our results showed that BtaF adhesin from *B. suis* is necessary for complete virulence of *B. suis* after both oral and intratracheal infection [70,71]. In line with these results, we assessed the immunogenic and protective potential of recombinant BtaF when administered intranasally with the mucosal adjuvant Bis-(3′,5′)-cyclic dimeric adenosine monophosphate (c-di-AMP) [71]. To this end, the trimeric form of the BtaF passenger domain fused at the C-terminus to the GCN4tri sequence to facilitate trimerization was successfully expressed and purified [71].

BALB/c mice intranasal immunization with BtaF plus c-di-AMP induced high levels of serum BtaF-specific IgG, IgA, IgG1, and IgG2a, and a mixed IgG1/IgG2a profile. In vitro, these serum antibodies reduced *B. suis* infection of human lung epithelial cells (A549 cell line), reduced bacterial binding to fetuin (a protein rich in sialic acid previously described as a ligand for BtaF), and enhanced *B. suis* phagocytosis by murine macrophages. In addition, immunization led to a significant production of specific IgA antibodies in the airways and in the gastrointestinal and genital mucosae.

BtaF immunization induced a systemic and pulmonary Th1 immune response, shown by the secretion of high levels of IFN-$\gamma$ by splenocytes and lung cells. Furthermore, depletion of CD4+ or CD8+ populations from spleen cells showed that CD4+ cells were responsible for IFN-$\gamma$ secretion. BtaF-vaccination also triggered the differentiation of specific CD4+ T cells to central memory cells in cervical lymph nodes, and differentiation of T cells to a Th17 profile in the spleen.

Mice vaccination with BtaF plus c-di-AMP demonstrated a high level of protection against *B. suis* oral infection, reducing the splenic burden of *B. suis* by 3.28 log. Unlike the protection achieved against oral infection, intranasal vaccination with BtaF failed to protect against respiratory infection with *B. suis* since no differences were observed in spleen or lung bacterial load between vaccinated and control mice after an intratracheal challenge.

In summary, despite extensive evidence supporting a role of *Brucella* adhesins in the infectious ability of this pathogen in vivo, only a few studies have assessed the protective efficacy of vaccination with adhesins. Such studies suggest that adhesins hold promise as appropriate antigens for vaccination against oral infection with *Brucella*, even though some protection against systemic infection might be attained.

## 6. Summary and Future Directions

For a long time, the process of *Brucella* adhesion to host cells has received little attention. Adhesins of *Brucella* spp. identified to date have been studied with varying degrees of depth. Some of them have been analyzed regarding their roles in the binding to components of the ECM and to different cell types as well as their in vivo role, while, in other cases, their functions have been only evaluated in vitro˙ Still, for various adhesins, it will be necessary to identify their ligands or cell receptors. The possible use of *Brucella* adhesins as vaccine candidates was tested only in two cases, and one of them showed encouraging results. Therefore, the in vivo role of many of the *Brucella* adhesins identified so far, and their possible applications as a basis for acellular vaccines, remains to be evaluated. It is expected that, in the future, new adhesins will be identified that mediate the initial adhesion of *Brucella* to the remarkable variety of cell types that this pathogen can invade. An interesting perspective will be to characterize the roles of the different adhesin variants from different species/strains and to determine whether they play a role in host preferences.

**Author Contributions:** Conceptualization, P.C.B. and A.Z. Literature review, critical analysis, and synthesis, M.G.B., G.S., M.C.F., F.M.G., P.C.B., and A.Z. Writing—original draft preparation, M.G.B., G.S., M.C.F., F.M.G., P.C.B., and A.Z. Supervision, P.C.B. and A.Z. All authors have read and agreed to the published version of the manuscript.

**Acknowledgments:** We are grateful to the staff of the BSL3 facility of INBIRS for expert assistance. We thank current and former members of our laboratories for all the contributions and helpful and inspiring discussions.

## References


1. Pappas, G.; Papadimitriou, P.; Akritidis, N.; Christou, L.; Tsianos, E.V. The new global map of human brucellosis. *Lancet Infect. Dis.* **2006**, *6*, 91–99. [CrossRef]
2. Traxler, R.M.; Lehman, M.W.; Bosserman, E.A.; Guerra, M.A.; Smith, T.L. A Literature Review of Laboratory-Acquired Brucellosis. *J. Clin. Microbiol.* **2013**, *51*, 3055–3062. [CrossRef] [PubMed]
3. Strausbaugh, L.J.; Berkelman, R.L. Human Illness Associated with Use of Veterinary Vaccines. *Clin. Infect. Dis.* **2003**, *37*, 407–414. [CrossRef] [PubMed]
4. Ashford, D.A.; Di Pietra, J.; Lingappa, J.; Woods, C.; Noll, H.; Neville, B.; Weyant, R.; Bragg, S.L.; Spiegel, R.A.; Tappero, J.; et al. Adverse events in humans associated with accidental exposure to the livestock brucellosis vaccine RB51. *Vaccine* **2004**, *22*, 3435–3439. [CrossRef]
5. Blasco, J.M.; Díaz, R. *Brucella melitensis* Rev-1 vaccine as a cause of human brucellosis. *Lancet* **1993**, *342*, 805. [CrossRef]
6. Vincent, P.; Joubert, L.; Prave, M. 2 occupational cases of brucellar infection after inoculation of B 19 vaccine. *Bull. Acad. Vet. Fr.* **1970**, *43*, 89–97.
7. Moreno, E. Retrospective and prospective perspectives on zoonotic brucellosis. *Front. Microbiol.* **2014**, *5*, 213. [CrossRef]
8. Buzgan, T.; Karahocagil, M.K.; Irmak, H.; Baran, A.I.; Karsen, H.; Evirgen, O.; Akdeniz, H. Clinical manifestations and complications in 1028 cases of brucellosis: A retrospective evaluation and review of the literature. *Int. J. Infect. Dis.* **2010**, *14*, e469–e478. [CrossRef]
9. Pourbagher, A.; Pourbagher, M.A.; Savas, L.; Turunc, T.; Demiroglu, Y.Z.; Erol, I.; Yalcintas, D. Epidemiologic, clinical, and imaging findings in brucellosis patients with osteoarticular involvement. *Am. J. Roentgenol.* **2006**, *187*, 873–880. [CrossRef]
10. Godfroid, J.; Cloeckaert, A.; Liautard, J.P.; Kohler, S.; Fretin, D.; Walravens, K.; Garin-Bastuji, B.; Letesson, J.J. From the discovery of the Malta fever's agent to the discovery of a marine mammal reservoir, brucellosis has continuously been a re-emerging zoonosis. *Vet. Res.* **2005**, *36*, 313–326. [CrossRef]
11. Baldi, P.C.; Giambartolomei, G.H. Pathogenesis and pathobiology of zoonotic brucellosis in humans. *OIE Rev. Sci. Tech.* **2013**, *32*, 117–125. [CrossRef] [PubMed]
12. Arenas, G.N.; Staskevich, A.S.; Aballay, A.; Mayorga, L.S. Intracellular trafficking of *Brucella abortus* in J774 macrophages. *Infect. Immun.* **2000**, *68*, 4255–4263. [CrossRef] [PubMed]
13. Price, R.E.; Templeton, J.W.; Smith, R.; Adams, L.G. Ability of mononuclear phagocytes from cattle naturally resistant or susceptible to brucellosis to control in vitro intracellular survival of *Brucella abortus*. *Infect. Immun.* **1990**, *58*, 879–886. [CrossRef]
14. Drazek, E.S.; Houng, H.S.; Crawford, R.M.; Hadfield, T.L.; Hoover, D.L.; Warren, R.L. Deletion of purE attenuates *Brucella melitensis* 16M for growth in human monocyte-derived macrophages. *Infect. Immun.* **1995**, *63*, 3297–3301. [CrossRef]
15. Pizarro-Cerdá, J.; Méresse, S.; Parton, R.G.; Van Der Goot, G.; Sola-Landa, A.; Lopez-Goñi, I.; Moreno, E.; Gorvel, J.P. *Brucella abortus* transits through the autophagic pathway and replicates in the endoplasmic reticulum of nonprofessional phagocytes. *Infect. Immun.* **1998**, *66*, 5711–5724. [CrossRef]
16. Detilleux, P.G.; Deyoe, B.L.; Cheville, N.F. Penetration and intracellular growth of *Brucella abortus* in nonphagocytic cells in vitro. *Infect. Immun.* **1990**, *58*, 2320–2328. [CrossRef] [PubMed]
17. Delpino, M.V.; Fossati, C.A.; Baldi, P.C. Proinflammatory response of human osteoblastic cell lines and Osteoblast-monocyte interaction upon infection with *Brucella* spp. *Infect. Immun.* **2009**, *77*. [CrossRef]
18. Scian, R.; Barrionuevo, P.; Giambartolomei, G.H.; De Simone, E.A.; Vanzulli, S.I.; Fossati, C.A.; Baldi, P.C.; Delpino, M.V. Potential role of fibroblast-like synoviocytes in joint damage induced by *Brucella abortus* infection through production and induction of matrix metalloproteinases. *Infect. Immun.* **2011**, *79*, 3619–3632. [CrossRef]
19. Fernández, A.G.; Ferrero, M.C.; Hielpos, M.S.; Fossati, C.A.; Baldi, P.C. Proinflammatory response of human trophoblastic cells to *Brucella abortus* infection and upon interactions with infected phagocytes. *Biol. Reprod.* **2016**, *94*, 48. [CrossRef] [PubMed]

20. Ferrero, M.C.; Bregante, J.; Delpino, M.V.; Barrionuevo, P.; Fossati, C.A.; Giambartolomei, G.H.; Baldi, P.C. Proinflammatory response of human endothelial cells to *Brucella* infection. *Microbes Infect.* **2011**, *13*, 852–861. [CrossRef]
21. Ferrero, M.C.; Fossati, C.A.; Baldi, P.C. Smooth *Brucella* strains invade and replicate in human lung epithelial cells without inducing cell death. *Microbes Infect.* **2009**, *11*, 476–483. [CrossRef]
22. Billard, E.; Cazevieille, C.; Dornand, J.; Gross, A. High Susceptibility of Human Dendritic Cells to Invasion by the Intracellular Pathogens *Brucella suis, B. abortus*, and *B. melitensis*. *Infect. Immun.* **2005**, *73*, 8418–8424. [CrossRef]
23. Delpino, M.V.; Barrionuevo, P.; Scian, R.; Fossati, C.A.; Baldi, P.C. *Brucella*-infected hepatocytes mediate potentially tissue-damaging immune responses. *J. Hepatol.* **2010**, *53*, 145–154. [CrossRef]
24. Ferrero, M.C.; Hielpos, M.S.; Carvalho, N.B.; Barrionuevo, P.; Corsetti, P.P.; Giambartolomei, G.H.; Oliveira, S.C.; Baldi, P.C. Key role of toll-like receptor 2 in the inflammatory response and major histocompatibility complex class ii downregulation in *Brucella abortus*-infected alveolar macrophages. *Infect. Immun.* **2014**, *82*, 626–639. [CrossRef]
25. Fernández, A.G.; Hielpos, M.S.; Ferrero, M.C.; Fossati, C.A.; Baldi, P.C. Proinflammatory response of canine trophoblasts to *Brucella canis* infection. *PLoS ONE* **2017**, *12*, e0186561. [CrossRef]
26. Sidhu-Muñoz, R.S.; Sancho, P.; Vizcaíno, N. Evaluation of human trophoblasts and ovine testis cell lines for the study of the intracellular pathogen *Brucella ovis*. *FEMS Microbiol. Lett.* **2018**, *365*. [CrossRef] [PubMed]
27. Castaneda-Roldan, E.I.; Avelino-Flores, F.; Dall'Agnol, M.; Freer, E.; Cedillo, L.; Dornand, J.; Giron, J.A. Adherence of *Brucella* to human epithelial cells and macrophages is mediated by sialic acid residues. *Cell. Microbiol.* **2004**, *6*, 435–445. [CrossRef]
28. Gomez, G.; Adams, L.G.; Rice-ficht, A.; Ficht, T.A. Host-*Brucella* interactions and the *Brucella* genome as tools for subunit antigen discovery and immunization against brucellosis. *Front. Cell. Infect. Microbiol.* **2013**, *3*, 1–15. [CrossRef]
29. Watarai, M.; Makino, S.I.; Fujii, Y.; Okamoto, K.; Shirahata, T. Modulation of *Brucella*-induced macropinocytosis by lipid rafts mediates intracellular replication. *Cell. Microbiol.* **2002**, *4*, 341–355. [CrossRef] [PubMed]
30. Pei, J.; Turse, J.E.; Ficht, T.A. Evidence of Brucella abortus OPS dictating uptake and restricting NF-k B activation in murine macrophages. *Microbes Infect.* **2008**, *10*, 582–590. [CrossRef]
31. Naroeni, A.; Porte, F. Role of cholesterol and the ganglioside GM1 in entry and short-term survival of Brucella suis in murine macrophages. *Infect. Immun.* **2002**, *70*, 1640–1644. [CrossRef]
32. Kim, S.; Watarai, M.; Suzuki, H.; Makino, S. Lipid raft microdomains mediate class A scavenger receptor-dependent infection of *Brucella abortus*. *Microb. Pathog.* **2004**, *37*, 11–19. [CrossRef]
33. Martín-Martín, A.I.; Vizcaíno, N.; Fernández-Lago, L. Cholesterol, ganglioside GM1 and class A scavenger receptor contribute to infection by *Brucella ovis* and *Brucella canis* in murine macrophages. *Microbes Infect.* **2010**, *12*, 246–251. [CrossRef]
34. Lapaque, N.; Moriyon, I.; Moreno, E.; Gorvel, J.P. *Brucella* lipopolysaccharide acts as a virulence factor. *Curr. Opin. Microbiol.* **2005**, *8*, 60–66. [CrossRef]
35. Manterola, L.; Guzmán-Verri, C.; Chaves-Olarte, E.; Barquero-Calvo, E.; de Miguel, M.-J.; Moriyón, I.; Grilló, M.-J.; López-Goñi, I.; Moreno, E. BvrR/BvrS-controlled outer membrane proteins Omp3a and Omp3b are not essential for *Brucella abortus* virulence. *Infect. Immun.* **2007**, *75*, 4867–4874. [CrossRef]
36. Martín-Martín, A.I.; Caro-Hernández, P.; Orduña, A.; Vizcaíno, N.; Fernández-Lago, L. Importance of the Omp25/Omp31 family in the internalization and intracellular replication of virulent *B. ovis* in murine macrophages and HeLa cells. *Microbes Infect.* **2008**, *10*, 706–710. [CrossRef] [PubMed]
37. Celli, J.; De Chastellier, C.; Franchini, D.M.; Pizarro-Cerda, J.; Moreno, E.; Gorvel, J.P. *Brucella* evades macrophage killing via VirB-dependent sustained interactions with the endoplasmic reticulum. *J. Exp. Med.* **2003**, *198*, 545–556. [CrossRef]
38. Comerci, D.J.; Martínez-Lorenzo, M.J.; Sieira, R.; Gorvel, J.P.; Ugalde, R.A. Essential role of the virB machinery in the maturation of the *Brucella abortus*-containing vacuole. *Cell. Microbiol.* **2001**, *3*, 159–168. [CrossRef]
39. Starr, T.; Child, R.; Wehrly, T.D.; Hansen, B.; Hwang, S.; López-Otin, C.; Virgin, H.W.; Celli, J. Selective subversion of autophagy complexes facilitates completion of the *Brucella* intracellular cycle. *Cell Host Microbe* **2012**, *11*, 33–45. [CrossRef]

40. Porte, F.; Naroeni, A.; Ouahrani-Bettache, S.; Liautard, J.P. Role of the *Brucella suis* lipopolysaccharide O antigen in phagosomal genesis and in inhibition of phagosome-lysosome fusion in murine macrophages. *Infect. Immun.* **2003**, *71*, 1481–1490. [CrossRef] [PubMed]
41. Kline, K.A.; Fälker, S.; Dahlberg, S.; Normark, S.; Henriques-Normark, B. Bacterial Adhesins in Host-Microbe Interactions. *Cell Host Microbe* **2009**, *5*, 580–592. [CrossRef]
42. Pizarro-Cerdá, J.; Cossart, P. Bacterial adhesion and entry into host cells. *Cell* **2006**, *124*, 715–727. [CrossRef]
43. Gerlach, R.G.; Hensel, M. Protein secretion systems and adhesins: The molecular armory of Gram-negative pathogens. *Int. J. Med. Microbiol.* **2007**, *297*, 401–415. [CrossRef]
44. Busch, A.; Waksman, G. Chaperone-usher pathways: Diversity and pilus assembly mechanism. *Philos. Trans. R. Soc. B Biol. Sci.* **2012**, *367*, 1112–1122. [CrossRef]
45. Giltner, C.L.; Nguyen, Y.; Burrows, L.L. Type IV Pilin Proteins: Versatile Molecular Modules. *Microbiol. Mol. Biol. Rev.* **2012**, *76*, 740–772. [CrossRef]
46. Saldaña, Z.; Xicohtencatl-Cortes, J.; Avelino, F.; Phillips, A.D.; Kaper, J.B.; Puente, J.L.; Girón, J.A. Synergistic role of curli and cellulose in cell adherence and biofilm formation of attaching and effacing *Escherichia coli* and identification of Fis as a negative regulator of curli. *Environ. Microbiol.* **2009**, *11*, 992–1006. [CrossRef]
47. Chapman, M.R.; Robinson, L.S.; Pinkner, J.S.; Roth, R.; Heuser, J.; Hammar, M.; Normark, S.; Hultgren, S.J. Role of *Escherichia coli* curli operons in directing amyloid fiber formation. *Science* **2002**, *295*, 851–855. [CrossRef]
48. Delepelaire, P. Type I secretion in gram-negative bacteria. *Biochim. Biophys. Acta Mol. Cell Res.* **2004**, *1694*, 149–161. [CrossRef]
49. Satchell, K.J.F. Structure and function of MARTX toxins and other large repetitive RTX proteins. *Annu. Rev. Microbiol.* **2011**, *65*, 71–90. [CrossRef]
50. Meuskens, I.; Saragliadis, A.; Leo, J.C.; Linke, D. Type V secretion systems: An overview of passenger domain functions. *Front. Microbiol.* **2019**, *10*, 1163. [CrossRef]
51. Babu, M.; Bundalovic-Torma, C.; Calmettes, C.; Phanse, S.; Zhang, Q.; Jiang, Y.; Minic, Z.; Kim, S.; Mehla, J.; Gagarinova, A.; et al. Global landscape of cell envelope protein complexes in *Escherichia coli*. *Nat. Biotechnol.* **2018**, *36*, 103–112. [CrossRef]
52. Kiessling, A.R.; Malik, A.; Goldman, A. Recent advances in the understanding of trimeric autotransporter adhesins. *Med. Microbiol. Immunol.* **2020**, *209*, 233–242. [CrossRef]
53. Bialer, M.G.; Ruiz-Ranwez, V.; Sycz, G.; Estein, S.M.; Russo, D.M.; Altabe, S.; Sieira, R.; Zorreguieta, A. MapB, the *Brucella suis* TamB homologue, is involved in cell envelope biogenesis, cell division and virulence. *Sci. Rep.* **2019**, *9*, 2158. [CrossRef]
54. Albenne, C.; Ieva, R. Job contenders: Roles of the β-barrel assembly machinery and the translocation and assembly module in autotransporter secretion. *Mol. Microbiol.* **2017**, *106*, 505–517. [CrossRef]
55. Van Ulsen, P.; ur Rahman, S.; Jong, W.S.P.; Daleke-Schermerhorn, M.H.; Luirink, J. Type V secretion: From biogenesis to biotechnology. *Biochim. Biophys. Acta Mol. Cell Res.* **2014**, *1843*, 1592–1611. [CrossRef]
56. Emsley, P.; Charles, I.G.; Fairweather, N.F.; Isaacs, N.W. Structure of *Bordetella pertussis* virulence factor P.69 pertactin. *Nature* **1996**, *381*, 90–92. [CrossRef]
57. Oberhettinger, P.; Leo, J.C.; Linke, D.; Autenrieth, I.B.; Schütz, M.S. The inverse autotransporter intimin exports its passenger domain via a hairpin intermediate. *J. Biol. Chem.* **2015**, *290*, 1837–1849. [CrossRef]
58. Leo, J.C.; Oberhettinger, P.; Schütz, M.; Linke, D. The inverse autotransporter family: Intimin, invasin and related proteins. *Int. J. Med. Microbiol.* **2015**, *305*, 276–282. [CrossRef]
59. Leibiger, K.; Schweers, J.M.; Schütz, M. Biogenesis and function of the autotransporter adhesins YadA, intimin and invasin. *Int. J. Med. Microbiol.* **2019**, *309*, 331–337. [CrossRef]
60. del Rocha-Gracia, R.C.; Castañeda-Roldán, E.I.; Giono-Cerezo, S.; Girón, J.A. *Brucella sp.* bind to sialic acid residues on human and animal red blood cells. *FEMS Microbiol. Lett.* **2002**, *213*, 219–224. [CrossRef]
61. Castañeda-Roldán, E.I.; Ouahrani-Bettache, S.; Saldaña, Z.; Avelino, F.; Rendón, M.A.; Dornand, J.; Girón, J.A. Characterization of SP41, a surface protein of *Brucella* associated with adherence and invasion of host epithelial cells. *Cell. Microbiol.* **2006**, *8*, 1877–1887. [CrossRef] [PubMed]
62. Sidhu-Muñoz, R.S.; Sancho, P.; Vizcaíno, N. Brucella ovis PA mutants for outer membrane proteins Omp10, Omp19, SP41, and BepC are not altered in their virulence and outer membrane properties. *Vet. Microbiol.* **2016**, *186*, 59–66. [CrossRef]

63. Czibener, C.; Ugalde, J.E. Identification of a unique gene cluster of *Brucella* spp. that mediates adhesion to host cells. *Microbes Infect.* **2012**, *14*, 79–85. [CrossRef]
64. Czibener, C.; Merwaiss, F.; Guaimas, F.; Del Giudice, M.G.; Serantes, D.A.R.; Spera, J.M.; Ugalde, J.E. BigA is a novel adhesin of *Brucella* that mediates adhesion to epithelial cells. *Cell. Microbiol.* **2016**, *18*, 500–513. [CrossRef]
65. Lopez, P.; Guaimas, F.; Czibener, C.; Ugalde, J.E. A genomic island in *Brucella* involved in the adhesion to host cells: Identification of a new adhesin and a translocation factor. *Cell. Microbiol.* **2020**, e13245. [CrossRef]
66. Eltahir, Y.; Al-Araimi, A.; Nair, R.; Autio, K.J.; Tu, H.; Leo, J.C.; Al-Marzooqi, W.; Johnson, E. Binding of *Brucella* protein, Bp26, to select extracellular matrix molecules. *BMC Mol. Cell Biol.* **2019**, *20*, 55. [CrossRef]
67. Posadas, D.M.; Ruiz-Ranwez, V.; Bonomi, H.R.; Martín, F.A.; Zorreguieta, A. BmaC, a novel autotransporter of *Brucella suis*, is involved in bacterial adhesion to host cells. *Cell. Microbiol.* **2012**, *14*, 965–982. [CrossRef]
68. Bialer, M.G.; Ferrero, M.C.; Delpino, M.V.; Ruiz-Ranwez, V.; Posadas, D.M.; Baldi, P.C.; Zorreguieta, A. Adhesive functions or pseudogenization of monomeric autotransporters in *Brucella* species. Unpublished.
69. Ruiz-Ranwez, V.; Posadas, D.M.; Van der Henst, C.; Estein, S.M.; Arocena, G.M.; Abdian, P.L.; Martín, F.A.; Sieira, R.; De Bolle, X.; Zorreguieta, A. BtaE, an Adhesin That Belongs to the Trimeric Autotransporter Family, Is Required for Full Virulence and Defines a Specific Adhesive Pole of *Brucella suis*. *Infect. Immun.* **2013**, *81*, 996–1007. [CrossRef]
70. Ruiz-Ranwez, V.; Posadas, D.M.; Estein, S.M.; Abdian, P.L.; Martin, F.A.; Zorreguieta, A. The BtaF trimeric autotransporter of *Brucella suis* is involved in attachment to various surfaces, resistance to serum and virulence. *PLoS ONE* **2013**, *8*, e79770. [CrossRef] [PubMed]
71. Muñoz González, F.; Sycz, G.; Alonso Paiva, I.M.; Linke, D.; Zorreguieta, A.; Baldi, P.C.; Ferrero, M.C. The BtaF Adhesin Is Necessary for Full Virulence During Respiratory Infection by *Brucella suis* and Is a Novel Immunogen for Nasal Vaccination Against Brucella Infection. *Front. Immunol.* **2019**, *10*, 1775. [CrossRef]
72. Vitry, M.A.; Mambres, D.H.; Deghelt, M.; Hack, K.; Machelart, A.; Lhomme, F.; Vanderwinden, J.M.; Vermeersch, M.; De Trez, C.; Pérez-Morga, D.; et al. *Brucella melitensis* invades murine erythrocytes during infection. *Infect. Immun.* **2014**, *82*, 3927–3938. [CrossRef] [PubMed]
73. Seco-Mediavilla, P.; Verger, J.M.; Grayon, M.; Cloeckaert, A.; Marín, C.M.; Zygmunt, M.S.; Fernández-Lago, L.; Vizcaíno, N. Epitope mapping of the *Brucella melitensis* BP26 immunogenic protein: Usefulness for diagnosis of sheep brucellosis. *Clin. Diagn. Lab. Immunol.* **2003**, *10*, 647–651. [CrossRef]
74. Bodelón, G.; Palomino, C.; Fernández, L.Á. Immunoglobulin domains in *Escherichia coli* and other enterobacteria: From pathogenesis to applications in antibody technologies. *FEMS Microbiol. Rev.* **2013**, *37*, 204–250. [CrossRef]
75. Posadas, D.M. Transport and Adhesion in *Brucella suis*: Role of a TolC Family Protein in the Eflux of Toxic Compounds and of Three Potential Adhesins Host in Colonization. Ph.D. Thesis, University of Buenos Aires, Buenos Aires, Argentina, 2010.
76. Jain, S.; Van Ulsen, P.; Benz, I.; Schmidt, M.A.; Fernandez, R.; Tommassen, J.; Goldberg, M.B. Polar localization of the autotransporter family of large bacterial virulence proteins. *J. Bacteriol.* **2006**, *188*, 4841–4850. [CrossRef]
77. Goldberg, M.B.; Barzu, O.; Parsot, C.; Sansonetti, P.J. Unipolar localization and ATPase activity of IcsA, a *Shigella flexneri* protein involved in intracellular movement. *J. Bacteriol.* **1993**, *175*, 2189–2196. [CrossRef]
78. Bandara, A.B.; Sriranganathan, N.; Schurig, G.G.; Boyle, S.M. Putative outer membrane autotransporter protein influences survival of *Brucella suis* in BALB/c mice. *Vet. Microbiol.* **2005**, *109*, 95–104. [CrossRef]
79. Sieira, R.; Bialer, M.G.; Roset, M.S.; Ruiz-Ranwez, V.; Langer, T.; Arocena, G.M.; Mancini, E.; Zorreguieta, A. Combinatorial control of adhesion of *Brucella abortus* 2308 to host cells by transcriptional rewiring of the trimeric autotransporter btaE gene. *Mol. Microbiol.* **2017**, *103*, 553–565. [CrossRef]
80. Sieira, R.; Arocena, G.M.; Bukata, L.; Comerci, D.J.; Ugalde, R.A. Metabolic control of virulence genes in *Brucella abortus*: HutC coordinates virB expression and the histidine utilization pathway by direct binding to both promoters. *J. Bacteriol.* **2010**, *192*, 217–224. [CrossRef]
81. Szczesny, P.; Lupas, A. Domain annotation of trimeric autotransporter adhesins—daTAA. *Bioinformatics* **2008**, *24*, 1251–1256. [CrossRef]
82. Linke, D. (University of Oslo, Oslo, Norway). Personal communication, 2018.
83. Eisenschenk, F.C.; Houle, J.J.; Hoffmann, E.M. Mechanism of serum resistance among *Brucella abortus* isolates. *Vet. Microbiol.* **1999**, *68*, 235–244. [CrossRef]

84. Fernandez-Prada, C.M.; Nikolich, M.; Vemulapalli, R.; Sriranganathan, N.; Boyle, S.M.; Schurig, G.G.; Hadfield, T.L.; Hoover, D.L. Deletion of wboA enhances activation of the lectin pathway of complement in Brucella abortus and *Brucella melitensis*. *Infect. Immun.* **2001**, *69*, 4407–4416. [CrossRef]
85. Mora-Cartín, R.; Chacón-Díaz, C.; Gutiérrez-Jiménez, C.; Gurdián-Murillo, S.; Lomonte, B.; Chaves-Olarte, E.; Barquero-Calvo, E.; Moreno, E. N -Formyl-Perosamine Surface Homopolysaccharides Hinder the Recognition of *Brucella abortus* by Mouse Neutrophils. *Infect. Immun.* **2016**, *84*, 1712–1721. [CrossRef]
86. Dotreppe, D.; Mullier, C.; Letesson, J.J.; De Bolle, X. The alkylation response protein AidB is localized at the new poles and constriction sites in *Brucella abortus*. *BMC Microbiol.* **2011**, *11*, 257. [CrossRef]
87. Hallez, R.; Mignolet, J.; Van Mullem, V.; Wery, M.; Vandenhaute, J.; Letesson, J.J.; Jacobs-Wagner, C.; De Bolle, X. The asymmetric distribution of the essential histidine kinase PdhS indicates a differentiation event in *Brucella abortus*. *EMBO J.* **2007**, *26*, 1444–1455. [CrossRef] [PubMed]
88. Brown, P.J.B.; De Pedro, M.A.; Kysela, D.T.; Van Der Henst, C.; Kim, J.; De Bolle, X.; Fuqua, C.; Brun, Y.V. Polar growth in the Alphaproteobacterial order Rhizobiales. *Proc. Natl. Acad. Sci. USA* **2012**, *109*, 1697–1701. [CrossRef]
89. Hallez, R.; Bellefontaine, A.F.; Letesson, J.J.; De Bolle, X. Morphological and functional asymmetry in α-proteobacteria. *Trends Microbiol.* **2004**, *12*, 361–365. [CrossRef]
90. Van Der Henst, C.; De Barsy, M.; Zorreguieta, A.; Letesson, J.J.; De Bolle, X. The *Brucella* pathogens are polarized bacteria. *Microbes Infect.* **2013**, *15*, 998–1004. [CrossRef]
91. Tomlinson, A.D.; Fuqua, C. Mechanisms and regulation of polar surface attachment in *Agrobacterium tumefaciens*. *Curr. Opin. Microbiol.* **2009**, *12*, 708–714. [CrossRef]
92. Merker, R.I.; Smit, J. Characterization of the Adhesive Holdfast of Marine and Freshwater Caulobacters. *Appl. Environ. Microbiol.* **1988**, *54*, 2078–2085. [CrossRef]
93. Selkrig, J.; Mosbahi, K.; Webb, C.T.; Belousoff, M.J.; Perry, A.J.; Wells, T.J.; Morris, F.; Leyton, D.L.; Totsika, M.; Phan, M.D.; et al. Discovery of an archetypal protein transport system in bacterial outer membranes. *Nat. Struct. Mol. Biol.* **2012**, *19*, 506–510. [CrossRef] [PubMed]
94. Heinz, E.; Selkrig, J.; Belousoff, M.J.; Lithgow, T. Evolution of the translocation and assembly module (TAM). *Genome Biol. Evol.* **2015**, *7*, 1628–1643. [CrossRef]
95. Davis, D.S.; Templeton, J.W.; Ficht, T.A.; Huber, J.D.; Angus, R.D.; Adams, L.G. *Brucella abortus* in Bison. II. Evaluation of strain 19 vaccination of pregnant cows. *J. Wildl. Dis.* **1991**, *27*, 258–264. [CrossRef] [PubMed]
96. Spink, W.W.; Hall, J.W.; Finstad, J.; Mallet, E. Immunization with viable *Brucella* organisms: Results of a safety test in humans. *Bull. World Health Organ.* **1962**, *26*, 409–419.
97. Pappagianis, D.; Elberg, S.S.; Crouch, D. Immunization against brucella infections: Effects of graded doses of viable attenuated *Brucella melitensis* in humans. *Am. J. Epidemiol.* **1966**, *84*, 21–31. [CrossRef]
98. Schurig, G.G.; Sriranganathan, N.; Corbel, M.J. Brucellosis vaccines: Past, present and future. *Vet. Microbiol.* **2002**, *90*, 479–496. [CrossRef]
99. Neutra, M.R.; Kozlowski, P.A. Mucosal vaccines: The promise and the challenge. *Nat. Rev. Immunol.* **2006**, *6*, 148–158. [CrossRef]
100. Pecha, B.; Low, D.; O'Hanley, P. Gal-Gal pili vaccines prevent pyelonephritis by piliated *Escherichia coli* in a murine model. Single-component Gal-Gal pili vaccines prevent pyelonephritis by homologous and heterologous piliated *E. coli* strains. *J. Clin. Investig.* **1989**, *83*, 2102–2108. [CrossRef]
101. Langermann, S.; Palaszynski, S.; Barnhart, M.; Auguste, G.; Pinkner, J.S.; Burlein, J.; Barren, P.; Koenig, S.; Leath, S.; Jones, C.H.; et al. Prevention of mucosal *Escherichia coli* infection by FimH-adhesin-based systemic vaccination. *Science* **1997**, *276*, 607–611. [CrossRef]
102. Samo, M.; Choudhary, N.R.; Riebe, K.J.; Shterev, I.; Staats, H.F.; Sempowski, G.D.; Leduc, I. Immunization with the *Haemophilus ducreyi* trimeric autotransporter adhesin DsrA with alum, CpG or imiquimod generates a persistent humoral immune response that recognizes the bacterial surface. *Vaccine* **2016**, *34*, 1193–1200. [CrossRef] [PubMed]
103. Comanducci, M.; Bambini, S.; Brunelli, B.; Adu-Bobie, J.; Aricò, B.; Capecchi, B.; Giuliani, M.M.; Masignani, V.; Santini, L.; Savino, S.; et al. NadA, a novel vaccine candidate of *Neisseria meningitidis*. *J. Exp. Med.* **2002**, *195*, 1445–1454. [CrossRef]

104. Hung, M.C.; Heckels, J.E.; Christodoulides, M. The adhesin complex protein (ACP) of *Neisseria meningitidis* is a new adhesin with vaccine potential. *MBio* **2013**, *4*, e00041-13. [CrossRef] [PubMed]
105. Al-Mariri, A.; Abbady, A.Q. Evaluation of the immunogenicity and the protective efficacy in mice of a DNA vaccine encoding SP41 from *Brucella melitensis*. *J. Infect. Dev. Ctries.* **2013**, *7*, 329–337. [CrossRef]

# 11

# The Effects of Silver Sulfadiazine on Methicillin-Resistant *Staphylococcus aureus* Biofilms

**Yutaka Ueda [1], Motoyasu Miyazaki [2], Kota Mashima [1], Satoshi Takagi [3], Shuuji Hara [4], Hidetoshi Kamimura [1] and Shiro Jimi [5,*]**

1 Department of Pharmacy, Fukuoka University Hospital, Fukuoka 814-0180, Japan; ueday@fukuoka-u.ac.jp (Y.U.); mashimakota210@fukuoka-u.ac.jp (K.M.); kamisan@fukuoka-u.ac.jp (H.K.)
2 Department of Pharmacy, Fukuoka University Chikushi Hospital, Fukuoka 818-8502, Japan; motoyasu@fukuoka-u.ac.jp
3 Departments of Plastic, Reconstructive and Aesthetic Surgery, Faculty of Medicine, Fukuoka University, Fukuoka 814-0180, Japan; stakagi@fukuoka-u.ac.jp
4 Department of Drug Informatics, Faculty of Pharmaceutical Sciences, Fukuoka University, Fukuoka 814-0180, Japan; harashu@fukuoka-u.ac.jp
5 Central Lab for Pathology and Morphology, Faculty of Medicine, Fukuoka University, Fukuoka 814-0180, Japan
* Correspondence: sjimi@fukuoka-u.ac.jp

**Abstract:** Methicillin-resistant *Staphylococcus aureus* (MRSA), the most commonly detected drug-resistant microbe in hospitals, adheres to substrates and forms biofilms that are resistant to immunological responses and antimicrobial drugs. Currently, there is a need to develop alternative approaches for treating infections caused by biofilms to prevent delays in wound healing. Silver has long been used as a disinfectant, which is non-specific and has relatively low cytotoxicity. Silver sulfadiazine (SSD) is a chemical complex clinically used for the prevention of wound infections after injury. However, its effects on biofilms are still unclear. In this study, we aimed to analyze the mechanisms underlying SSD action on biofilms formed by MRSA. The antibacterial effects of SSD were a result of silver ions and not sulfadiazine. Ionized silver from SSD in culture media was lower than that from silver nitrate; however, SSD, rather than silver nitrate, eradicated mature biofilms by bacterial killing. In SSD, sulfadiazine selectively bound to biofilms, and silver ions were then liberated. Consequently, the addition of an ion-chelator reduced the bactericidal effects of SSD on biofilms. These results indicate that SSD is an effective compound for the eradication of biofilms; thus, SSD should be used for the removal of biofilms formed on wounds.

**Keywords:** biofilm; MRSA; silver ion; silver sulfadiazine; wound infections

---

## 1. Introduction

Biofilms (BFs) are a cause of chronic infections. Several types of symbiotic bacteria, such as *Staphylococcus aureus* and *Pseudomonas aeruginosa*, colonize our body and form BFs [1–3]. Methicillin-resistant *S. aureus* (MRSA) causes soft-tissue infections, indwelling catheter-associated infections, bacteremia, endocarditis, and osteomyelitis. Approximately 80% of chronic wound infections are attributed to bacteria or BFs [1]. BFs produce a subpopulation of drug-resistant cells called persister cells [4–7]. The BF matrix predominantly contains extracellular polysaccharides [8,9], and interacts with other molecules, including quorum-sensing signaling molecules/autoinducers, polypeptides, lectins, lipids, and extracellular DNA [10–12]. Specific molecules targeting BFs and effective drugs for BF eradication have not been identified yet. Therefore, the BF itself becomes a serious exacerbation factor in antimicrobial resistance.

Silver sulfadiazine (SSD) has been used as an exogenous antimicrobial agent since the 1960s [13,14], and is used on partial and full thickness burns to prevent infection [15,16]. It is registered on the World Health Organization's List of Essential Medicines [17]. SSD possesses broad-spectrum antibacterial activity, reacting nonspecifically to Gram-negative and Gram-positive bacteria, causing distortion of the cell membrane and inhibition of DNA replication [14,18–20]. Nevertheless, common side effects of SSD include pruritus and pain at the site of administration [21], as well as decreased white blood cell counts, allergic reactions, bluish-gray skin discoloration, and liver inflammation [22–24]. However, the Cochrane systematic review (2010) [25] did not recommend the use of SSD because of insufficient evidence on whether silver-containing dressings or topical agents promote wound healing or prevent wound infection. In addition, specific antibiotics are currently obtainable.

To overcome these shortcomings, new nanomedicine technologies using SSD have been adopted to enhance its antibacterial activity [26]. Unfortunately, SSD cytotoxicity was also enhanced [27]. SSD is poorly soluble and has limited penetration through intact skin [28,29]. When it comes in contact with body fluids, free sulfadiazine (SD) can be absorbed systemically, metabolized in the liver [30], and excreted in urine [31].

Notably, the effects of SSD on BFs have been widely studied [32,33]; however, their mechanisms of action have not been investigated yet. In this study, we investigated the essential mechanisms underlying the antibacterial action of SSD, especially against BFs, using SSD elements, silver and SD, and analyzed the effects of the compounds on a clinical strain of BF-forming MRSA.

## 2. Materials and Methods

### 2.1. Preparation of BF Chips

ATCC BAA-2856 (OJ-1) [34–36], a high-BF-forming strain of MRSA, was used. BF chips were prepared as previously described [36,37]. In brief, one colony grown on tryptic soy agar (TSA) was inoculated in tryptic soy broth (TSB) and grown at 37 °C until the optical density (OD) = 0.57 ($\lambda$ = 578 nm). A 1000-fold diluted bacterial solution was used for BF formation in the culture, and a plastic sheet for an overhead projector (3M Japan Ltd., Tokyo, Japan) was used as the substrate. Sterilized plastic sheets (1 × 8 cm) were immersed in bacterial solution and incubated to obtain a mature and uniform BF. After incubation, the sheets were washed 3 times with 10 mL of 0.01 M phosphate buffered saline solution (pH 7.4) to remove planktonic cells. Plastic sheets cut into 1 × 1 cm pieces termed "BF chips" were primarily used in the experiments unless otherwise stated.

### 2.2. Determination of BF Mass and Number of Live Bacteria

BF mass using crystal violet (CV) stain and live bacteria number by colony forming unit (CFU) counts were determined using a BF sheet (1 × 8 cm) in 10 mL media. CV-positive BF mass was measured according to a previously reported method [38] with some modifications. In brief, BFs stained with CV (Sigma-Aldrich, Tokyo, Japan) were eluted in 1 mL of 30% acetic acid in an RIA tube (Eiken Chemical Co., Ltd., Tokyo, Japan). Absorbance ($\lambda$ = 570 nm) was measured using a spectrometer (GENESYS 10S VIS, Thermo Scientific, LMS, Tokyo, Japan). Bacteria on BF sheets were dissociated using an ultrasonic generator (Sonifier 250; Branson Ultrasonics, Emerson Japan, Ltd., Kanagawa, Japan), and colony-forming units (CFUs) were assessed.

### 2.3. Antibacterial Effects of Compounds on BFs

Ethylene oxide-gas sterilized silver nitrate (FUJIFILM Wako Pure Chemical Corporation, Osaka, Japan), SSD (ALCARE Co., Ltd., Tokyo, Japan), and SD (FUJIFILM Wako Pure Chemical Corporation) were freshly prepared in TSB at a maximum concentration of 11,200 μM and serially diluted to 43.75 μM. After dipping the BF chip in 5-mL TSB-containing tubes in the presence of different concentrations of $AgNO_3$, SSD, and SD, the tubes were incubated at 37 °C for 24 h. Because SSD and SD were insoluble in liquid, cultures were continuously and gently mixed on a horizontal-rotation shaker (G10: New Brunswick Scientific, New Brunswick, NJ, USA) at 120 rpm.

The minimum antibacterial concentrations of compounds for bacteria derived from BF were determined by methods described previously [37] with some modifications (Figure S1). After incubating the bacteria for 24 h with different compounds at different concentrations, the culture tubes were kept at 4 °C for 1 h in order for the compounds to sediment, before the turbidity in the supernatant was measured ($\lambda = 578$ nm). The values were used to assess the minimum inhibitory concentration for planktonic cells derived from BFs (bMIC). Next, 10 µL of medium from each previously used tube for the bMIC analysis, was blotted on a $1 \times 1$ cm filter paper on TSA, and then they were incubated overnight at 37 °C. The colonies that grew around the paper were used to assess the minimum bactericidal concentration for planktonic cells derived from BF (bMBC). BF chips used for the bMIC analysis were directly placed on TSA and incubated overnight at 37 °C, and the minimum BF eradication concentration (MBEC) was determined.

In some studies, ethylenediamine-*N,N,N′,N′*-tetraacetic acid (EDTA) (FUJIFILM Wako Pure Chemical Corporation) (pH 7.4) at a concentration of 680 µM was used to chelate ions including silver.

### *2.4. Measurement of Liberated Silver Ions in the Media*

After BF chips were incubated in TSB at 37 °C for 24 h in the presence of different concentrations of $AgNO_3$ and SSD, media were centrifuged (EX-136: TOMY SEIKO Co. Ltd., Tokyo, Japan) at 3000 rpm for 10 min, and filtered with a 0.45-µm filter unit (Merck Millipore, Darmstadt, Germany) to remove bacteria and SSD aggregates. Ionized silver in the media was diluted 4 times with water and was quantified with AGT-131 using a NI-Ag kit (range: 0.01–0.25 ppm Ag-ion/mg (mL), Japan Ion Co., Tokyo, Japan).

### *2.5. SSD Inducing Direct/Indirect Bactericidal Effects on BFs*

To avoid direct contact of insoluble SSD with BFs, a closed chamber with a semipermeable membrane (Intercell S well chamber: KURABO INDUSTRIES LTD, Osaka, Japan) was used along with a syringe gasket to close the lid. Different concentrations of SSD in 500 µL TSB were injected into the chamber, which rested on 4.5 mL of TSB with the BF chip in a tube. The tube was cultured at 120 rpm at 37 °C for 24 h. For the control, 500 µL of SSD was directly added into TSB with the BF chip, and the chamber was filled with 500 µL TSB floated on the medium. After incubation, bMIC, bMBC, and MBEC were determined.

### *2.6. Quantification of the SD Attachment on BFs*

The SD amount was quantified using the diazo-coupling reaction with *N,N*-diethyl-*N′* 1-naphthylethylenediamine oxalate, also known as Tsuda's reagent (FUJIFILM Wako Pure Chemical Corporation) [39]. BFs formed on the surface of plastic tubes after 24 h incubation at 37 °C were washed 3 times with phosphate buffered saline. Freshly prepared SD was added at concentrations between 43.75 and 11,200 µM in 3 mL of TSB in a tube and was incubated at 37 °C for 1 h. After incubation, BFs were washed 3 times to remove any unbound SD on the BFs, and 500 µL of 1 N HCl was added to the tubes, from which 70 µL of the solution was mixed with 20 µL of 10% sodium nitrite (FUJIFILM Wako Pure Chemical Corporation) and reacted for 2 min on ice. Then, 100 µL of 2.5% ammonium amidosulfate (FUJIFILM Wako Pure Chemical Corporation) was added and reacted for 1 min, and 100 µL of Tsuda's reagent was added to the solution and mixed. The solutions in a 96-well plate (Becton, Dickinson and Company, Franklin Lakes, NJ, USA) were measured ($\lambda = 550$ nm) using a microplate reader (Model680; Bio-Rad Laboratories, Inc., Hercules, CA, USA). The SD calibration curve was also prepared (Figure S2).

### 2.7. Morphological Analysis of the Effects of $AgNO_3$, SSD, and SD on BFs

Plastic sheets in TSB with or without bacteria were incubated at 37 °C for 24 h. After incubation, the sheets were placed in media containing $AgNO_3$, SSD, and SD at a concentration of 2800 μM and incubated at 37 °C for 24 h. After incubation, the sheets were fixed in 5% formalin (pH 7.4), and stained with crystal violet. Another set of unstained sheets was placed in the air for more than one week to be turned black due to the sulfur-oxidized silver properties.

### 2.8. Ethics Approval

All methods involving bacterial handling were performed in accordance with the relevant guidelines and regulations under the Fukuoka University's experiment regulations.

### 2.9. Data and Statistical Analysis

Results from two different experimental groups initially underwent a distribution analysis using the F-test, before the Student's $t$-test or Mann–Whitney $U$ test were performed. $p$ values $< 0.05$ were considered to denote statistical significance. Sample numbers and repeated experiments are indicated in the legends of the figures and tables. Data are expressed as mean ± standard error.

## 3. Results

### 3.1. Antibacterial Effects of $AgNO_3$, SSD, and SD

Antibacterial effects of $AgNO_3$, SSD, and SD on BF chips were obtained (Table 1): bMIC for planktonic bacteria from BFs: 700 μM, 700 μM, >5600 μM; bMBC for planktonic bacteria from BFs 2800 μM, 1400 μM, >5600 μM; and MBEC for bacteria in BFs: >5600 μM, 2800 μM, >5600 μM, respectively. Results showed that SD had no antibacterial effects, but SSD alone was effective in targeting BFs.

**Table 1.** Minimum antimicrobial concentration (bMIC) and minimum bactericidal concentration (bMBC) for planktonic bacteria from BFs and bacteria in BFs (minimum BF eradication concentration (MBEC)).

| Compounds | | 0 | 175 | 350 | 700 | 1400 | 2800 | 5600 (μM) |
|---|---|---|---|---|---|---|---|---|
| $AgNO_3$ | bMIC | + | + | + | − | − | − | − |
| | bMBC | + | + | + | + | +/− | − | − |
| | MBEC | + | + | + | + | + | + | + |
| SSD | bMIC | + | + | + | − | − | − | − |
| | bMBC | + | + | + | −/+ | − | − | − |
| | MBEC | + | + | + | +/− | −/+ | − | − |
| SD | bMIC | + | + | + | + | + | + | + |
| | bMBC | + | + | + | + | + | + | + |
| | MBEC | + | + | + | + | + | + | + |

The antimicrobial effects of different concentrations of the compounds were determined using the BF chip method described in the Materials and Methods. The results of independent replicated examinations were evaluated: "−": 0/6 cases, "−/+":2/6 cases, "+/−":4/6 cases, "+": 6/6 cases. N size = 6; number of experimental replicates = 2.

### 3.2. Effects of $AgNO_3$, SSD, and SD on Viable Cells in BFs

Viable cell numbers on the BF chip were analyzed (Figure 1) using a CFU assay technique after collecting all bacteria from the BF chip. After 24-h incubation with $AgNO_3$ in different concentrations, bacterial growth was all arrested to some extent, but a dose-dependent suppression effect was not found. In contrast, SSD suppressed CFU values in a dose-dependent manner, especially at concentrations higher than 2800 μM, and the values reached approximately 1/100 of the initial density (0 h) and approximately 1/100,000 of cell density after 24 h of incubation. However, no effects on cell growth were found in any of the SD exposures.

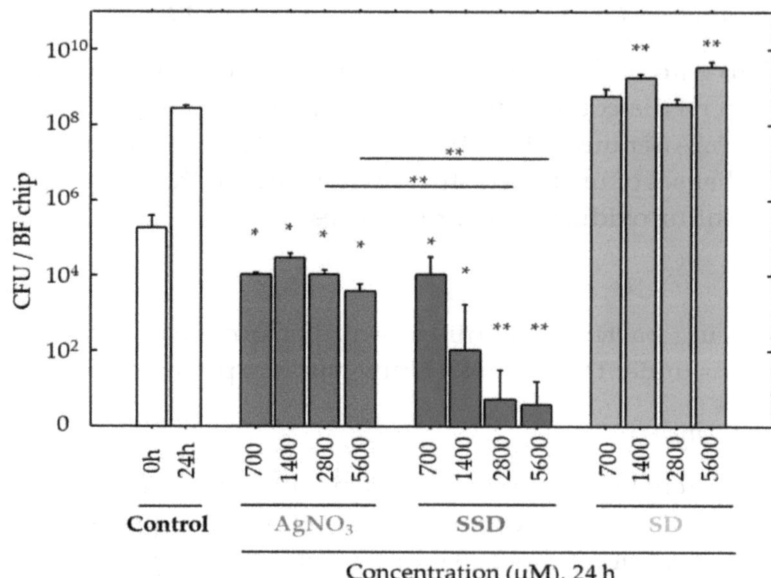


**Figure 1.** Colony forming unit (CFU) values for bacteria in biofilms (BFs). CFU values before (0 h) and after 24 h incubation at 37 °C in the vehicle alone (Control) and with $AgNO_3$, silver sulfadiazine (SSD), and sulfadiazine (SD) at different concentrations. Data are presented as the mean ± SE. N size = 5; number of experimental replicates = 2. *: $p < 0.05$, **: $p < 0.01$ vs. control after 24-h incubation with different compounds.

*3.3. Silver Ion Release in the Culture with SSD*

Antibacterial effects of SSD could be due to the ionized silver's action. As such, silver ions liberated in the cultures with $AgNO_3$ and SSD were measured. In both cases, liberated silver ions increased dose-dependently (Figure 2). Silver ions in the culture with $AgNO_3$ demonstrated a linear increase pattern, while those in the culture with SSD showed a logarithmic increase pattern. Significant differences were noted in the samples with concentrations greater than 700 μM (Figure 2). In the cultures with $AgNO_3$ and SSD at the concentration of 700 μM, silver ions increased to 5.91 μM and 4.64 μM, respectively. At the maximum concentration (11,200 μM), the silver ion concentration in the culture with $AgNO_3$ was three times higher than that in SSD culture (Figure 2).

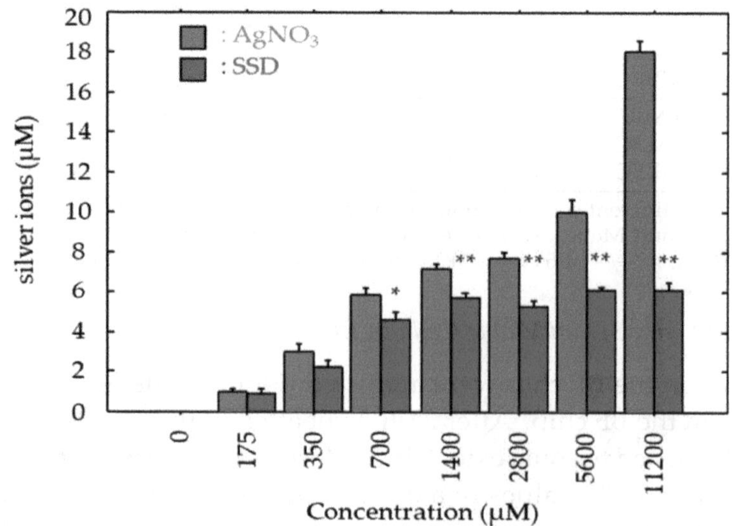


**Figure 2.** Medial ionized silver concentrations. After 24-h incubation at 37 °C in the presence of BF chips with $AgNO_3$ and SSD, medial ionized silver concentration was determined. Data are presented as the mean ± SE. N size = 3; number of experimental replicates = 2. *: $p < 0.05$ and **: $p < 0.01$ vs. $AgNO_3$.

### 3.4. Direct/Indirect Bactericidal Effect of SSDs on BFs

We hypothesized that the antibacterial effects of SSD could be due to its direct attachment to the BFs. To examine this mechanism, SSD was confined to a chamber, and a similar culture study was carried out. The bMIC in the culture with the chambered SSD was 1400 μM, which was two times greater than that in the culture with direct addition of SSD (Table 2). In terms of the bactericidal effects, the bMBC increased to >5600 μM, which was greater in the chambered SSD condition compared to that by direct SSD addition. Similarly, the MBEC value also increased to >5600 μM, which was greater in the chambered SSD compared to that in the direct SSD condition (Table 2). Therefore, the direct attachment of SSD to the BFs was necessary for initiating its antibacterial effects.

**Table 2.** Summary of antimicrobial threshold values.

| | bMIC | (+EDTA) | bMBC | (+EDTA) | MBEC | (+EDTA) |
|---|---|---|---|---|---|---|
| <Direct> | | | | | | |
| $AgNO_3$ (μM) | 700 | (700) | 2800 | (2800) | >5600 | (>5600) |
| SSD (μM) | 700 | (1400) | 1400 | (2800) | 2800 | (>5600) |
| SSD chambered (μM) | 1400 | (ND) | >5600 | (ND) | >5600 | (ND) |

ND: not determined.

### 3.5. Effects of EDTA on $AgNO_3$ and SSD-Induced Antibacterial Activity

EDTA is a chelator for silver ions [40]. To analyze the involvement of silver ions in the antibacterial effects, EDTA at a concentration of 640 μM was used. In the cultures with $AgNO_3$, EDTA did not affect bMIC, bMBC, and MBEC values (700 μM, 2800 μM, and >5600 μM, respectively, Table 2). However, all the values in the cultures with SSD were increased by the addition of EDTA (700 μM → 1400 μM, 1400 μM → 2800 μM, and 2800 μM → >5600 μM, respectively, Table 2).

### 3.6. Effects of Compounds on BF Eradication

The BF eradication effects of the compounds were analyzed at a dose of 2800 μM, an effective concentration for SSD (Table 2). The results of BF mass, CFU and morphological analysis are shown in Figure 3a–c, respectively. The BF eradication effects were shown to be SSD > $AgNO_3$ > SD, and SD had no effect. Silver-containing compounds, including $AgNO_3$ and SSD, displayed BF eradication activities, where SSD showed the greatest effects.

Given that SSD is chemically synthesized with SD and silver, we examined the BF eradication effects of a mixture of SD and $AgNO_3$ at a concentration of 2800 μM (Figure 3a,c). The values of BF mass and live bacteria number in the mixture were lower than those in SD alone, equivalent to those in $AgNO_3$ alone, but were still significantly higher than those in SSD, suggesting that SSD was still the most effective at eradicating BFs.

### 3.7. SD Deposition to BFs

In the control without BFs, the addition of any concentration of SD to the media only resulted in a consistently low level of SD being deposited on the plastic surface (Figure 3d). However, in the BF tubes, SD was deposited on the BFs in a dose-dependent manner, and prominent deposition started once concentrations of SD were greater than 1400 μM (Figure 3d).

### 3.8. BF Morphology

After exposure to the different compounds, the BFs stained with crystal violet are shown in the left panels of Figure 3b, where the overall purple staining intensity among the groups follows the next scheme: Control > SD > $AgNO_3$ > SSD. String-like structures (i.e., thickened BFs) found in the

Control group remained in the SD and $AgNO_3$ groups, and such structures were scarcely detected in the SSD group.

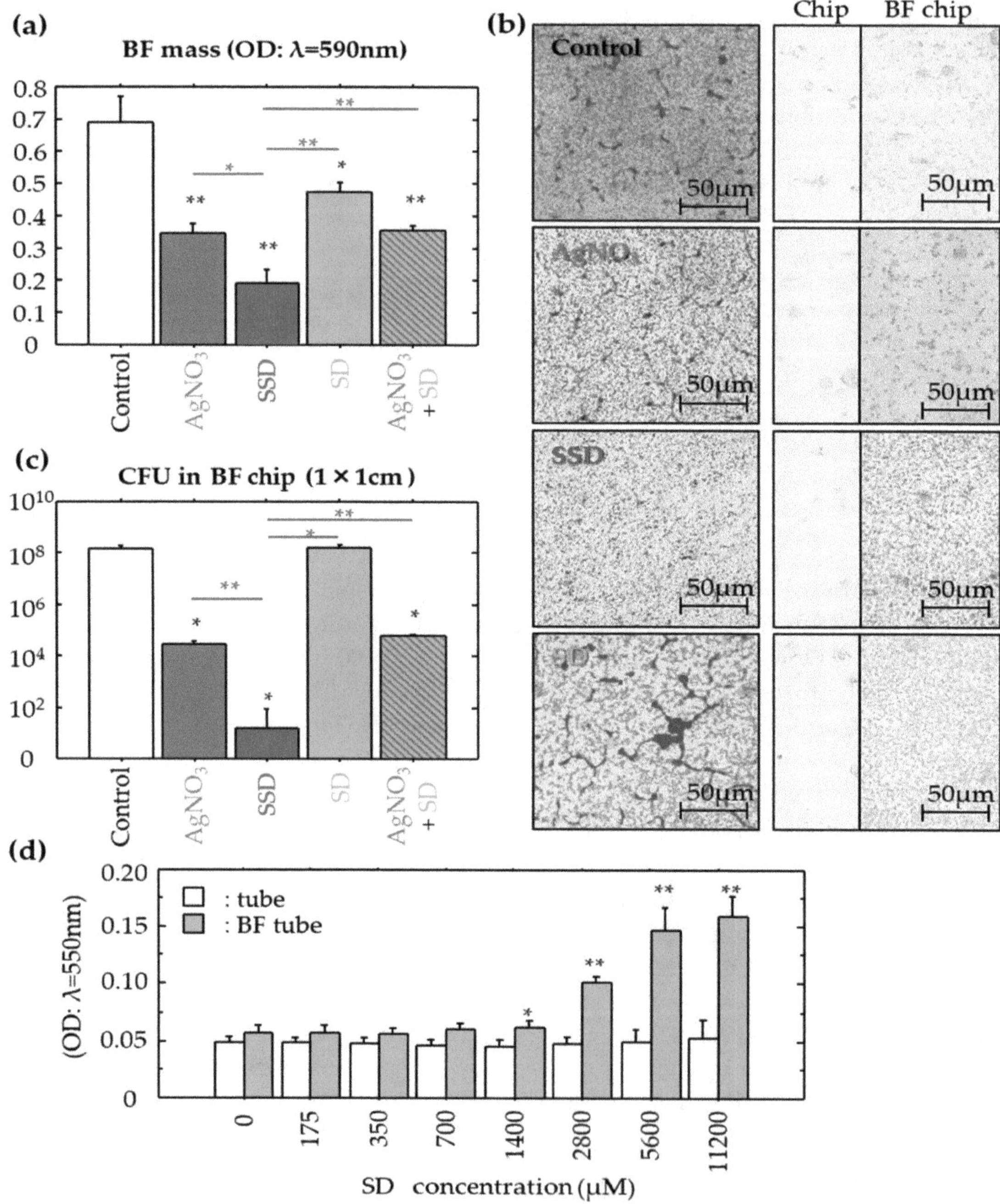


**Figure 3.** Effects of the compounds on mature BFs formed on plastic chips and SD deposition on BF chips. The effects of the compounds (2800 μM) after 24-h incubation at 37 °C on BF masses (**a**), morphology (**b**) and CFU values (**c**) were determined. Data are presented as the mean ± SE. N size = 5; number of experimental replicates = 2. *: $p < 0.05$ and **: $p < 0.01$ vs. negative control shown in white, and vs. SSD shown in red. In morphology, silver was deposited on the plastic chips with/without matured BFs after 24 h incubation at 37 °C with/without the different compounds. Sulfurized silvers in air are shown as black elements. Objective lens: ×20. In SD quantification, SD deposited on the BFs formed in plastic tubes after 24 h incubation at 37 °C was determined according to the method described in the Materials and Methods (**d**). Data are presented as the mean ± SE. N size = 5; number of experimental replicates = 2. *: $p < 0.05$ and **: $p < 0.01$ vs. BF-free control.

### *3.9. Localization of Silver on BFs*

Oxidized silver turns black in color. No black spots were found on the plastic substrate even after exposure to all compounds, nor were there spots on the BFs following SD exposure (Figure 3b, right panels). In contrast, fine small black spots accumulated on the BFs after exposure to $AgNO_3$ and SSD. Moreover, SSD induced denser accumulation of black spots.

## 4. Discussion

BFs are formed by bacteria, which contain massive extracellular polysaccharides/exopolymeric substances, and acquire an antibiotic-tolerant nature, which is explained by the decreased drug penetration [41] and the appearance of dormant cells (i.e., persister cells) [42]. Thus, the eradication of BFs has become a difficult task. In this study, we examined the effect of SSD on MRSA-formed BFs. The SSD concentration in the medical drugs, namely Silvadene, etc., is of 1% (10,000 μg/mL), which could be diluted with tissue fluid exuded from wounds after application. The highest concentration used in this study was 4000 μg/mL (0.4%: 11,200 μM), which is slightly lower than that of the medical drug.

The bactericidal actions of silver can be explained through three different mechanisms: (1) the production of dissolved oxygen-derived reactive oxygen species by its catalytic activity [43,44]; (2) the cross-linkage with silver at the sites of hydrogen bonding between the double strands of DNA [44,45]; and (3) the inhibition of enzyme activities by intracellular silver ions [44,46]. SD, a sulfonamide, inhibits intracellular folate metabolism in bacteria, resulting in proliferation arrest. During the investigation of silver-containing agents, SD has been selected amongst different compounds due to its high and broad bactericidal effects [14,18–20]. We showed here that SSD is effective against MRSA, especially in the BF state, but SD itself had no effect.

Among the compounds tested, SSD significantly decreased BF mass and live bacteria number in BFs, which was a considerably greater difference than that of $AgNO_3$. This phenomenon was supported by silver deposition on the BF chip. We also examined the efficacy of the simultaneous addition of $AgNO_3$ and SD instead of the SSD. These effects are equivalent to those of $AgNO_3$ alone, but are significantly lower than SSD, indicating that the molecular form of SSD is important to induce a silver-related BF eradication. In contrast, it was reported that SSD was ineffective for MRSA BFs formed on a polycarbonate filter using a novel in vitro model (colony/drip-flow reactor) [32]. However, its exposure time was quite short (15 min) to induce an eradication reaction as compared with the present study (24 h).

In contrast, SSD and $AgNO_3$ induced a similar antibacterial potency in bMIC and bMBC, both of which were detected in the planktonic state derived from the BFs. In accordance with their threshold levels, silver ion concentrations reached more than 5 μM. In any case, such concentration could be necessary for growth inhibition and the killing of planktonic bacteria. However, in the range of effective doses, silver ions were always generated at higher levels in $AgNO_3$ conditions compared to those in SSD conditions. This reflects the contradiction of the bactericidal effects of SSD. The results suggest that some factor(s) other than simple diffusion of silver ion may be involved.

The MBEC level is a threshold for bacterial killing concentration in BFs. It was detectable only in SSD rather than $AgNO_3$ or SD, which was also confirmed by the CFU assay. It clearly showed that $AgNO_3$ could not efficiently kill the bacteria in BFs, and its effect remained within the level of growth inhibition. On the other hand, SSD strongly and dose-dependently depressed the live cell number in matured BFs, in which the live cell density at a concentration of 700 μM (the equivalent level of bMIC) reached 1/10,000 of the control after 24 h of incubation, and, at a concentration of 2800 μM (the equivalent level of MBEC), the number was 1/300 of the initial cell density and 1/300,000 of the control after 24-h incubation. These results suggest that the medial silver ion concentration is not a direct influencing factor for the SSD.

To clarify its mechanism of action, SSD was confined in a sealed chamber, by which SSD was constrained and bacteria could not penetrate beyond the membrane, but ionized silver could pass freely. As a result, silver ion concentrations in the chambered condition were lower at 700 and 1400 μM SSD, as compared with non-chambered conditions, but, in higher concentrations more than 2800 μM SSD, they became similar levels (Figure S3). Upon confinement, SSD-induced bMBC and MBEC levels were no longer seen. Our study also showed that SD itself had the property to bind to the BFs. Although, the exact binding mechanisms of SSD on BFs are still unknown, the direct attachment of SSD to BFs is crucial to induce bactericidal effects. Therefore, if the SSD binding site on BFs is identified in the near future, this would lead to the discovery of target molecules of the BFs formed by MRSA.

To validate the silver ion's role in the antibacterial properties of SSD, EDTA was used at a concentration of 680 μM, which was about 40 to 100 times greater than that of the liberated silver ions in the media with $AgNO_3$ and SSD. With the addition of EDTA, no changes in bMIC, bMBC, and MBEC were found in the culture with $AgNO_3$. In contrast, the addition of EDTA increased in all of the values in the culture with SSD. The mechanisms of antibacterial action between $AgNO_3$ and SSD are unknown; however, their mode of reaction might be different. Silver ions from $AgNO_3$ could be easily bound by negatively charged components in media as compared to SSD, and a bound–liberation cycle may be repeated. Moreover, the silver holding capacity in SSD may be greater than that in $AgNO_3$, by which SSD could reach the BFs. This mechanism may act especially on the BFs due to the selective adhesion of SD on BFs, which was confirmed by the greater deposition of silver on the BF chip incubated with SSD, whereas sites of the deposition on BFs could not be identified. It therefore seems likely that SSD may act on the bacteria settled in the BFs. After deposition of SSD, silver ions could be released in a micro environment in BFs, by which the opportunities for bacteria to kill could be increased, resulting in severe distortions of BF structure (Figure 3b).

As a limitation of the present study, one MRSA strain was only used. In future study, we will use different clinical strains, such as low-BF formers and high-BF formers in our collection [47].

## 5. Conclusions

SSD is a reliable anti-bacterial agent, and thus has recently been used as a coating material for indwelling catheters [48,49]. However, the mechanisms of action of SSD on BFs are still unclear. This study is the first to elucidate the mechanisms behind the efficacy of SSD on BFs. Therefore, it is possible that SSD preferably binds to BFs, and then it releases silver ions, by which bacteria settled in the BFs under a micro-environment are killed (Figure 4). In the future, further molecular levels of investigation are needed to identify the SSD binding site on BFs. Additionally, a systematic investigation of the use of SSD in BF-infected wounds should be performed in clinical practices rather than for infection prevention.

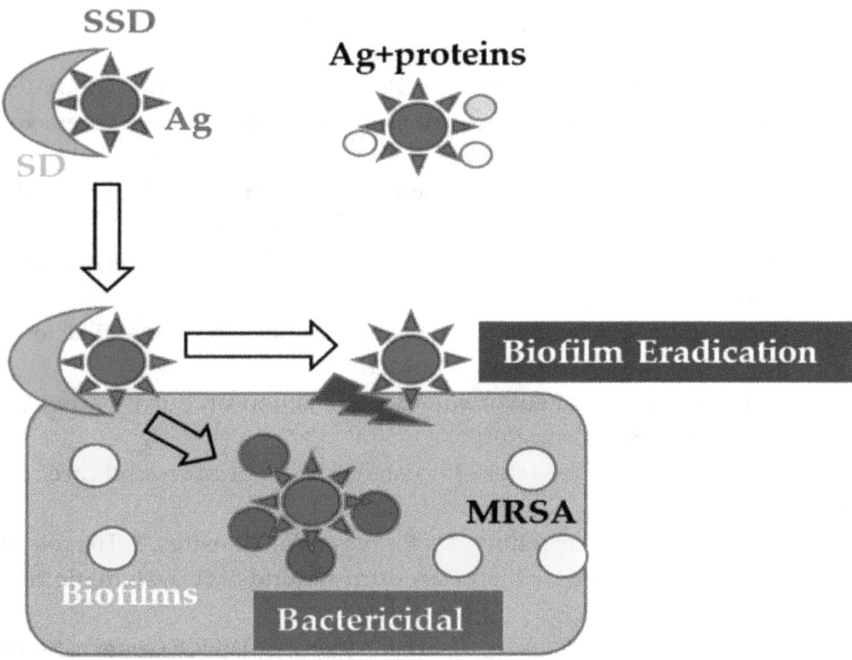


**Figure 4.** Working hypothesis of the action of SSD on biofilms formed by MRSA.

**Author Contributions:** All experimental designs were created by S.J. and Y.U. Bacterial experiments were carried out by Y.U., M.M., K.M., and S.H. Statistical analyses were performed by Y.U., S.T., S.H., and H.K. The manuscript was prepared by Y.U., M.M., and S.J. and revised by S.J., K.M., S.T., S.H., and H.K. All authors have read and agreed to the published version of the manuscript.

**Acknowledgments:** The authors thank Yui Ohshiro, Arisa Nita, Kaede Munekata, and Yuka Kawamura (Faculty of Pharmaceutical Science, Fukuoka University, Fukuoka, Japan) for their excellent technical assistance. We are also thankful for the research support and the valuable discussion with Takashi Kubo, ALCARE Co., Ltd., Tokyo, Japan.

## References

1. Kiedrowski, M.R.; Horswill, A.R. New approaches for treating staphylococcal biofilm infections. *Ann. N. Y. Acad. Sci.* **2011**, *1241*, 104–121. [CrossRef] [PubMed]
2. Otto, M. Staphylococcal infections: Mechanisms of biofilm maturation and detachment as critical determinants of pathogenicity. *Annu. Rev. Med.* **2013**, *64*, 175–188. [CrossRef] [PubMed]
3. Abdulhaq, N.; Nawaz, Z.; Zahoor, M.A.; Siddique, A.B. Association of biofilm formation with multi drug resistance in clinical isolates of. *EXCLI J.* **2020**, *19*, 201–208. [CrossRef] [PubMed]
4. Evans, R.C.; Holmes, C.J. Effect of vancomycin hydrochloride on *Staphylococcus epidermidis* biofilm associated with silicone elastomer. *Antimicrob. Agents Chemother.* **1987**, *31*, 889–894. [CrossRef]
5. Stewart, P.S. Mechanisms of antibiotic resistance in bacterial biofilms. *Int. J. Med. Microbiol.* **2002**, *292*, 107–113. [CrossRef]
6. Singh, R.; Ray, P.; Das, A.; Sharma, M. Role of persisters and small-colony variants in antibiotic resistance of planktonic and biofilm-associated Staphylococcus aureus: An in vitro study. *J. Med. Microbiol.* **2009**, *58*, 1067–1073. [CrossRef]
7. Garcia, L.G.; Lemaire, S.; Kahl, B.C.; Becker, K.; Proctor, R.A.; Denis, O.; Tulkens, P.M.; Van Bambeke, F. Antibiotic activity against small-colony variants of *Staphylococcus aureus*: Review of in vitro, animal and clinical data. *J. Antimicrob. Chemother.* **2013**, *68*, 1455–1464. [CrossRef]
8. Flemming, H.C. The perfect slime. *Colloids Surf. B: Biointerfaces* **2011** *86*, 251–259. [CrossRef]

9. Dragoš, A.; Kovács, Á. The Peculiar Functions of the Bacterial Extracellular Matrix. *Trends Microbiol.* **2017**, *25*, 257–266. [CrossRef]
10. Roche, F.M.; Downer, R.; Keane, F.; Speziale, P.; Park, P.W.; Foster, T.J. The N-terminal A domain of fibronectin-binding proteins A and B promotes adhesion of Staphylococcus aureus to elastin. *J. Biol. Chem.* **2004**, *279*, 38433–38440. [CrossRef]
11. Kline, K.A.; Fälker, S.; Dahlberg, S.; Normark, S.; Henriques-Normark, B. Bacterial adhesins in host-microbe interactions. *Cell Host Microbe* **2009**, *5*, 580–592. [CrossRef] [PubMed]
12. Lee, J.; Zhang, L. The hierarchy quorum sensing network in *Pseudomonas aeruginosa*. *Protein Cell* **2015**, *6*, 26–41. [CrossRef] [PubMed]
13. Carr, H.S.; Wlodkowski, T.J.; Rosenkranz, H.S. Silver sulfadiazine: In vitro antibacterial activity. *Antimicrob. Agents Chemother.* **1973**, *4*, 585–587. [CrossRef] [PubMed]
14. Fox, C.L.; Modak, S.M. Mechanism of silver sulfadiazine action on burn wound infections. *Antimicrob. Agents Chemother.* **1974**, *5*, 582–588. [CrossRef]
15. Percival, S.L.; Bowler, P.G.; Russell, D. Bacterial resistance to silver in wound care. *J. Hosp. Infect.* **2005**, *60*, 1–7. [CrossRef]
16. Heyneman, A.; Hoeksema, H.; Vandekerckhove, D.; Pirayesh, A.; Monstrey, S. The role of silver sulphadiazine in the conservative treatment of partial thickness burn wounds: A systematic review. *Burns* **2016**, *42*, 1377–1386. [CrossRef]
17. WHO. *World Health Organization Model List of Essential Medicines*; WHO: Geneva, Switzerland, 2019.
18. Hoffmann, S. Silver sulfadiazine: An antibacterial agent for topical use in burns. A review of the literature. *Scand. J. Plast. Reconstr. Surg.* **1984**, *18*, 119–126. [CrossRef]
19. Fuller, F.W.; Parrish, M.; Nance, F.C. A review of the dosimetry of 1% silver sulfadiazine cream in burn wound treatment. *J. Burn Care Rehabil.* **1994**, *15*, 213–223. [CrossRef]
20. Miller, A.C.; Rashid, R.M.; Falzon, L.; Elamin, E.M.; Zehtabchi, S. Silver sulfadiazine for the treatment of partial-thickness burns and venous stasis ulcers. *J. Am. Acad. Dermatol.* **2012**, *66*, e159–e165. [CrossRef]
21. Fuller, F.W. The side effects of silver sulfadiazine. *J. Burn Care Res.* **2009**, *30*, 464–470. [CrossRef]
22. Muller, M.J.; Hollyoak, M.A.; Moaveni, Z.; Brown, T.L.; Herndon, D.N.; Heggers, J.P. Retardation of wound healing by silver sulfadiazine is reversed by Aloe vera and nystatin. *Burns* **2003**, *29*, 834–836. [CrossRef]
23. Chaby, G.; Viseux, V.; Poulain, J.F.; De Cagny, B.; Denoeux, J.P.; Lok, C. Topical silver sulfadiazine-induced acute renal failure. *Ann. Dermatol. Vénéréologie* **2005**, *132*, 891–893. [CrossRef]
24. Abedini, F.; Ahmadi, A.; Yavari, A.; Hosseini, V.; Mousavi, S. Comparison of silver nylon wound dressing and silver sulfadiazine in partial burn wound therapy. *Int. Wound J.* **2013**, *10*, 573–578. [CrossRef]
25. Carter, M.J.; Tingley-Kelley, K.; Warriner, R.A. Silver treatments and silver-impregnated dressings for the healing of leg wounds and ulcers: A systematic review and meta-analysis. *J. Am. Acad. Dermatol.* **2010**, *63*, 668–679. [CrossRef] [PubMed]
26. Adhya, A.; Bain, J.; Ray, O.; Hazra, A.; Adhikari, S.; Dutta, G.; Ray, S.; Majumdar, B.K. Healing of burn wounds by topical treatment: A randomized controlled comparison between silver sulfadiazine and nano-crystalline silver. *J. Basic Clin. Pharm.* **2014**, *6*, 29–34. [CrossRef]
27. Liu, X.; Gan, H.; Hu, C.; Sun, W.; Zhu, X.; Meng, Z.; Gu, R.; Wu, Z.; Dou, G. Silver sulfadiazine nanosuspension-loaded thermosensitive hydrogel as a topical antibacterial agent. *Int. J. Nanomed.* **2019**, *14*, 289–300. [CrossRef] [PubMed]
28. Dellera, E.; Bonferoni, M.C.; Sandri, G.; Rossi, S.; Ferrari, F.; Del Fante, C.; Perotti, C.; Grisoli, P.; Caramella, C. Development of chitosan oleate ionic micelles loaded with silver sulfadiazine to be associated with platelet lysate for application in wound healing. *Eur. J. Pharm. Biopharm.* **2014**, *88*, 643–650. [CrossRef]
29. Kumar, P.M.; Ghosh, A. Development and evaluation of silver sulfadiazine loaded microsponge based gel for partial thickness (second degree) burn wounds. *Eur. J. Pharm. Sci.* **2017**, *96*, 243–254. [CrossRef]
30. Winter, H.R.; Unadkat, J.D. Identification of cytochrome P450 and arylamine N-acetyltransferase isoforms involved in sulfadiazine metabolism. *Drug Metab. Dispos.* **2005**, *33*, 969–976. [CrossRef]
31. Nouws, J.F.; Firth, E.C.; Vree, T.B.; Baakman, M. Pharmacokinetics and renal clearance of sulfamethazine, sulfamerazine, and sulfadiazine and their N4-acetyl and hydroxy metabolites in horses. *Am. J. Vet. Res.* **1987**, *48*, 392–402.

32. Agostinho, A.M.; Hartman, A.; Lipp, C.; Parker, A.E.; Stewart, P.S.; James, G.A. An in vitro model for the growth and analysis of chronic wound MRSA biofilms. *J. Appl. Microbiol.* **2011**, *111*, 1275–1282. [CrossRef] [PubMed]
33. Lemire, J.A.; Kalan, L.; Bradu, A.; Turner, R.J. Silver oxynitrate, an unexplored silver compound with antimicrobial and antibiofilm activity. *Antimicrob. Agents Chemother.* **2015**, *59*, 4031–4039. [CrossRef] [PubMed]
34. Makino, T.; Jimi, S.; Oyama, T.; Nakano, Y.; Hamamoto, K.; Mamishin, K.; Yahiro, T.; Hara, S.; Takata, T.; Ohjimi, H. Infection mechanism of biofilm-forming Staphylococcus aureus on indwelling foreign materials in mice. *Int. Wound J.* **2015**, *12*, 122–131. [CrossRef] [PubMed]
35. Haraga, I.; Abe, S.; Jimi, S.; Kiyomi, F.; Yamaura, K. Increased biofilm formation ability and accelerated transport of *Staphylococcus aureus* along a catheter during reciprocal movements. *J. Microbiol. Methods* **2017**, *132*, 63–68. [CrossRef]
36. Ueda, Y.; Mashima, K.; Miyazaki, M.; Hara, S.; Takata, T.; Kamimura, H.; Takagi, S.; Jimi, S. Inhibitory effects of polysorbate 80 on MRSA biofilm formed on different substrates including dermal tissue. *Sci. Rep.* **2019**, *9*, 3128. [CrossRef]
37. Jimi, S.; Miyazaki, M.; Takata, T.; Ohjimi, H.; Akita, S.; Hara, S. Increased drug resistance of meticillin-resistant *Staphylococcus aureus* biofilms formed on a mouse dermal chip model. *J. Med. Microbiol.* **2017**, *66*, 542–550. [CrossRef]
38. Bendouah, Z.; Barbeau, J.; Hamad, W.A.; Desrosiers, M. Use of an in vitro assay for determination of biofilm-forming capacity of bacteria in chronic rhinosinusitis. *Am. J. Rhinol.* **2006**, *20*, 434–438. [CrossRef]
39. Negoro, H. Colorimetric Estimation of PAS Using Tsuda Reagent. *Yakugaku Zasshi-J. Pharm. Soc. Jpn.* **1951**, *71*, 209–210. [CrossRef]
40. Saran, L.; Cavalheiro, E.; Neves, E.A. New aspects of the reaction of silver(I) cations with the ethylenediaminetetraacetate ion. *Talanta* **1995**, *42*, 2027–2032. [CrossRef]
41. Flemming, H.C.; Wingender, J.; Szewzyk, U.; Steinberg, P.; Rice, S.A.; Kjelleberg, S. Biofilms: An emergent form of bacterial life. *Nat. Rev. Microbiol.* **2016**, *14*, 563–575. [CrossRef]
42. Harms, A.; Maisonneuve, E.; Gerdes, K. Mechanisms of bacterial persistence during stress and antibiotic exposure. *Science* **2016**, *354*. [CrossRef] [PubMed]
43. Lansdown, A.B. Silver I: Its antibacterial properties and mechanism of action. *J. Wound Care* **2002**, *11*, 125–130. [CrossRef] [PubMed]
44. Marx, D.E.; Barillo, D.J. Silver in medicine: The basic science. *Burns* **2014**, *40*, S9–S18. [CrossRef]
45. Russell, A.D.; Hugo, W.B. Antimicrobial activity and action of silver. *Prog. Med. Chem.* **1994**, *31*, 351–370. [CrossRef] [PubMed]
46. Slawson, R.M.; Lee, H.; Trevors, J.T. Bacterial interactions with silver. *Biol. Met.* **1990**, *3*, 151–154. [CrossRef]
47. Oyama, T.; Miyazaki, M.; Yoshimura, M.; Takata, T.; Ohjimi, H.; Jimi, S. Biofilm-Forming Methicillin-Resistant Staphylococcus aureus Survive in Kupffer Cells and Exhibit High Virulence in Mice. *Toxins* **2016**, *8*, 198. [CrossRef]
48. de Sousa, J.K.T.; Haddad, J.P.A.; de Oliveira, A.C.; Vieira, C.D.; Dos Santos, S.G. In vitro activity of antimicrobial-impregnated catheters against biofilms formed by KPC-producing Klebsiella pneumoniae. *J. Appl. Microbiol.* **2019**, *127*, 1018–1027. [CrossRef]
49. Mohseni, M.; Shamloo, A.; Aghababaie, Z.; Afjoul, H.; Abdi, S.; Moravvej, H.; Vossoughi, M. A comparative study of wound dressings loaded with silver sulfadiazine and silver nanoparticles: In vitro and in vivo evaluation. *Int. J. Pharm.* **2019**, *564*, 350–358. [CrossRef]

# 12

# Role of SrtA in Pathogenicity of *Staphylococcus lugdunensis*

**Muzaffar Hussain [1], Christian Kohler [2] and Karsten Becker [1,2,*]**

[1] Institute of Medical Microbiology, University Hospital Münster, 48149 Münster, Germany; muzaffa@uni-muenster.de

[2] Friedrich Loeffler-Institute of Medical Microbiology, University Medicine Greifswald, 17475 Greifswald, Germany; christian.kohler@med.uni-greifswald.de

* Correspondence: karsten.becker@med.uni-greifswald.de

**Abstract:** Among coagulase-negative staphylococci (CoNS), *Staphylococcus lugdunensis* has a special position as causative agent of aggressive courses of infectious endocarditis (IE) more reminiscent of IEs caused by *Staphylococcus aureus* than those by CoNS. To initiate colonization and invasion, bacterial cell surface proteins are required; however, only little is known about adhesion of *S. lugdunensis* to biotic surfaces. Cell surface proteins containing the LPXTG anchor motif are covalently attached to the cell wall by sortases. Here, we report the functionality of *Staphylococcus lugdunensis* sortase A (SrtA) to link LPXTG substrates to the cell wall. To determine the role of SrtA dependent surface proteins in biofilm formation and binding eukaryotic cells, we generated SrtA-deficient mutants (Δ*srtA*). These mutants formed a smaller amount of biofilm and bound less to immobilized fibronectin, fibrinogen, and vitronectin. Furthermore, SrtA absence affected the gene expression of two different adhesins on transcription level. Surprisingly, we found no decreased adherence and invasion in human cell lines, probably caused by the upregulation of further adhesins in Δ*srtA* mutant strains. In conclusion, the functionality of *S. lugdunensis* SrtA in anchoring LPXTG substrates to the cell wall let us define it as the pathogen's housekeeping sortase.

**Keywords:** *Staphylococcus lugdunensis*; sortase A; surface proteins; LPXTG

---

## 1. Introduction

As known for *Staphylococcus aureus* and several coagulase-negative staphylococci (CoNS), *Staphylococcus lugdunensis* is also part of the human microbiota colonizing miscellaneous skin surface habitats [1–4]. Infections due to this opportunistic pathogen resemble those caused by *S. aureus* rather than "classical" CoNS infections [5]. In particular, aggressive and highly destructive courses of native and prosthetic valve infectious endocarditis (IE) have been reported [6]. Despite its clinical impact, only a few factors contributing to the pathogenicity of *S. lugdunensis* have been described, including a fibrinogen binding protein (Fbl), a von Willebrand-factor binding protein (vWbl), and a multifunctional autolysin (AtlL) [7–11].

To initiate invasive infections, staphylococcal cells irrespective of the species must adhere to cells of the host tissue or to the extracellular matrix (ECM) [12]. This complex multifactorial process is mediated by strong interactions of cell wall-anchored (CWA) proteins with host structures. In *S. aureus*, four distinct categories of CWA proteins have been identified, and the microbial surface components recognizing adhesive matrix molecules (MSCRAMMs) were elucidated as the largest group [13,14]. CWA proteins are characterized by a sorting signal containing a carboxy-terminal LPXTG anchor motif [15]. In *S. aureus* and many other Gram-positive species, the LPXTG-sorting signal is cleaved by the membrane-bound transpeptidase sortase. Sortase A (SrtA) of *S. aureus* is a 206 amino acid peptide that catalyzes a transpeptidation process [16–19] and is considered the housekeeping sortase [20]. The cleavage between the threonine and glycine residues is followed by an amide linkage between the carboxyl group of threonine and the amino group of a pentaglycine cross-bridge of the cell membrane-attached peptidoglycan [16,19]. Mutants lacking the *srtA* gene are defective in the cell wall anchoring of LPXTG proteins [16,17,21]. Thus, cells are unable to anchor cell surface proteins of *S. aureus* important for the adherence to eukaryotic cell structures. In consequence, Δ*srtA* mutants were attenuated to establish infections as shown in a murine model [22]. In addition to SrtA, a further *S. aureus* sortase isoform, sortase B (SrtB), has been described [23]. More recently, a SrtA-deficient mutant of *S. lugdunensis* was significantly less virulent than the parental strain in a rat IE model [24]. Therefore, the use of SrtA inhibitors might represent a promising anti-virulence therapy strategy to disrupt the anchoring of bacterial surface proteins, which are critical for the pathogen´s adherence to the host.

Here, we report the functionality of *S. lugdunensis* SrtA to anchor LPXTG substrates to the cell wall and defined it as the pathogen´s housekeeping sortase. Generating *srtA*-deficient mutants, we determined the role of *S. lugdunensis* SrtA-dependent surface proteins in biofilm formation and invasion of eukaryotic cells. Finally, we confirmed the influence of a functional SrtA on the gene expression of further LPXTG proteins.

## 2. Materials and Methods

### 2.1. Bacterial Strains and Culture Media

Lysogeny broth (LB) or agar were used for cultivation of *Escherichia coli* and staphylococci were cultivated in tryptic soy broth (TSB) or agar (TSA) (Difco, Detroit, Detroit, MI, USA), brain heart infusion (BHI) broth or agar (Merck, Darmstadt, Darmstadt, Germany) and Mueller Hinton (MH) broth or agar (Mast, Merseyside, Bootle, UK). Antibiotics were added to MH agar in appropriate amounts (ampicillin, 100 µg/mL, Sigma; erythromycin, 10 µg/mL Serva; and chloramphenicol, 10 µg/mL, Serva, Heidelberg, Germany) for selection of resistance in *E. coli* or *S. lugdunensis*. All bacterial strains and plasmids used in this study are presented in Table 1.

**Table 1.** Bacterial strains used in this study.

| Strains | Relevant Genotype or Plasmid | Properties | Reference or Source |
|---|---|---|---|
| | *S. lugdunensis* strains | | |
| Sl48 | | Clinical isolate | Germany [b] |
| Sl44 | | Clinical isolate | Germany [b] |
| Mut7 | Sl48 *srtA::EmR* | Sl48 deficient in sortase-A | This study |
| Mut47 | Sl44 *srtA::EmR* | Sl44 deficient in sortase-A | This study |
| SL241 | | Clinical isolate | Germany [b] |
| SL253 | | Clinical isolate | Germany [b] |
| | *S. aureus* strain | | |
| *S. aureus* Cowan 1 (ATCC 12598) | | Reference isolate from septic arthritis | ATCC |
| | *E. coli* strains | | |
| DH5α | *supE44ΔlacU169 (ϕ80 lacZΔM15) hsdR17 recA1 end A1 gyrA96 thi-1 relA1* | Cloning host | Stratagene |
| TG1 | *supE hsdΔ5 thiΔ(lac-proAB) F′(traD36 proAB⁺ lacI^q lacZΔM15)* | Cloning host | Stratagene |
| DH5α (pBT37) | pBT9*atlL::Em*$^R$ | Shuttle vector pBT9 containing *atlL::Em*$^R$ | This study |
| | Eukaryotic strains | | |
| EA.hy 926 cells | | | [25] |
| A549 fibroblast | | | [26] |
| Human bladder carcinoma cell line 5637 | | | [27] |

[b] kindly provided by F. Szabados and S. Gatermann (Bochum, Germany).

### 2.2. Growth Characteristics

Bacteria were grown in BHI overnight followed by dilution of the culture in 100 mL fresh BHI in 500 mL flask to an optical density 578 nm ($OD_{578}$) of 0.05. Flasks were incubated at 160 rpm at 37 °C and the $OD_{578}$ were determined hourly for a period of 10 h followed by final sampling after 24 h to establish growth curves. For growth experiments in little volumes, bacteria were grown in 5 mL BHI in a glass tube for 18 h at 37 °C with shaking at 160 rpm. Next day, $OD_{578}$ were measured against un-inoculated BHI after vortex of growth.

### 2.3. Characterization of Agglutination

The Pastorex Staph Plus (Sanofi Diagnostic Pasteur, S.A., Marnes la Coquette, France), a rapid agglutination test for the simultaneous detection of the *S. aureus* fibrinogen affinity antigen (clumping factor), protein A, and capsular polysaccharides were used to differentiate wild type and mutant strains.

### 2.4. Biofilm Formation

Bacteria were grown in 5 mL BHI, TSB, BHI + 0.5% glucose or TSB + 0.5% glucose in glass test tubes for 8 h at 37 °C with shaking at 160 rpm. Afterwards, the cells were diluted 1:100 with the same type of fresh medium and 100 μL were added to wells in quadruplicate in a 96 well microtiterplate. Plates were incubated at 37 °C without shaking overnight. Next day bacteria were removed, plates were washed with PBS, and the biofilm bacteria mix were fixed with ice cold methanol for 10 min at −20 °C. After washing once with PBS, the adhered biofilm bacteria were stained with crystal violet for 10 min at room temperature. Excess stain was removed by washing 3x with PBS. The adhered bacteria were brought into solution by the addition of 100 μL of 35% acetic acid to each well. Finally, the plates were read at $OD_{595}$ in the iMark Microplate Reader (Bio Rad).

### 2.5. DNA Manipulations and Transformations

Staphylococcal cells were lysed with recombinant lysostaphin (20 U/mL, Applied Micro, New York, NY, USA). Genomic DNA isolated by using QIAamp DNA Mini Blood Kit (Qiagen, Hilden, Germany) and plasmid DNA were prepared using the Qiagen Plasmid Mini kit (Qiagen). DNA fragments were isolated from agarose gels using the QIAEX II Gel extraction kit (Qiagen).

### 2.6. Construction of a srtA-Deficient Mutant

The method for the generation of a *srtA* lacking mutant was same as described before [7]. In brief, a primer set 1 (SrtA1FH, and SrtA1RE,) and primer set 2 (SrtA2FE, and SrtAl2RB,) were used to amplify PCR products of 809 and 793 bp, respectively, from genomic DNA of *S. lugdunensis* strain Sl48. The PCR products were ligated into the shuttle vector pBT9 and the ligation mixes were transformed into *E. coli* TG1 cells followed by incubation on LB plates containing ampicillin. A clone containing *srtA* gene as an insert was designated as pBT*srtA*. The restriction analysis and sequence data of clone pBT*srtA* confirmed *srtA* gene as an insert. A 1,479 bp *ermB* fragment was PCR-amplified using primers ermbF, and ermbR from the plasmid pEC4 containing the staphylococcal transposon Tn551 in a ClaI restriction site [28]. The *ermB* primers were designed from NCBI accession # AF239773. The *ermB* cassette was restricted with EcoRI and ligated into the EcoRI-restricted pBT*srtA*. The freshly prepared *E. coli* TG1 cells were transformed with the ligation mix and cultivated on ampicillin- and erythromycin-containing LB plates. Clones with the plasmid conferring resistance to both antibiotics were designated as pBT*srtA*E and were selected for further work. For the construction of the *srtA* allelic replacement mutant, protoplasts of *S. lugdunensis* strains Sl44 and Sl48 transformed with pBT*srtA*E as described by Palma et al. [29]. Chloramphenicol-sensitive and erythromycin-resistant colonies were detected by replica plating protocol onto plates containing chloramphenicol or erythromycin at 37 °C overnight. Clones that were sensitive for chloramphenicol and resistant to erythromycin were designated Mut7 (wild-type strain Sl44) or Mut47 (Wild-type strain Sl48) and were selected for further analyses. Correct insertion of the *ermB* was confirmed by using srtA1FH, srtA2RB, ermBF and ermBR primers.

## 2.7. The Complementation of a srtA-Deficient Mutant

A PCR product of *srtA* gene including the ribosomal binding site was PCR amplified using primer set (SrtA1FB and SrtA2RH) from genomic DNA of *S. lugdunensis* Sl48. The PCR product was restricted with HindIII and BamHI and ligated into the pCU1 plasmid. Freshly prepared *E. coli* TG1 cells were transformed with ligation mix. A plasmid containing *srtA* as an insert was designated pCU*srtA*. The isolated plasmid pCU*srtA* transformed by protoplast method into Mut7 and Mut47. Transformants were grown on TSA plates containing 10 μg of chloramphenicol and 10 μg of erythromycin per mL. Clones expressing SrtA were designated as Mut7C and Mut47C.

## 2.8. Cell Protein Preparations

Briefly, bacteria were grown overnight in BHI at 37 °C with shaking (160 rpm) and cells were harvested by centrifugation and washed twice with phosphate buffered saline. Then the cell surface proteins from PBS washed bacterial cells pellets were extracted by following methods: In the first method, bacterial cell surface proteins were generated by the LiCl method as described earlier [30]. In a second method we used the hydroxylamine hydrochloride method as described by Ton-That et al. [17]. In brief, bacteria were grown in two tubes containing 5 mL BHI overnight at 37 °C and 160 rpm. After centrifugation of the cells (4000 rpm 4 °C for 10 min), 1.5 mL of 50 mM Tris HCl pH 7.0 was added to the control pellet. To the other pellet we added 1.5 mL 100 mM hydroxylamine HCl in 50 mM Tris HCl pH 7.0. Both sets were stirred at 600 rpm at 37 °C in a heating block for 1 h. After centrifugation of the cells (13,000 rpm 4 °C for 10 min, the proteins in both supernatants were precipitated by addition of TCA to a final concentration of 10%. After centrifugation of the precipitated proteins (13,000 rpm 4 °C for 10 min), 200 μL of a 2× standard Laemmli SDS Page Sample buffer were added to the pellets and heated at 95 °C for 3 min. Samples were ready to load on SDS page gel. Whole cell lysates were prepared by the suspensions of cell pellets solved in 50 mM Tris/HCl pH 8.0 containing 15 mM NaCl. Lysostaphin (20 U/mL) and protease inhibitor cocktail (Sigma) were added and the mixture was incubated at 37 °C for 30 min. Then DNase was added to break DNA and, after centrifugation, liquid supernatants were used as whole cell lysate. For the generation of cell wall, cytoplasmic, and membrane proteins, *S. lugdunensis* were grown overnight in BHI at 37 °C with shaking (160 rpm) and cells were harvested by centrifugation and washed twice with phosphate buffered saline, and adjusted to an optical density at 600 nm ($OD_{600}$) of 20. A 1-mL portion of the bacterial suspension was pelleted and resuspended in 200 μL of digestion buffer (50 mM Tris-HCl, 20 mM $MgCl_2$, 30% [wt/vol] raffinose; pH 7.5) containing complete mini-EDTA-free protease inhibitors (Roche). Cell wall proteins were solubilized by digestion with lysostaphin (500 μg/mL) at 37 °C for 30 min. Protoplasts were harvested by centrifugation (5000× *g*, 15 min) and the supernatant was retained as the cell wall fraction. Protoplast pellets were washed once in digestion buffer, sedimented (5000× *g*, 15 min), and resuspended in ice-cold lysis buffer (50 mM Tris-HCl [pH 7.5]) containing protease inhibitors and DNase (80 μg/mL). Protoplasts were lysed on ice by vortexing. Complete lysis was confirmed by phase-contrast microscopy. The membrane fractions were obtained by centrifugation at 18,500× *g* for 1 h at 4 °C. The pellets (membrane fraction) were washed once and resuspended in ice-cold lysis buffer. Cell wall fractions and protoplast suspensions were centrifuged under the same conditions and the pellets were resuspended in 200 μL of lysis buffer. The liquid supernatant from protoplast suspension retained as cytosolic fraction.

## 2.9. SDS-PAGE and Ligand Overlay Analysis

The prepared cell surface proteins were separated in SDS-PAGE mini gel approaches. For Western ligand blot analysis, fibronectin (Fn) (Chemicon, Temecula, CA, USA), fibrinogen (Fg) (Calbiochem, San Diego, CA, USA), collagen type I (Cn) (Sigma; Sigma product #7774) or vitronectin (Vn) were purified by the method of Yatohgo et al. [28] and labeled with biotin (Roche, Mannheim, Germany). The cell surface proteins separated by SDS-PAGE were transferred electrophoretically (Trans-blot SD, Bio-Rad, Munich, Germany) onto nitrocellulose membranes (Schleicher and Schüll, Dassel, Germany) and were blocked with 3% BSA (bovine serum albumin fraction V, Sigma). The nitrocellulose membranes with blotted proteins were exposed to biotinylated ligands, treated with avidin alkaline phosphatase and subsequently bands were revealed using NBT (Nitrotetrazolium Blue chloride) and BCIP (5-Brom-4-chlor-3-indoxylphosphat) as recommended by the manufacturer´s protocol (Bio-Rad).

### 2.10. *Expression of Recombinant Sortase-A*

The *srtA* gene was PCR amplified from genomic DNA of *S. lugdunensis* using the primer set srtAF and srtAR. The *srtA* gene was ligated into the pQE30 expression vector and the vector was transformed into *E. coli* TG1 cells. *E. coli* TG1 cells were cultivated in LB medium plus 100 μg/mL ampicillin at 37 °C and 150 rpm and IPTG (Isopropyl-β-D-thiogalactopyranosid, Sigma) 1 mM per mL was added at $OD_{578}$ 0.5 to induce the expression of SrtA. *E. coli* cells were lysed by lysozyme (1 mg/mL lysis buffer) and SrtA was purified in a single step under native conditions using Ni-nitrilotriacetic acid (NINTA) resin column according to the manufacturer's recommendation (Qiagen).

### 2.11. *Polyclonal Antibodies*

Polyclonal antibodies against the recombinant SrtA were raised commercially (Genosphere Biotechnologies, Paris, France) in two rabbits by applying standard procedures of 70 days with 4 immunizations. The IgG fraction from the crude antiserum was obtained on protein-A column (Pierce, Rockford, IL, USA). In IgG preparations, naturally occurring anti-staphylococcal antibodies were removed by mixing the sera with 10 volumes of LiCl cell surface extracts from strain Mut47, which does not produce the SrtA, and immune-complexes were partially removed by centrifugation at 14,000 rpm at 4 °C for 15 min. For control Western immunoblotting, cell proteins of the wild type were separated on a SDS-Page, transferred onto a nitrocellulose membrane (Schleicher and Schuell, Dassel, Germany), and probed with anti-SrtA sera. Bound rabbit immunoglobulin G was detected with alkaline phosphatase-conjugated goat anti-rabbit immunoglobulins (Dako Diagnostika, Hamburg, Germany) with NBT/BCIP color reaction (Bio-Rad).

### 2.12. *ELISA Adherence assays*

The employed ELISA adherence assays of staphylococci to ECM proteins and host cells were done as described earlier [7].

### 2.13. *Cell Culture and Flow Cytometric Invasion Assay*

For flow cytometric invasion assay, three types of host cells were used (Table 1). The human endothelial cell line, Ea.hy926 [31], was grown in DMEM (Sigma-Aldrich) supplemented with 10% fetal calf serum (FCS).The fibroblasts (human fetal lung cells) were cultivated as described earlier [31,32]. Another epithelial cell line, the human bladder carcinoma cell line 5637 (ATCC HTB-9™), which secretes several functionally active cytokines, was also used in this study and maintained as described by Quentmeier et al. [33]. Fluorescein-5-isothiocyanate (FITC)-labeling of bacterial cells was performed as described elsewhere [34,35] and labeled cells were used within 24 h. The flow cytometric invasion assay was performed as described previously with minor modifications, such as the addition of propidium iodide just before samples were analyzed on a FacsCALIBUR™ (BD Bioscience) [36]. Results were normalized according to the mean fluorescence intensity of the respective bacterial preparation, as determined by flow cytometry. The invasiveness of the laboratory *S. aureus* strain Cowan I was set as 100% and the results are shown as means ± SEM of three independent experiments performed in duplicates.

## 2.14. Sortase Inhibitor PVS

5 mL BHI in a glass tube was inoculate with *S. lugdunensis* and grown overnight at 37 °C with 160 rpm. Next day, cultures were diluted to an $OD_{595}$ of 0.01. 180 µL of diluted culture was mixed with 20 µL of 10× concentrated phenyl vinyl sulfone (PVS) in a microtiter plate well. Plates were incubated at 37 °C for 18 h. Growth was determined in a Biorad reader at absorbance 595 nm. The lowest concentration that inhibited the cell growth was considered to be the MIC. For determination of effects of PVS on binding of bacteria to Fn and Fg, microtiter plate wells were coated with Fn and Fg separately. Overnight cultures were diluted to an $OD_{595}$ of 0.01. 180 µL of diluted culture were mixed with 20 µL of concentrated phenyl vinyl sulfone (PVS) to the respective concentrations in a microtiter plate well. Plates were incubated at room temperature for 1 h. Adherence was determined in a Biorad reader at absorbance of 595 nm.

## 2.15. Hydroxylaminolysis of LPXTG Peptide

Reactions were performed in 260 µL volume containing 50 mM Hepes buffer pH 7.5, 5 mM $CaCl_2$, and 10 µM LPETG fluorescent labeled peptides (DABCYL-LPETG-EDANS). The mixture was heated at 95 °C for 5 min. Then 10 µL of staphylococcal LiCl-extracts or 10 µM recombinant sortase-A, 5 µM of the sortase inhibitor p-hydroxymecuribenzoic acid (pHMB), or 10 mM DTT were added. Reactions were incubated for 1 h at 37 °C and then analyzed fluorometrically at 350 nm for excitations and 495 nm for recordings. Experiments were performed in triplicates.

## 2.16. Real-Time Reverse-Transcription PCR (qtRT-PCR)

For RNA isolation, *S. lugdunensis* was grown in BHI for 18 h at 37 °C with 160 rpm. After bacterial RNA isolation, real-time amplification and transcripts quantification was done as described earlier [37]. Primer sequences are given in Table 2.

**Table 2.** Primer sequences used in this study.

| Primer | Sequence (5′–3′) | Reference |
|---|---|---|
| srtA1FH | AAAAAGCTTTAAGAAAGCTAAAAAAATGACATAGTTG | This study |
| srtA1RE | AAAGAATTCCTCCAATAATGGTCATCAATTGGTTTGTCC | This study |
| srtA2FE | AAGAATTCTATTTATAGCAGAACAGATTAAATAATTGTAG | This study |
| srtA2RB | AAAGGATCCCATCTGAGTCAA GACTACTAGCAAGTGG | This study |
| Ery-EF, | ATATATCGATTAGGGACCTCTTTAGC | [28] |
| Ery-ER | ATATATCGATATCATGAGTATTGTCCG | [28] |
| SrtA1FH | AAAAAGCTTTAAGAAAGCTAAAAAAATGACATAGTTG | This study |
| SrtA2RB | AAAGGATCCCATCTGAGTCAAGACTACTAGCAAGTGG | This study |
| srtAF | CTCGGATCCAAACCTCATATTGATAGTTATTTACATGAC | This study |
| srtAR | CTCGGTACCTTATTTAATCTGTTCTGCTATAAATATTTTACGC | This study |
| RTFblF | GAAGCAACAACGCAGAACAA | [38] |
| RTFblR | TGCTTGTGCCTCGCTATTTA | [38] |
| RT16SF | CAGCTCGTGTCGTGAGATGT | [38] |
| RT16SR | TAGCACGTGTGTAGCCCAAA | [38] |
| RTvWbF | GGACCAGGTGAAGGTGATGT | This study |
| RTvWbR | GCCGCTGATTTTCGTGTAAT | This study |

### 2.17. Statistical Analysis

Using GraphPad Instat3, the statistical significance of the results was analyzed by ANOVA in combination with Bonferroni´s post test (compare all pairs of column or compare selected pairs of column) or Dunnett´s post test (compare all columns vs. control). Differences with $p$ values $\leq 0.05$ were considered as significant and are indicated with asterisks: * ($p \leq 0.05$), ** ($p \leq 0.01$), and *** ($p \leq 0.001$).

## 3. Results

### 3.1. Sortase A-Dependent Proteins

A comparative bioinformatic analysis of publicly available *S. lugdunensis* genomes identified the presence of only 11 sortase-A dependent proteins (Table 3) [39]. All 11 MSCRAMMs were found to be highly conserved in the *S. lugdunensis* strains as genomic data from strain HKU09-01, M23590, VCU139 and N920143 revealed [39]. However, there exist only minor differences in the number of repeats within the stalk regions that in turns affect the length of mature proteins [39].

**Table 3.** Known *S. lugdunensis* putative sortase-A mediated cell surface proteins with relevant properties [39].

| Genetic Identifiers (GN) | Annotation | Cleavage Motif | Size (aa) | Predicted Protein Size (kDa) | NCBI BLAST Hit (Protein [1], Strain [2], Length [3]) |
|---|---|---|---|---|---|
| SLUG_00890<br>SLGD_00061 | IsdB | LPATG | 690 | 76.9 | Surface protein SasI, HKU09-01 |
| SLUG_00930<br>SLGD_00065<br>SasE | IsdJ | LPNTG | 646 | 71.5 | Cell surface protein IsdA, HKU09-01<br>LPXTG cell wall surface anchor protein, M23590 |
| SLUG_02990<br>SLGD_00301 | SlsF | LPASG | 659 | 73.4 | Predicted cell-wall-anchored protein SasF, HKU09-01 |
| SLUG_03480<br>SLGD_00351 | SlsA | LPDTG | 1930 | 207.3 | Cell wall associated biofilm protein, HKU09-01, 3799 |
| SLUG_03490<br>SEVCU139_1800<br>SLGD_00352 | SlsD [4] | LPATG | 1619 | 175.8 | Putative serine-aspartate repeat protein F, VCU139, 2190<br>Putative uncharacterized protein, HKU09-01, 1136 |
| SLUG_03850<br>SLGD_00389<br>HMPREF0790_1688 | Slsc | LPETG | 190 | 21 | LPXTG protein, HKU09-01<br>Cell wall surface anchor family protein, M23590, 196 |
| SLUG_04710<br>SEVCU139_1680<br>SLGD_00473 | SlsE | LPETG | 3459 | 364 | Gram-positive signal peptide protein, VCU139, 2988<br>Hypothetical membrane protein, HKU09-01, 3232 |
| SLUG_04760<br>SLGD_00478 | SlsB | LPNTG | 277 | 30.6 | Putative uncharacterized protein, HKU09-01 |
| SLUG_22400<br>SLGD_02322<br>bca PE | SlsG | LPDTG | 2079 | 222.1 | Putative uncharacterized protein, HKU09-01, 2886<br>C protein alpha-antigen, VCU139, 2031 |

**Table 3.** *Cont.*

| Genetic Identifiers (GN) | Annotation | Cleavage Motif | Size (aa) | Predicted Protein Size (kDa) | NCBI BLAST Hit (Protein [1], Strain [2], Length [3]) |
|---|---|---|---|---|---|
| SLUG_16350<br>HMPREF0790_0533<br>SLGD_01633 | Fbl | LPKTG | 881 | 94.2 | Clumping factor A, M23590, 857<br>Clumping factor A (fragment), VCU139, 688<br>Methicillin-resistant surface protein, HKU09-01, 701 |
| SLUG_23290<br>SLGD_02429 | vWbF | LPETG | 1869 | 209.4 | Von Willebrand factor-binding protein, HKU09-01, 2194 |

[1] Different designation as given in annotation column. [2] Strain different from N920143. Annotation, cleavage motif, size (aa), and predicted protein size (kDa) columns are based on strain N920143. [3] Length if different from size (aa) column. [4] The *slsD* contains a nonsense codon located just 5′ to the region encoding LPXTG.

### 3.2. Alignment of Sortase A Sequences

In silico analyses identified homologs of *S. aureus srtA* in published genomes of *S. lugdunensis*. A nucleotide NCBI BLAST search identified SLGD_00472 (strain HKU09-01), *srtA* (strain N920143 and M23590), and SEVCU139_1681 (strain VCU139) as homologs of staphylococcal *srtA*. A ClustalW alignment of amino acid sequences was performed for pairwise comparison of SrtA between the four *S. lugdunensis* strains (N920143, HKU09-01, M23590, VCU139) and *S. aureus* SrtA. The SrtA sequences of four strains of *S. lugdunensis* (N920143, HKU09-01, M23590, VCU139) showed an intra-alignment score of 98–100%. The identity of SrtA sequences from four *S. lugdunensis* strains to *S. aureus* Newman was between 76–77% and revealed significant similarities. We identified conserved amino acid residues in the region corresponding to the calcium binding cleft of the *S. aureus* SrtA, such as the three glutamate residues Glu105, Glu108 and Glu166 as well as the aspartate residue Asp112, which is also highly conserved in the SrtA sequences of the *S. lugdunensis* strains (Figure 1) [40].

```
S.lug-N920143       MRKWTNQLMTIIGVILILVAIYLFAKPHIDSYLHDKDNSDKIENYDKTVSKANKNDK--- 57
S.lug-HKU09-01      MRKWTNQLMTIIGVILILVAIYLFAKPHIDSYLHDKDNSDKIENYDKTVSKANKNDK--- 57
S.lug-M23590        MRKWTNQLMTIIGVILILVAIYLFAKPHIDSYLHDKDNSDKIENYDKTVSKANKNDK--- 57
S.lug-VCU139        MKKWTNQLMTIIGVILILVAIYLFAKPHIDSYLHDKDNSDKIENYDKIVSKANKNDK--- 57
S.aur-Newman        MKKWTNRLMTIAGVVLILVAAYLFAKPHIDNYLHDKDKDEKIEQYDKNVKEQASKDKKQQ 60
                    *:****:**** **:***** *********.******:.:***:*** *.:  .:**

S.lug-N920143       --PTIPKDKAEMAGYLRIPDADINEPVYPGPATPEQLNRGVSFAEEQESLDDQNIAIAGH 115
S.lug-HKU09-01      --PTIPKDKAEMAGYLRIPDADINEPVYPGPATPEQLNRGVSFAEEQESLDDQNIAIAGH 115
S.lug-M23590        --PTIPKDKAEMAGYLRIPDADINEPVYPGPATPEQLNRGVSFAEEQESLDDQNIAIAGH 115
S.lug-VCU139        --PTIPKDKAEMAGYLRIPDADINEPVYPGPATPEQLNRGVSFAEEQESLDDQNIAIAGH 115
S.aur-Newman        AKPQIPKDKSKVAGYIEIPDADIKEPVYPGPATPEQLNRGVSFAEENESLDDQNISIAGH 120
                      * *****:::***:.******:**********************:********:****

S.lug-N920143       TYIGRPHYQFTNLKAAKKGSKVYFKVGNETREYKMTTIRDVNPDEIDVLDEHRGDKNRLT 175
S.lug-HKU09-01      TYIGRPHYQFTNLKAAKKGSKVYFKVGNETREYKMTTIRDVNPDEIDVLDEHRGDKNRLT 175
S.lug-M23590        TYIGRPHYQFTNLKAAKKGSKVYFKVGNETREYKMTTIRDVDPDEIDVLDEHRGDKNRLT 175
S.lug-VCU139        TYIGRPHYQFTNLKAAKKGSKVYFKVGNETREYKMTTIRDVNPDEIDVLDEHRGDKNRLT 175
S.aur-Newman        TFIDRPNYQFTNLKAAKKGSMVYFKVGNETRKYKMTSIRDVKPTDVGVLDEQKGKDKQLT 180
                    *:*.**:************* **********:****:****.* ::.****::*..::**

S.lug-N920143       LITCDDYNEKTGVWEKRKIFIAEQIK 201
S.lug-HKU09-01      LITCDDYNEKTGVWEKRKIFIAEQIK 201
S.lug-M23590        LITCDDYNEKTGVWEKRKIFIAEQIK 201
S.lug-VCU139        LITCDDYNEKTGVWEKRKIFIAEQIK 201
S.aur-Newman        LITCDDYNEKTGVWEKRKIFVATEVK 206
                    ********************:* ::*
```

**Figure 1.** Multiple sequence alignments of SrtA sequences of four strains of *S. lugdunensis* (N920143, HKU09-01, M23590, and VCU139) and *S. aureus* Newman using CLUSTAL 2.1. The highly conserved amino acids glutamate (E) and aspartate (D) involved in binding of calcium ions are highlighted in yellow. Stars denote highly conserved amino acids.

*3.3. Generation of Sortase-A Mutants*

To generate knockout mutants of *srtA*, we used the homologous recombination method leading to an insertion of *ermB* cassette into the *srtA* genes of *S. lugdunensis* strains Sl48 and Sl44 (Figure 2A). The *srtA* deficient mutants Mut7 and Mut47 were analyzed by PCR amplification using primers that anneal to *ermB* or to sequences flanking the *srtA* gene. The *ermB* gene could be amplified from the chromosomal DNA of the *srtA* deficient mutants Mut7 and Mut47, but not from the isogenic parent strains Sl48 and Sl44 respectively. Amplifications with primers specific for sequences flanking *srtA* revealed the insertion of *ermB* into the *srtA* genes of strains Mut7 and Mut47. The final double cross-over *srtA::ermB* allelic replacement in the Δ*srtA* mutant was confirmed by PCR amplification. Figure 2 shows the map of *srtA* as well as the positions of the primer pairs and the expected sizes of the amplified PCR products. The expected PCR product size with primers SrtA1F and SrtA2R was 2.2 Kb for the wild type and is 3.1 kb for the mutant. The observed sizes for the increased PCR products could only be generated from the Δ*srtA* mutants with primers SrtA1F and SrtA2R, but not with DNA of the wild type strains. In addition, amplifications of the *ermB* cassette were only successful with the Δ*srtA* mutant, but not with wild type strain DNAs using primers Ery-FE and Ery-RE which confirmed the allelic replacement of the wild type *srtA* gene by *srtA::ermB* (Figure 2). Results are exemplary shown for the Sl48 parental and Mut47 mutant strain.

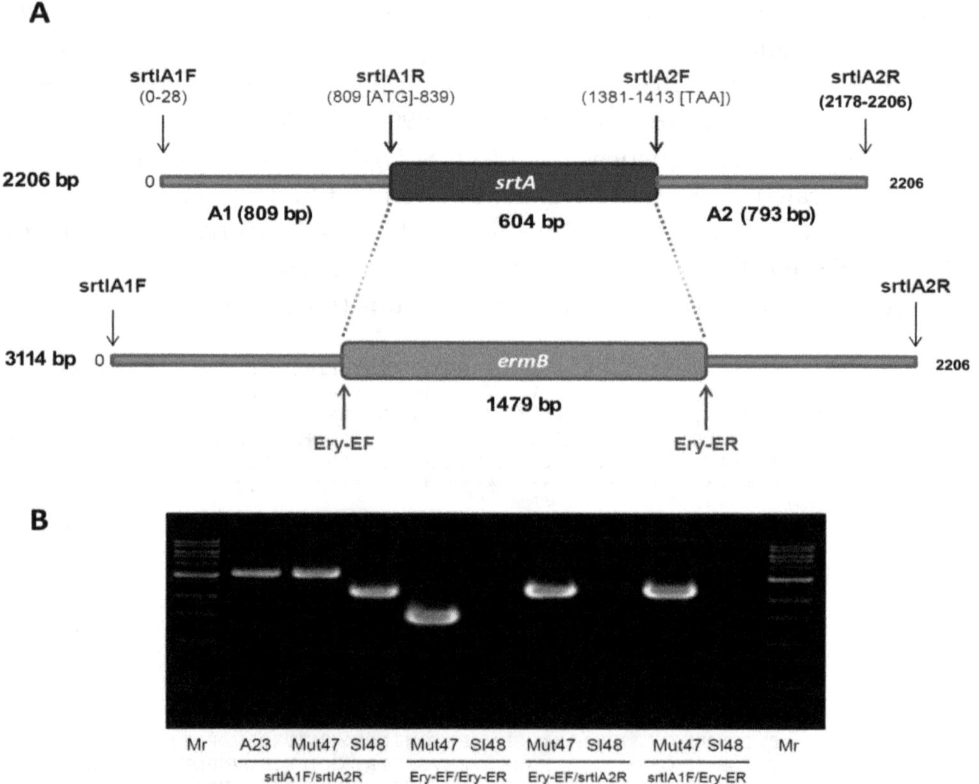


**Figure 2.** Genetic map of srtA and primer positions for PCRs, and results of PCR confirming the gene depletion of *srtA* in *S. lugdunensis*. (**A**) The gene loci of *srtA* in *S. lugdunensis* are shown before and after the recombination processes. In addition, the gene loci of *srtA* in *S. lugdunensis* are shown before and after the recombination processes and the positions of primer binding during the PCR experiments. The primer sequences are given in Table 3. (**B**) PCR products from genomic DNA of *S. lugdunensis* wild-type strain (Sl48), Δ*srtA* mutant strain (Mut47) and plasmid DNA of pBT*srtA* (A23) were separated by 1% agarose gels. Higher molecular bands (PCR with primer pair srtlA1F and srtlA2R) and the occurrence of corresponding bands to PCR experiments with primer pairs Ery-EF/Ery-ER, Ery-EF/srtlA2R and srtlA1F/Ery-ER confirming the deletion of *srtA* and following substitution by the *ermB* gene in the genome *S. lugdunensis* Δ*srtA* mutant strain Mut47.

### *3.4. Characterisation of the Surface Proteins by SDS Page and Western Blot Experiments*

First, we verified the absence of the sortase A protein in mutant strains and confirmed their complementation with plasmid coded *srtA*. Specific polyclonal antibodies rose in rabbits against recombinant sortase-A lacking the N-terminal signal peptide, revealed sortase-A as a polypeptide of 26 kDa in staphylococcal extracts. Immunoblotting with anti-SrtA antibodies failed to detect the SrtA in the whole cell extract from mutant strains, but complementation of the mutant strain with plasmid pCUSrtA encoded sortase-A restored the appearance of the sortase-A (Figure 3A). As a control, the transformation of the mutant strain with empty vector DNA did not affect the expression of the SrtA (data not shown). Together, these results confirmed the successful depletion of the *srtA* gene from the genome of *S. lugdunensis*. The SrtA is described as a membrane bound protein. Therefore, an attempt was made to localize SrtA within *S. lugdunensis* cells. For this purpose, *S. lugdunensis* cultivated in BHI broth medium was fractionated into the extracellular fraction, cell wall fraction, cytosol fraction and membrane digest. SrtA was detected in the membrane digest (Figure 3B). Strongest signal was observed for the membrane fractions. Only weak signals were found for the other protein fractions. It shows that SrtA of *S. lugdunensis* is also a membrane bound protein.

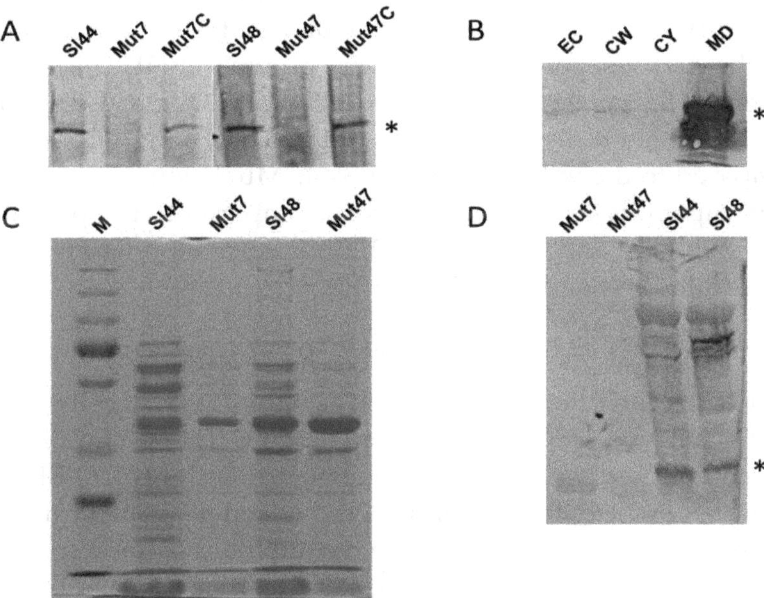


**Figure 3.** Western blot and SDS page experiments of *S. lugdunensis* wild type, Δ*srtA* mutant and complemented mutant strains showing the presence and localization of SrtA and consequences of a *srtA* deletion for the surface proteome pattern. (**A**) Western blots showing the absence of SrtA in the mutant strains but specific signals in wild types and complemented mutants. Whole cell extracts of wild type, mutant and complemented mutant strains were separated via SDS page and blotted. The Western blots were probed with specific polyclonal antibodies rose in rabbits against recombinant SrtA (anti-SrtA). Asterisk mark the specific SrtA band. (**B**) Localization of SrtA within *S. lugdunensis* cells. For this purpose, *S. lugdunensis* were grown in BHI broth medium. Cells were fractionated into the extracellular fraction (EC), cell wall fraction (CW), cytosol fraction (CY), and membrane digest fraction (MD). Proteins were separated by SDS page. After blotting, SrtA was detected in the different fractions with anti-SrtA antibodies. Asterisk mark the specific SrtA band. (**C**) Analyses of cell surface proteins of the wild-type strains and the mutant strains. The cell surface proteins were extracted in 0.1 M hydroxylamine hydrochloride. The protein extracts were separated on SDS-Page and stained with Coomassie Blue. (**D**) Cell surface protein fractions of wild type and mutant strains were separated via SDS page. After blotting, the Western blots were probed with a mixture of antibodies raised in rabbits against formaldehyde fixed whole cells of *S. lugdunensis*. Asterisk mark the specific SrtA band. M—page ruler, Sl44 and Sl48—wild type strains, Mut7 and Mut47—Δ*srtA* mutant strains, Mut7C and Mut47C—complemented Δ*srtA* mutant strains.

### 3.5. The Surface Proteomes of the Wild-Type and Mutant Strains Differ in the Absence of the SrtA

The cell surface protein expression patterns were determined by extracting the whole cells in 0.1 M hydroxylamine hydrochloride. In Coomassie Blue-stained SDS-PAGEs, the wild type and the mutant strains showed similar but non-identical patterns of proteins. In general, gels revealed less protein concentrations and protein bands for the mutants strains compared to the wild type isolates. In all strains, lower molecular mass proteins were detected in the both wild type and mutant strains, but higher molecular mass proteins were dominant in the wild type strains (Figure 3C). The results were confirmed by Western immunoblots probed with a mixture of antibodies raised in rabbits against formaldehyde fixed whole cells of *S. lugdunensis*. This experiment revealed a lack of bands for several proteins of the cell surface protein extracts of the Δ*srtA* mutant strains, although these bands were present in wild type strains (Figure 3D).

### 3.6. Determination of Growth and Biofilm Formation of the ΔsrtA Mutants

The parental strains of *S. lugdunensis* Sl44/Sl48 and Δ*srtA* mutants Mut7/Mut47 showed nearly the same growth rate for 24 h when cultivated in 100 mL of BHI in 500 mL flask (Figure 4A). Furthermore, in little cultivation volumes of only 5 mL in 14 mL glass tubes, no difference in growth could be observed (data not shown). Next, we determined the biofilm formation of all strains. *S. lugdunensis* wild type strains and the Δ*srtA* isogenic mutants were grown in BHI, TSB, and were supplemented with or without additional 0.5% (*w*/*v*) glucose. The addition of 0.5% glucose to TSB resulted in a strong biofilm formation but it was not observed in BHI with glucose. Mut7 produced 70% less biofilm in TSB plus glucose compared to the wild type strain Sl44. Mut47 showed only a decrease of around 25% biofilm formation capability in TSB plus glucose compared to the parental strain Sl48. In the complemented mutants Mut47C and Mut47C, the biofilm formation was completely restored to the levels of the wild-type strains Sl44 and Sl48 (Figure 4B).

### 3.7. Recombinant SrtA As Well As Cell Extracts Catalyzes Hydroxylaminolysis

To verify whether our recombinant SrtA as well as native SrtA of cell extracts still had catalytic activities, we incubated them with LPXTG fluorescent peptides (DABCYL-LPETG-EDANS) and $CaCl_2$. An increase in fluorescence is observed when cleavage of DABCYL-LPETG-EDANS peptide separates the fluorophore from Dabcyl which in turn confirmed the well described catalytic mechanism of the sortase. First, we tested the catalytic activity of the recombinant sortase SrtA from *S. lugdunensis* in this experimental setup. We observed an increase in fluorescence caused by SrtA which confirmed the cleavage of the DABCYL-LPETG-EDANS peptide (Figure 5A). As hydroxylaminolysis of LPXTG peptides depends on the sulfhydryl of the sortase, an addition of the known sortase inhibitor pHMB at least strongly abolished enzymatic activity of SrtA. However, the enzymatic activity was restored when DTT was added. We next tested cell lysates containing the native SrtA of the wild type strain Sl48, the mutant Mut47, the complemented mutant Mut47C and compare them with the recombinant SrtA in one experiment (Figure 5B). We observed an increase in fluorescence intensity when cell extracts of Sl48, Mut47C and the recombinant SrtA were incubated with the LPXTG peptide in presence of $CaCl_2$. As expected, only low activity was measured for cell extracts of the mutant strain Mut47 and the control (Figure 5B). Comparable results were found in cell extracts of the Sl44 wild type, the SrtA deficient mutant Mut7, and the Mut7C complemented mutant strains.

### 3.8. Agglutination Test and Adherence of ΔsrtA Mutants to Immobilized Fibronectin (Fn), Fibrinogen (Fg), and Vitronectin (Vn)

Clumping factor (ClfA) of *S. aureus* is a SrtA substrate and constitutes an important virulence factor. It binds to the C-terminus of the $\gamma$-chain of fibrinogen and allowed the adhesion to different eukaryotic cell types. Besides, the ClfA is used as an antigen for rapid diagnostic identification based on latex-agglutination test systems. *S. lugdunensis* also possesses a clumping factor (Fbl) with an amino acid identity of 58% to ClfA of *S. aureus*. Hence, we applied the Pastorex Staph Plus (Sanofi Diagnostic Pasteur, S.A., Marnes la Coquette, France) rapid agglutination test for simultaneous detection of the fibrinogen affinity antigen (clumping factor) combined with the detection of protein A and capsular polysaccharides of *S. aureus*. The WT strains showed positive agglutination reactions, but the Δ*srtA* strains did not reveal any agglutination (Figure 6A). It suggested the absence of Fbl in mutant strains, but we verified this result and measured the adherence of the Δ*srtA* mutants to different ECM proteins via ELISA assays. All strains were investigated for binding to immobilized Fn, Fg, and Vn. Compared to wild-type strain Sl44, the adherence of the Δ*srtA* mutant Mut7 was significantly decreased to immobilized Fg, Fn, and Vn (Figure 6B). The binding of complemented mutant Mut7C was indistinguishable to the parent level binding capacity. For the mutant strain Mut47, the results differed a little bit. We observed only a strong decrease in binding Fg, but the binding to Fn- and Vn-coated surfaces of Mut47 decreased less and not to a significant extent. However, binding of complemented mutant Mut47C to Fg, Fn and Vn was restored to the same extent as the wild type strain Sl48 (Figure 6B).

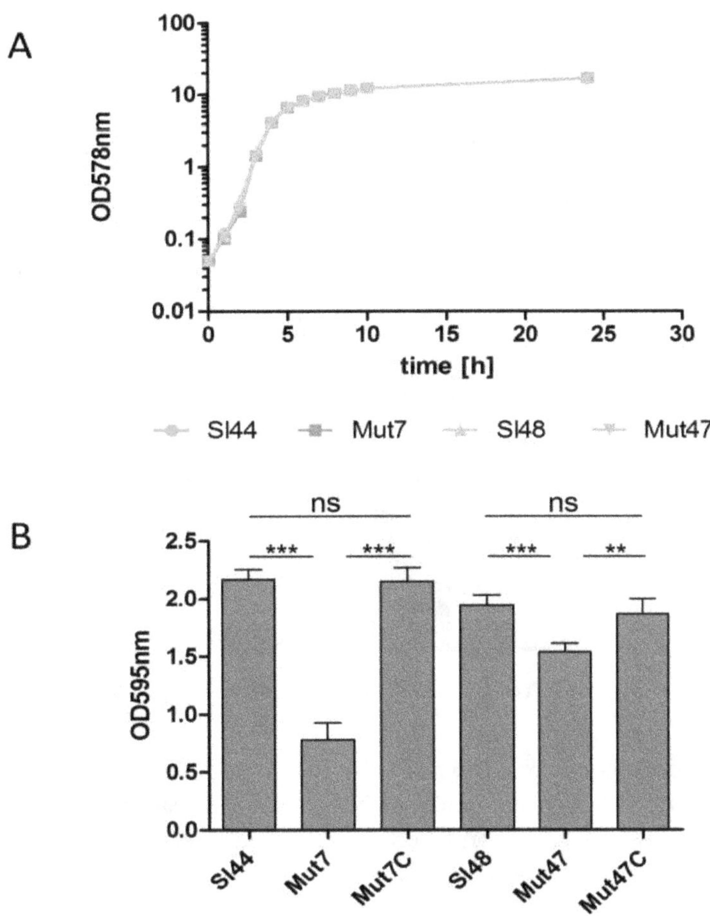


**Figure 4.** Growth and biofilm formation of *S. lugdunensis* wild-type strains (Sl44, Sl48) and Δ*srtA* mutant strains (Mut7, Mut47). (**A**) Bacterial growth was monitored for 24 h. Bacteria were cultivated in 100 mL BHI in 500 mL flask at 37 °C under permanent agitation. The experiment was done on triplicates. One representative experiment is shown. (**B**) Biofilm-forming capacities of the *S. lugdunensis* wild type (Sl44 and Sl48), Δ*srtA* mutant (Mut7 and Mut47) and complemented mutant (Mut7C and Mut47C) strains were assessed by a quantitative biofilm assay performed in microtiter plates applying crystal violet and determination of the $OD_{595nm}$. Results are shown as the mean of five independent experiments with the standard deviation (SD). Statistical analyses were performed using one-way ANOVA with Bonferroni multiple comparisons posttest (** $p < 0.01$; *** $p < 0.001$). ns—not significant.

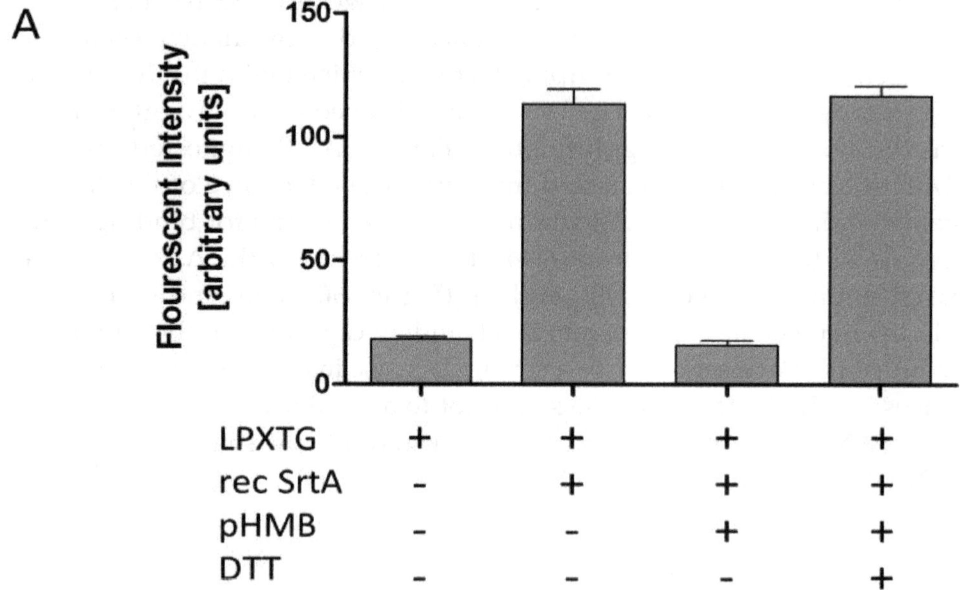


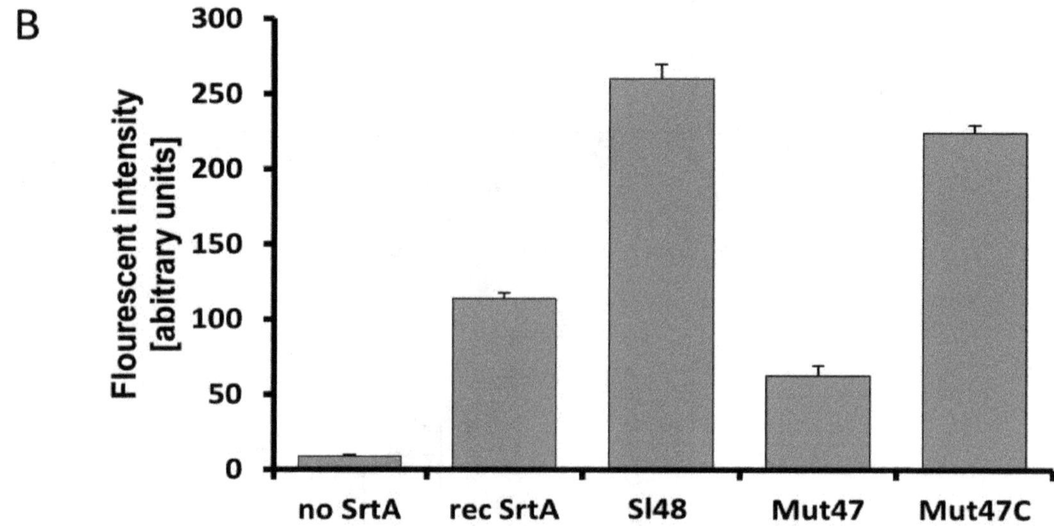


**Figure 5.** Hydroxylaminolysis of LPXTG peptide by recombinant SrtA (**A**) and whole *S. lugdunensis* cells (**B**). (**A**) Recombinant SrtA was incubated with the sorting substrate Dabcyl-QALPETGEE-Edans (LPXTG), and peptide cleavage was monitored as an increase in fluorescence. The reactions were influenced by the addition of pHMB (inhibitor of sortase activity) and by DTT. Presence and absence of the respective substance is shown as "+" or "-", respectively. (**B**) Same assay using the LPXTG substrate, recombinant SrtA, *S. lugdunensis* cell extracts from the wild-type strain Sl48, Δ*srtA* mutant Mut47 and complemented Mut47C. All results are shown as the mean of three independent experiments with the standard deviation (SD).

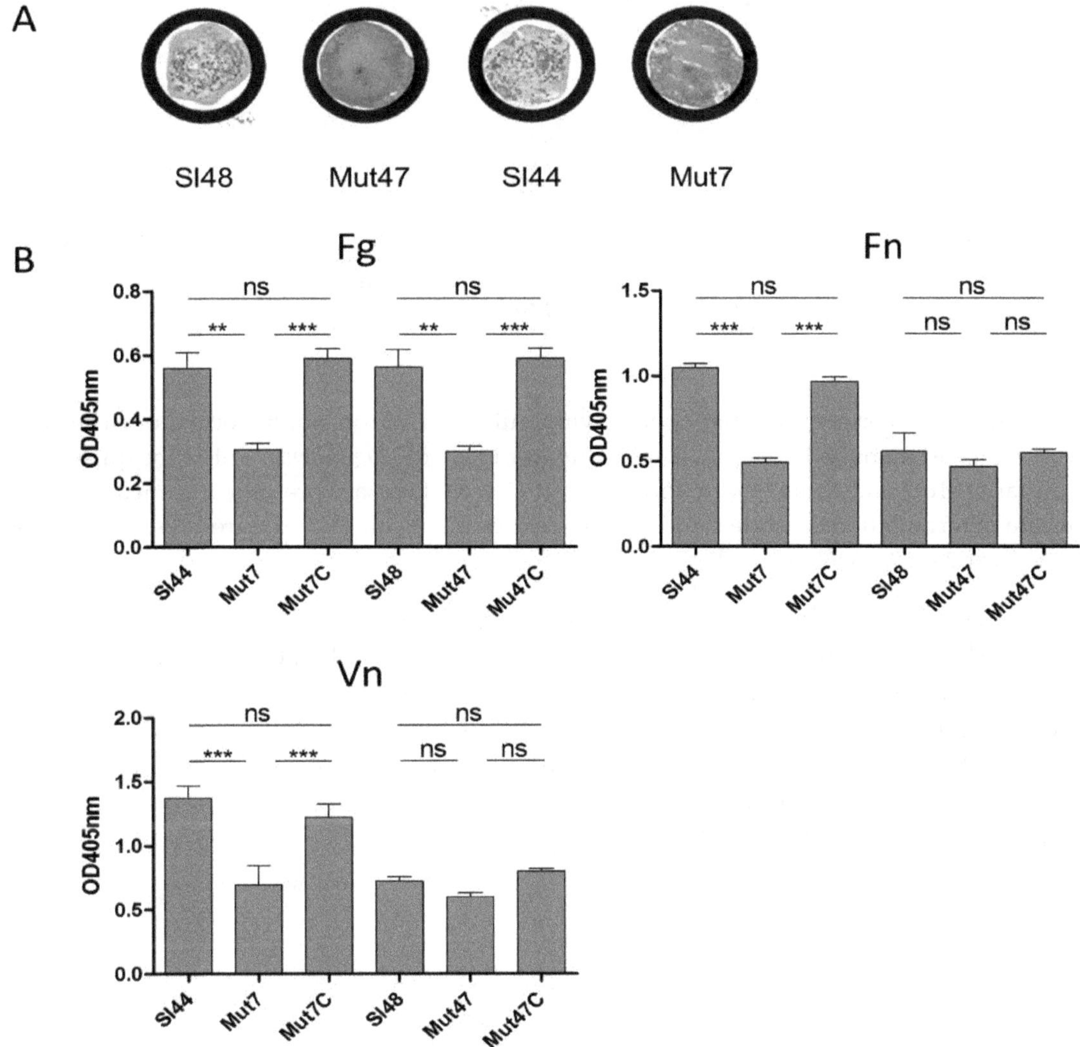


**Figure 6.** Pastorex staph plus agglutination test and binding of *S. lugdunensis* strains to ECM proteins. (**A**): Pastorex staph plus test of *S. lugdunensis* wild-type strains (Sl44, Sl48) and Δ*srtA* mutant strains (Mut7, Mut47) grown overnight on blood agar plates. Material was mixed in one drop of Pastorex reagent on a Pastorex disposable card. Results were recorded as positive on visual agglutination of wild-type strains and as negative for both mutants showing no agglutination. (**B**): Binding of *S. lugdunensis* wild-type strains (Sl44 and Sl48), Δ*srtA* mutants (Mut7 and Mut47), and complemented mutant strains (Mut7C and Mut47C) to immobilized fibrinogen (Fg), fibronectin (Fn), and vitronectin (Vn) assessed by ELISA adherence assays. Results are shown as the mean of four independent experiments with the standard deviation (SD). Statistical analyses were performed using one-way ANOVA with Bonferroni multiple comparisons posttest (** $p < 0.01$; *** $p < 0.001$). ns—not significant.

### *3.9. Sortase A Inhibition Resulted in Decreased Biofilm Formation and Binding to Fg and Fn*

The function of sortase A could be blocked by different classes of sortase A inhibitors like berberine chloride (BBCl), phenyl vinyl sulfone (PVS) and pHMB as shown before in Section 3.7 [41]. In our study, we used the well-known SrtA blocking reagent PVS. Our experiments with PVS showed MIC's of about 8-12 mM for strains Sl48, Sl44 and the corresponding Δ*srtA* mutants (Figure 7).

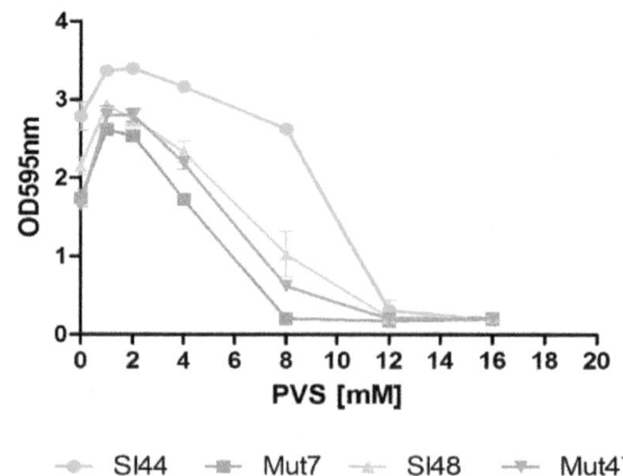


**Figure 7.** Effect of the sortase A inhibitor phenyl vinyl sulfone (PVS) on biofilm formation. The minimum inhibitory concentrations (MIC) of the sortase inhibitor PVS of *S. lugdunensis* wild type (SL44 and Sl48), Δ*srtA* mutant (Mut7 and Mut47) were found about 12 mM. Data are presented as mean adsorptions of triplicate determinations. Single representative experiments out of three are presented. Error bars show standard deviations (SD).

The sortase A inhibition by PVS appeared to be highly effective because at 8 mM concentration we observed a reduction in biofilm formation of strains Sl48 and Sl44, but also for the corresponding Δ*srtA* mutants. We further determined the effect of PVS on the binding capability of *S. lugdunensis* to Fn- and Fg-coated surfaces (Figure 8). A PVS treatment decreased the ability of *S. lugdunensis* to adhere to Fn- and Fg-coated surfaces. PVS at a concentration of 12 mM abolishes the binding to Fn of the parental strains Sl48 and Sl44 and the Δ*srtA* mutants Mut47 and Mut7. The effective PVS concentration to stop binding to Fg is about 16–20 mM (Figure 8).

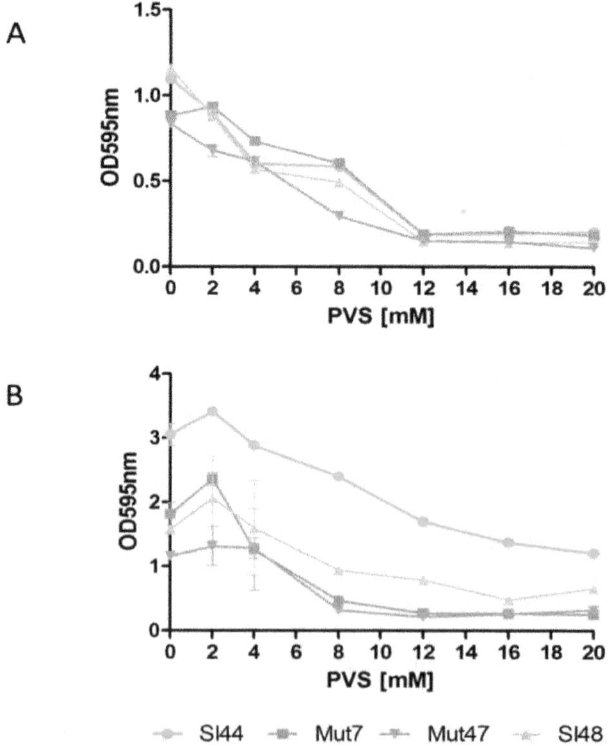


**Figure 8.** Effect of PVS on binding of *S. lugdunensis* to Fn- (**A**) or Fg- (**B**) coated wells. Overnight cultures were diluted to an $OD_{595}$ of 0.01. 180 μL of diluted culture were mixed with 20 μL of concentrated phenyl vinyl sulfone (PVS) to the respective concentrations in a microtiter plate well. Plates were incubated at room temperature for 1 h. Adherence was determined in a Biorad reader at absorbance of 595 nm. Data are presented as mean adsorptions of triplicate determinations. Single representative experiments out of three are presented. Error bars show standard deviations (SD).

### 3.10. Hydroxylamine HCl Treatment Decrease Binding to Immobilized Fg and Fn

A hydroxylamine treatment causes a formation of the C-terminal threonine hydroxamate of surface proteins, which are thereby released into the culture medium (17). Therefore, a hydroxylamine exposure of *S. lugdunensis* should result in a decreased adherence to Fg, Fn or other eukaryotic proteins. Here, we incubated different wild type strains (Sl44, Sl48, Sl253 and Sl241) with different concentrations of hydroxalamine HCL and found that 20 mM hydroxylamine HCl tendentially reduced the ability to adhere to Fn (Figure 9A). The reduction of Fg adherence was much lower than that observed for adherence to Fn (Figure 9B). In addition, we observed strain specific differences in binding Fg and Fn.

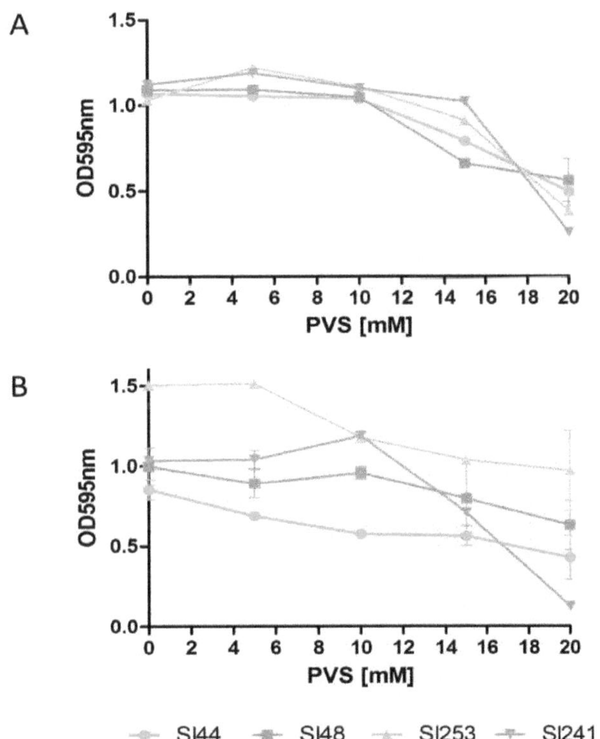


**Figure 9.** Effect of hydroxylamine HCl on binding of *S. lugdunensis* wild type strains to Fn (**A**) or Fg (**B**) coated wells. Overnight cultures were diluted to an $OD_{595}$ of 0.01. 180 μL of diluted culture were mixed with 20 μL of concentrated hydroxylamine hydrochloride to the respective concentrations in a microtiter plate well. Plates were incubated at room temperature for 1 h. Adherence was determined in a Biorad reader at absorbance of 595 nm. Data are presented as mean adsorptions of triplicate determinations. Single representative experiments out of three are presented. Error bars show standard deviations (SD).

### 3.11. Adherence and Invasion to Eucaryotic Cell Lines

LPXTG motif cell surface proteins are important for the adherence of staphylococci to eukaryotic cells. As shown above, Δ*srtA* mutants showed decreased adherence to different ECM proteins. Therefore we tested the adherence of the wild type, mutant, and complemented strains to three different eukaryotic cell lines. Surprisingly, the Δ*srtA* mutant Mut47 showed only slight reduced adherence to confluent epithelial cells, fibroblast, and 5637 cells. However, another Δ*srtA* mutant Mut7 bound more to the above mention three cell lines. However, these differences were not statistically significant (Figure 10).

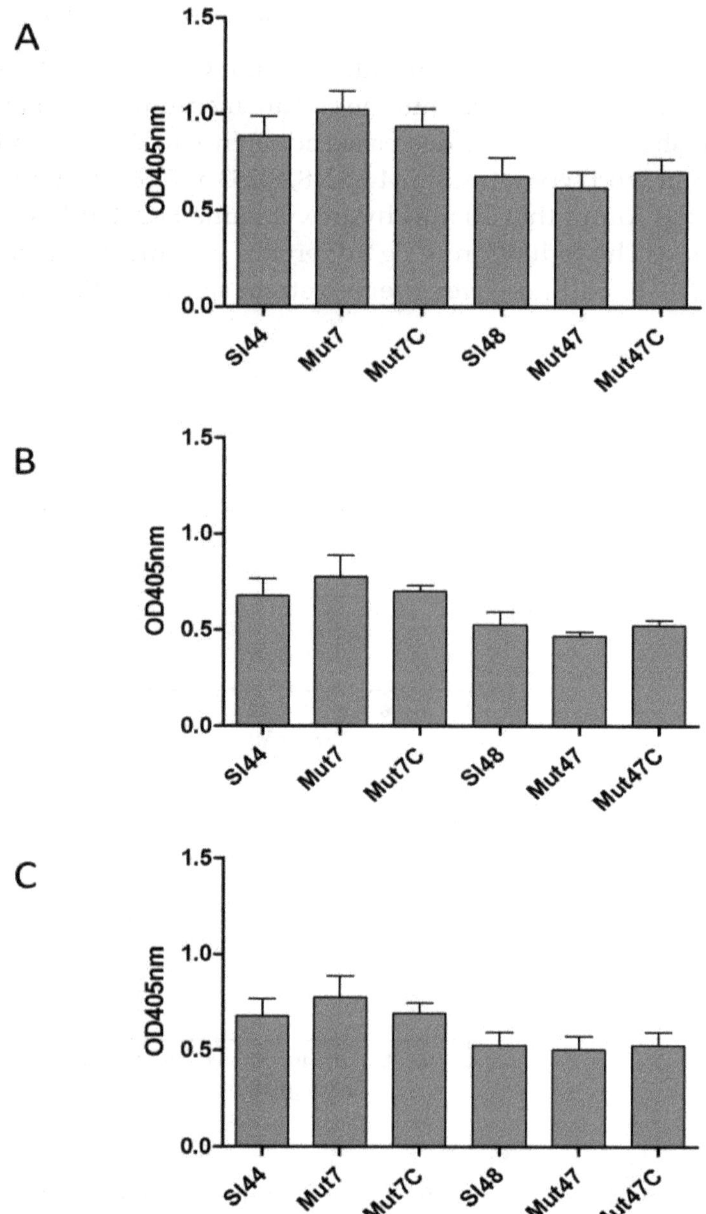


**Figure 10.** Adherence of *S. lugdunensis* wild-type (Sl44 and Sl48), Δ*srtA* mutant (Mut7 and Mut47) and complemented mutant (Mut7C and Mut47C) strains to immobilized human host cells ((**A**): endothelial cell line EA.hy926, (**B**): fetal lung A549 fibroblasts and (**C**): urinary bladder carcinoma cell line 5637 (ATCC HTB-9™)) assessed by ELISA adherence assays. Results are shown as the mean of at three independent experiments with the standard deviation (SD). Statistical analyses were performed using one-way ANOVA with Bonferroni multiple comparisons posttest, but the differences were not significant.

Next, we analyzed the invasion to the same eukaryotic cell lines. In experiments with FITC-labeled wild type and mutant strains, we observed a significant increase in invasion of both mutant strains to 5637 cells, Ea.hy926, and fibroblast cells (Figure 11). In addition, we observed that the Mut7 strain reached same invasion levels as the other mutant Mut47, but the wild type Sl44 revealed a comparable low capability to invade these cell lines. However, complemented strains showed the same invasion behavior as the respective wild type strains.

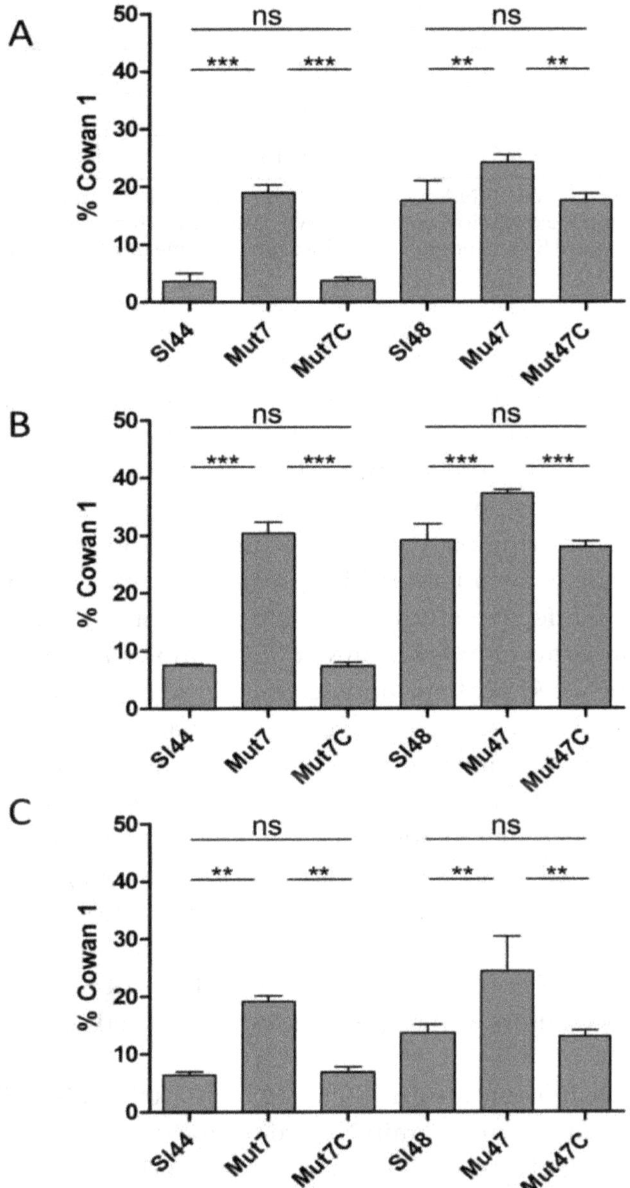


**Figure 11.** Internalization of FITC-labelled *S. lugdunensis* wild-type (Sl44 and Sl48), Δ*srtA* mutant (Mut7 and Mut47) and complemented mutant (Mut7C and Mut47C) strains by human host cells ((**A**): endothelial cell line EA.hy926, (**B**): fetal lung A549 fibroblasts and (**C**): urinary bladder carcinoma cell line 5637 (ATCC HTB-9™)) assessed by flow cytometry and computed in relation to *S. aureus* Cowan 1. Results are shown as the mean of three independent experiments with the standard deviation (SD). Statistical analyses were performed using one-way ANOVA with Bonferroni multiple comparisons posttest (** $p < 0.01$; *** $p < 0.001$). ns—not significant.

### *3.12. SrtA Influence the Gene Expression of Further Adhesins*

As shown for Δ*srtA* mutants of other species [42], we finally measured the influence of a *srtA* gene deletion on the gene expression of two further LPXTG proteins. Using real-time PCR, expression of specific mRNA for the LPXTG-bearing adhesin proteins Fbl and vWbF were compared in wild type strains Sl44, Sl48 and their respective Δ*srtA* mutants Mut7 and Mut47. The RT-PCR analyses of *fbl* and *vwbF* genes showed a modest up-regulation of about 2 fold in both Δ*srtA* mutant backgrounds (Table 4), verifying results observed for sortase mutants of other species [42].

**Table 4.** Real-time quantification of *fbl* and *vWbF*genes.

| Strain | fbl [a] | vWbF [a] |
|---|---|---|
| Mut47 vs. Sl48 | 1.88 ± 0.12 | 2.35 ± 0.25 |
| Mut7 vs. Sl44 | 1.99 ± 0.18 | 1.68 ± 0.12 |

[a] Relative levels of *fbl* and *vWbF* specific RNA in *S. lugdunensis srtA* deletion mutants were compared to wild-type strains. The fold changes in gene expression of *fbl* and *vWbF* are shown for the Δ*srtA* mutants Mut47 and Mut7 relative to wild type strains Sl48 and Sl44. Above given values represent mean ± SD of three independent experiments performed in triplicate.

## 4. Discussion

The LPXTG-anchored proteins are covalently anchored to the cell surface of many bacteria and play key roles as virulence factors for the establishment of bacterial infections [43]. The executive enzyme, the sortase, is essential for the functional assembly of cell surface virulence factors and, hence, important for the pathogenesis of staphylococci and in particular *S. aureus* [22,44]. Mutations in the *srtA* gene result in non-anchoring of 19 surface proteins to the cell envelope [23]. Consequently, *S. aureus* sortase mutants are defective in assembling surface proteins and are highly attenuated in the pathogenesis shown in animal infection studies [23]. However, only a few reports are available on the properties of *S. lugdunensis* in terms of interaction with ECM molecules or host tissues, as well as on its possession of the LPXTG-anchored proteins and secretable expanded repertoire adhesive molecules (SERAMs). *S. lugdunensis* strains were shown to bind to varying extents to collagen, fibronectin, vitronectin, laminin, fibrinogen, thrombospondin, plasminogen, and human IgG [45]. A comparative bioinformatic analysis of publicly available *S. lugdunensis* genomes identified the presence of 11 sortase A dependent proteins (Table 3) [39]. In the published genome sequence of *S. lugdunensis* strain N920143, Heilbronner et al. (2011) identified *fbl, vWbF, isdB, isdJ* and seven more genes coding for proteins with a LPXTG motif [39]. In the present study, we report the characterization of *S. lugdunensis* housekeeping SrtA and its role in LPXTG motif proteins' cell wall anchoring. To address the role of SrtA to adhere to different ECMs and cell lines, stable and defined Δ*srtA* mutants of *S. lugdunensis* were generated by allelic replacement as described before and respective complemented mutant strains were generated to exclude any polar effects [30].

The SrtA from four *S. lugdunensis* strains (N920143, HKU09-01, M23590, and VCU139) showed identity of 76-77% to the SrtA of *S. aureus* strain Newman, and a very high identity of 98-100% among themselves. Specific residues identified in the region corresponding to the described calcium binding cleft in the *S. aureus* SrtA were also highly conserved in the amino acid sequence of the *S. lugdunensis* SrtA. In our study, the recombinant SrtA as well as whole cell extracts of *S. lugdunensis* could catalyze the hydrolysis of a fluorescent quenched polypeptide carrying the LPXTG motif, which is characteristic of sortases of other species (Figure 5) [16,23]. In addition, it was observed that the *S. lugdunensis* SrtA activity also required calcium ions for hydroxyaminolysis of LPXTG peptide, as reported previously for SrtA of *S. aureus* [40]. Therefore, we propose a quite similar catalytic mechanism as shown for SrtA from *S. aureus* [16,17]. These assumption was approved by further phenotypic analyzes showing an essential role for SrtA in anchoring LPXTG proteins, as Δ*srtA* mutants differ in their cell surface proteome from that of the wild type strains, showing the absence of several proteins in the Δ*srtA* mutants (Figure 3). It's therefore not surprising that we found a reduced biofilm formation and a reduced adherence to Fn, Fg, and Vn in the Δ*srtA* mutants (Figures 4B and 6). Similar results were shown for *S. aureus* as the inhibition of SrtA activity caused loss of binding to Fn, Fg, and IgG and reduced biofilm formation [46,47]. Furthermore, these results are consistent with our observations that the Δ*srtA* mutants showed no agglutination in the Pastorex test.

The presence of an active sortase enzyme in *S. aureus* was found essential for the adherence to host eukaryotic cell [46]. In a previous study, the Δ*srtA* mutants of *S. aureus* and *Lactococcus salivarius* showed significant reductions in adhesion to different epithelial cell lines compared to the WT strain [48,49]. Here in this report, the Δ*srtA* mutant Mut47 showed only slightly reduced adherence to confluent epithelial cells, fibroblast, and 5637 cells, whereas another Δ*srtA* mutant Mut7 bound to above mention cell lines (Figure 10). Obviously, the differences were not statistically significant, which was surprising because formerly we observed significant lower adherences to Fn, Fg, and Vn. As shown previously, the autolysin AtlL of *S. lugdunensis* was identified as a major adhesin being crucially involved in the internalization process to cells [47]. Here, we showed a fairly strong imbalance in the proteome of the cell surface compartment in the mutant strains. These probably make AtlL more available to the cell surface, which might compensate for the loss of other LPXTG adhesins in case of Mut47 and eventually enhance, to a limited extent, the binding of Mut7 to eukaryotic cells. On the other hand, we cannot exclude the absence of the respective LPXTG counterpart proteins on the cell surfaces of the eukaryotic cell lines under our experimental conditions. This rather would lead to indistinguishable results between the adherence of wild types and Δ*srtA* mutant to cell lines, similar to what we have observed here.

We further observed that SrtA is not only involved in the covalent binding of LPXTG motif proteins to cell wall, but also influenced the transcriptional regulation of at least two adhesin genes *fbl* and *vWbF*. RT-PCR results showed that expression of *fbl* and *vWbI* genes were upregulated at least 2 fold in the Δ*srtA* mutants relative to their wild type strains (Table 4). This is in agreement with the previous finding that mutations in the *srtA* gene result in the upregulated expression of cell surface proteins in an oral *Streptococcus* strain [42]. Nobbs et al. (2007) reported a significant upregulation of the expression of adhesin genes like *sspA/B*, *cshA/B*, and *fbpA* in Δ*srtA* deficient mutants relative to their wild type strains of *S. gordonii* [50]. Moreover, Hall et al. (2019) recently described a quality control mechanism that monitors the processing of LPXTG adhesins by SrtA via measuring the left C-terminal cleaved LPXTG proteins (C-peps) which stayed membrane located and are recognized by a previously uncharacterized intramembrane two-component system (TCS) [51]. Prevention of C-peps generation de-repressed this TCS and resulted in increased expression of further adhesins compensating for the loss of LXPTG adhesins [51]. Since TCS is conserved in streptococci but not in staphylococci, the same C-pep-driven regulatory circuit is unlikely in *S. lugdunensis*, but similar regulation processes cannot be ruled out and might be an explanation for the upregulation of *fbl* and *vWbF*. Thus, the comparable high adherence to host cells of Δ*srtA* mutants might also be a result of the upregulation of other adhesins in the mutant strains and is object of future research. However, detailed in vitro infection studies are necessary to understand the impact of the single LPXTG proteins of *S. lugdunensis* to the respective adherence mechanisms.

A sudden inhibition of the SrtA activity could disrupt the pathogenesis of bacterial infections [22]. The surface-protein anchoring is sensitive to sulfhydryl-modifying reagents like methanethiosulfonates, berberine chloride, (Z)-3-(2,5-dimethoxyphenyl)-2-(4-methoxyphenyl)acrylonitrile or PVS under in vitro conditions [17,42,52]. Here, the SrtA inhibitor PVS reduced the biofilm formation and also reduced the adherence to Fg and Fn (Figures 7 and 8). Further, a hydroxylamine HCl treatment, releasing LPXTG proteins from the cell wall, was found to greatly reduce the adherence of four wild type strains of *S. lugdunensis* to immobilized Fg and Fn (Figure 9). This clearly showed the important role of the LPXTG proteins in *S. lugdunensis* for the binding to some ECM proteins, and confirmed the data observed for *S. aureus* [23,53]. The inhibition of the sortase activity may provide a novel approach for the treatment of infections with staphylococci and a development of such inhibitors could complement the current dependence on antibiotics. Therefore, SrtA is an attractive target to attenuate virulence and hamper infections. In addition, the Δ*srtA* mutant of *S. lugdunensis* showed significant defects in virulence including reduced bacteremia, reduced bacterial spreading to the kidneys and reduced size/density of endocardial vegetations [24]. This highlights the importance of LPXTG-anchored proteins in the pathogenesis of this germ. These and our results suggest that several surface proteins probably act in concert to promote the adhesion to different host structures and enable a survival in the host [24]. Sortase A mutants of *S. aureus* displayed strongly reduced virulence in various infection models including experimental sepsis, highlighting the impact of properly displayed surface proteins during infection [54,55]. Therefore, the Δ*srtA* mutants were recently discussed as vaccine candidates due to the attenuated but still immune-stimulating phenotype [56].

In conclusions, SrtA was found as the housekeeping sortase of *S. lugdunensis* like in other Gram-positive bacteria and in particular in *S. aureus*. Disrupting the presentation of surface proteins by gene deletion or blocking the activity of SrtA via different SrtA inhibitors could therefore effectively cause a reduction in binding to ECM proteins and could probably disrupt the pathogenesis of bacterial infections and promote bacterial clearance by the host. Although some strain-specific differences may exist, SrtA could be an attractive target to attenuate the virulence of *S. lugdunensis*.

**Author Contributions:** Conceptualization, M.H. and K.B.; methodology, M.H.; software, M.H.; validation, M.H., C.K. and K.B.; formal analysis, M.H.; investigation, M.H.; resources, M.H.; data curation, M.H.; writing—original draft preparation, M.H., C.K.; writing—review and editing, C.K., K.B.; visualization, M.H., C.K.; supervision, K.B.; project administration, M.H.; funding acquisition, K.B. All authors have read and agreed to the published version of the manuscript.

**Acknowledgments:** The paper is dedicated to Georg Peters (1951–2018).

## References

1. Becker, K.; Heilmann, C.; Peters, G. Coagulase-negative staphylococci. *Clin. Microbiol. Rev.* **2014**, *27*, 870–926. [CrossRef]
2. Sundqvist, M.; Bieber, L.; Smyth, R.; Kahlmeter, G. Detection and identification of *Staphylococcus lugdunensis* are not hampered by use of defibrinated horse blood in blood agar plates. *J. Clin. Microbiol* **2010**, *48*, 1987–1988. [CrossRef]
3. van der Mee-Marquet, N.; Achard, A.; Mereghetti, L.; Danton, A.; Minier, M.; Quentin, R. *Staphylococcus lugdunensis* infections: High frequency of inguinal area carriage. *J. Clin. Microbiol.* **2003**, *41*, 1404–1409. [CrossRef]
4. Kaspar, U.; Kriegeskorte, A.; Schubert, T.; Peters, G.; Rudack, C.; Pieper, D.H.; Wos-Oxley, M.; Becker, K. The culturome of the human nose habitats reveals individual bacterial fingerprint patterns. *Environ. Microbiol.* **2016**, *18*, 2130–2142. [CrossRef]
5. Frank, K.L.; Del Pozo, J.L.; Patel, R. From clinical microbiology to infection pathogenesis: How daring to be different works for *Staphylococcus lugdunensis*. *Clin. Microbiol. Rev.* **2008**, *21*, 111–133. [CrossRef]
6. Anguera, I.; Del, R.A.; Miro, J.M.; Matinez-Lacasa, X.; Marco, F.; Guma, J.R.; Quaglio, G.; Claramonte, X.; Moreno, A.; Mestres, C.A.; et al. *Staphylococcus lugdunensis* infective endocarditis: Description of 10 cases and analysis of native valve, prosthetic valve, and pacemaker lead endocarditis clinical profiles. *Heart* **2005**, *91*, e10. [CrossRef]
7. Hussain, M.; Steinbacher, T.; Peters, G.; Heilmann, C.; Becker, K. The adhesive properties of the *Staphylococcus lugdunensis* multifunctional autolysin AtlL and its role in biofilm formation and internalization. *Int. J. Med. Microbiol.* **2015**, *305*, 129–139. [CrossRef]
8. Marlinghaus, L.; Becker, K.; Korte, M.; Neumann, S.; Gatermann, S.G.; Szabados, F. Construction and characterization of three knockout mutants of the *fbl* gene in *Staphylococcus lugdunensis*. *APMIS* **2012**, *120*, 108–116. [CrossRef]
9. Nilsson, M.; Bjerketorp, J.; Wiebensjo, A.; Ljungh, A.; Frykberg, L.; Guss, B. A von Willebrand factor-binding protein from *Staphylococcus lugdunensis*. *FEMS Microbiol. Lett.* **2004**, *234*, 155–161. [CrossRef]
10. Nilsson, M.; Bjerketorp, J.; Guss, B.; Frykberg, L. A fibrinogen-binding protein of *Staphylococcus lugdunensis*. *FEMS Microbiol. Lett.* **2004**, *241*, 87–93. [CrossRef]
11. Donvito, B.; Etienne, J.; Denoroy, L.; Greenland, T.; Benito, Y.; Vandenesch, F. Synergistic hemolytic activity of *Staphylococcus lugdunensis* is mediated by three peptides encoded by a non-agr genetic locus. *Infect. Immun.* **1997**, *65*, 95–100. [CrossRef]
12. Heilmann, C. Adhesion mechanisms of staphylococci. *Adv. Exp. Med. Biol.* **2011**, *715*, 105–123. [CrossRef]
13. Foster, T.J.; Geoghegan, J.A.; Ganesh, V.K.; Hook, M. Adhesion, invasion and evasion: The many functions of the surface proteins of *Staphylococcus aureus*. *Nat. Rev. Microbiol.* **2014**, *12*, 49–62. [CrossRef]

14. Patti, J.M.; Höök, M. Microbial adhesins recognizing extracellular matrix macromolecules. *Curr. Opin. Cell. Biol.* **1994**, *6*, 752–758. [CrossRef]
15. Navarre, W.W.; Schneewind, O. Proteolytic cleavage and cell wall anchoring at the LPXTG motif of surface proteins in gram-positive bacteria. *Mol. Microbiol.* **1994**, *14*, 115–121. [CrossRef]
16. Mazmanian, S.K.; Liu, G.; Ton-That, H.; Schneewind, O. *Staphylococcus aureus* sortase, an enzyme that anchors surface proteins to the cell wall. *Science* **1999**, *285*, 760–763. [CrossRef]
17. Ton-That, H.; Liu, G.; Mazmanian, S.K.; Faull, K.F.; Schneewind, O. Purification and characterization of sortase, the transpeptidase that cleaves surface proteins of *Staphylococcus aureus* at the LPXTG motif. *Proc. Natl. Acad. Sci. USA* **1999**, *96*, 12424–12429. [CrossRef]
18. Perry, A.M.; Ton-That, H.; Mazmanian, S.K.; Schneewind, O. Anchoring of surface proteins to the cell wall of *Staphylococcus aureus*. III. Lipid II is an in vivo peptidoglycan substrate for sortase-catalyzed surface protein anchoring. *J. Biol. Chem.* **2002**, *277*, 16241–16248. [CrossRef]
19. Ruzin, A.; Severin, A.; Ritacco, F.; Tabei, K.; Singh, G.; Bradford, P.A.; Siegel, M.M.; Projan, S.J.; Shlaes, D.M. Further evidence that a cell wall precursor [C(55)-MurNAc-(peptide)-GlcNAc] serves as an acceptor in a sorting reaction. *J. Bacteriol.* **2002**, *184*, 2141–2147. [CrossRef]
20. Spirig, T.; Weiner, E.M.; Clubb, R.T. Sortase enzymes in Gram-positive bacteria. *Mol. Microbiol.* **2011**, *82*, 1044–1059. [CrossRef]
21. Oshida, T.; Sugai, M.; Komatsuzawa, H.; Hong, Y.M.; Suginaka, H.; Tomasz, A. A *Staphylococcus aureus* autolysin that has an N-acetylmuramoyl-L-alanine amidase domain and an endo-beta-N-acetylglucosaminidase domain: Cloning, sequence analysis, and characterization. *Proc. Natl. Acad. Sci. USA* **1995**, *92*, 285–289. [CrossRef]
22. Mazmanian, S.K.; Liu, G.; Jensen, E.R.; Lenoy, E.; Schneewind, O. *Staphylococcus aureus* sortase mutants defective in the display of surface proteins and in the pathogenesis of animal infections. *Proc. Natl. Acad. Sci. USA* **2000**, *97*, 5510–5515. [CrossRef]
23. Mazmanian, S.K.; Ton-That, H.; Su, K.; Schneewind, O. An iron-regulated sortase anchors a class of surface protein during *Staphylococcus aureus* pathogenesis. *Proc. Natl. Acad. Sci. USA* **2002**, *99*, 2293–2298. [CrossRef]
24. Heilbronner, S.; Hanses, F.; Monk, I.R.; Speziale, P.; Foster, T.J. Sortase A promotes virulence in experimental *Staphylococcus lugdunensis* endocarditis. *Microbiology* **2013**, *159*, 2141–2152. [CrossRef]
25. Edgell, C.J.; McDonald, C.C.; Graham, J.B. Permanent cell line expressing human factor VIII-related antigen established by hybridization. *Proc. Natl. Acad. Sci. USA* **1983**, *80*, 3734–3737. [CrossRef]
26. Giard, D.J.; Aaronson, S.A.; Todaro, G.J.; Arnstein, P.; Kersey, J.H.; Dosik, H.; Parks, W.P. In vitro cultivation of human tumors: Establishment of cell lines derived from a series of solid tumors. *J. Natl. Cancer Inst.* **1973**, *51*, 1417–1423. [CrossRef]
27. Fogh, J.; Wright, W.C.; Loveless, J.D. Absence of HeLa cell contamination in 169 cell lines derived from human tumors. *J. Natl. Cancer Inst.* **1977**, *58*, 209–214. [CrossRef]
28. Khan, S.A.; Novick, R.P. Terminal nucleotide sequences of Tn551, a transposon specifying erythromycin resistance in *Staphylococcus aureus*: Homology with Tn3. *Plasmid* **1980**, *4*, 148–154. [CrossRef]
29. Palma, M.; Nozohoor, S.; Schennings, T.; Heimdahl, A.; Flock, J.I. Lack of the extracellular 19-kilodalton fibrinogen-binding protein from *Staphylococcus aureus* decreases virulence in experimental wound infection. *Infect. Immun.* **1996**, *64*, 5284–5289. [CrossRef]
30. Hussain, M.; Becker, K.; von Eiff, C.; Schrenzel, J.; Peters, G.; Herrmann, M. Identification and characterization of a novel 38.5-kilodalton cell surface protein of *Staphylococcus aureus* with extended-spectrum binding activity for extracellular matrix and plasma proteins. *J. Bacteriol.* **2001**, *183*, 6778–6786. [CrossRef]
31. Tuchscherr, L.; Löffler, B.; Buzzola, F.R.; Sordelli, D.O. *Staphylococcus aureus* adaptation to the host and persistence: Role of loss of capsular polysaccharide expression. *Future Microbiol.* **2010**, *5*, 1823–1832. [CrossRef]
32. Hussain, M.; Haggar, A.; Heilmann, C.; Peters, G.; Flock, J.I.; Herrmann, M. Insertional inactivation of Eap in *Staphylococcus aureus* strain Newman confers reduced staphylococcal binding to fibroblasts. *Infect. Immun.* **2002**, *70*, 2933–2940. [CrossRef]
33. Quentmeier, H.; Zaborski, M.; Drexler, H.G. The human bladder carcinoma cell line 5637 constitutively secretes functional cytokines. *Leuk. Res.* **1997**, *21*, 343–350. [CrossRef]
34. Sinha, B.; Francois, P.; Que, Y.A.; Hussain, M.; Heilmann, C.; Moreillon, P.; Lew, D.; Krause, K.H.; Peters, G.; Herrmann, M. Heterologously expressed *Staphylococcus aureus* fibronectin-binding proteins are sufficient for invasion of host cells. *Infect. Immun.* **2000**, *68*, 6871–6878. [CrossRef]

35. Juuti, K.M.; Sinha, B.; Werbick, C.; Peters, G.; Kuusela, P.I. Reduced adherence and host cell invasion by methicillin-resistant *Staphylococcus aureus* expressing the surface protein Pls. *J. Infect. Dis.* **2004**, *189*, 1574–1584. [CrossRef]
36. Sinha, B.; Francois, P.P.; Nüße, O.; Foti, M.; Hartford, O.M.; Vaudaux, P.; Foster, T.J.; Lew, D.P.; Herrmann, M.; Krause, K.H. Fibronectin-binding protein acts as *Staphylococcus aureus* invasin via fibronectin bridging to integrin $\alpha_5\beta_1$. *Cell. Microbiol.* **1999**, *1*, 101–117. [CrossRef]
37. Hussain, M.; Schafer, D.; Juuti, K.M.; Peters, G.; Haslinger-Loffler, B.; Kuusela, P.I.; Sinha, B. Expression of Pls (plasmin sensitive) in *Staphylococcus aureus* negative for *pls* reduces adherence and cellular invasion and acts by steric hindrance. *J. Infect. Dis.* **2009**, *200*, 107–117. [CrossRef]
38. Pinsky, B.A.; Samson, D.; Ghafghaichi, L.; Baron, E.J.; Banaei, N. Comparison of real-time PCR and conventional biochemical methods for identification of *Staphylococcus lugdunensis*. *J. Clin. Microbiol.* **2009**, *47*, 3472–3477. [CrossRef]
39. Heilbronner, S.; Holden, M.T.; van Tonder, A.; Geoghegan, J.A.; Foster, T.J.; Parkhill, J.; Bentley, S.D. Genome sequence of *Staphylococcus lugdunensis* N920143 allows identification of putative colonization and virulence factors. *FEMS Microbiol. Lett.* **2011**, *322*, 60–67. [CrossRef]
40. Hirakawa, H.; Ishikawa, S.; Nagamune, T. Design of $Ca^{2+}$-independent *Staphylococcus aureus* sortase A mutants. *Biotechnol. Bioeng.* **2012**, *109*, 2955–2961. [CrossRef]
41. Kim, S.H.; Shin, D.S.; Oh, M.N.; Chung, S.C.; Lee, J.S.; Oh, K.B. Inhibition of the bacterial surface protein anchoring transpeptidase sortase by isoquinoline alkaloids. *Biosci. Biotechnol. Biochem.* **2004**, *68*, 421–424. [CrossRef] [PubMed]
42. Maresso, A.W.; Schneewind, O. Sortase as a target of anti-infective therapy. *Pharmacol. Rev.* **2008**, *60*, 128–141. [CrossRef] [PubMed]
43. Que, Y.A.; Francois, P.; Haefliger, J.A.; Entenza, J.M.; Vaudaux, P.; Moreillon, P. Reassessing the role of *Staphylococcus aureus* clumping factor and fibronectin-binding protein by expression in *Lactococcus lactis*. *Infect. Immun.* **2001**, *69*, 6296–6302. [CrossRef]
44. Jonsson, I.M.; Mazmanian, S.K.; Schneewind, O.; Verdrengh, M.; Bremell, T.; Tarkowski, A. On the role of *Staphylococcus aureus* sortase and sortase-catalyzed surface protein anchoring in murine septic arthritis. *J. Infect. Dis.* **2002**, *185*, 1417–1424. [CrossRef] [PubMed]
45. Paulsson, M.; Petersson, A.C.; Ljungh, A. Serum and tissue protein binding and cell surface properties of *Staphylococcus lugdunensis*. *J. Med. Microbiol.* **1993**, *38*, 96–102. [CrossRef]
46. Oh, K.B.; Oh, M.N.; Kim, J.G.; Shin, D.S.; Shin, J. Inhibition of sortase-mediated *Staphylococcus aureus* adhesion to fibronectin via fibronectin-binding protein by sortase inhibitors. *Appl. Microbiol. Biotechnol.* **2006**, *70*, 102–106. [CrossRef]
47. Sibbald, M.J.; Yang, X.M.; Tsompanidou, E.; Qu, D.; Hecker, M.; Becher, D.; Buist, G.; van Dijl, J.M. Partially overlapping substrate specificities of staphylococcal group A sortases. *Proteomics* **2012**, *12*, 3049–3062. [CrossRef]
48. Weidenmaier, C.; Kokai-Kun, J.F.; Kristian, S.A.; Chanturiya, T.; Kalbacher, H.; Gross, M.; Nicholson, G.; Neumeister, B.; Mond, J.J.; Peschel, A. Role of teichoic acids in *Staphylococcus aureus* nasal colonization, a major risk factor in nosocomial infections. *Nat. Med.* **2004**, *10*, 243–245. [CrossRef]
49. van Pijkeren, J.P.; Canchaya, C.; Ryan, K.A.; Li, Y.; Claesson, M.J.; Sheil, B.; Steidler, L.; O'Mahony, L.; Fitzgerald, G.F.; van Sinderen, S.; et al. Comparative and functional analysis of sortase-dependent proteins in the predicted secretome of *Lactobacillus salivarius* UCC118. *Appl. Environ. Microbiol.* **2006**, *72*, 4143–4153. [CrossRef]
50. Nobbs, A.H.; Vajna, R.M.; Johnson, J.R.; Zhang, Y.; Erlandsen, S.L.; Oli, M.W.; Kreth, J.; Brady, L.J.; Herzberg, M.C. Consequences of a sortase A mutation in *Streptococcus gordonii*. *Microbiology* **2007**, *153*, 4088–4097. [CrossRef]
51. Hall, J.W.; Lima, B.P.; Herbomel, G.G.; Gopinath, T.; McDonald, L.; Shyne, M.T.; Lee, J.K.; Kreth, J.; Ross, K.F.; Veglia, G.; et al. An intramembrane sensory circuit monitors sortase A-mediated processing of streptococcal adhesins. *Sci. Signal.* **2019**, *12*. [CrossRef] [PubMed]
52. Sudheesh, P.S.; Crane, S.; Cain, K.D.; Strom, M.S. Sortase inhibitor phenyl vinyl sulfone inhibits *Renibacterium salmoninarum* adherence and invasion of host cells. *Dis. Aquat. Organ.* **2007**, *78*, 115–127. [CrossRef] [PubMed]
53. Ton-That, H.; Schneewind, O. Anchor structure of staphylococcal surface proteins. IV. Inhibitors of the cell wall sorting reaction. *J. Biol. Chem.* **1999**, *274*, 24316–24320. [CrossRef] [PubMed]

54. Jonsson, I.M.; Mazmanian, S.K.; Schneewind, O.; Bremell, T.; Tarkowski, A. The role of *Staphylococcus aureus* sortase A and sortase B in murine arthritis. *Microbes Infect.* **2003**, *5*, 775–780. [CrossRef]
55. Weiss, W.J.; Lenoy, E.; Murphy, T.; Tardio, L.; Burgio, P.; Projan, S.J.; Schneewind, O.; Alksne, L. Effect of srtA and srtB gene expression on the virulence of *Staphylococcus aureus* in animal models of infection. *J. Antimicrob. Chemother.* **2004**, *53*, 480–486. [CrossRef] [PubMed]
56. Kim, H.K.; Kim, H.Y.; Schneewind, O.; Missiakas, D. Identifying protective antigens of *Staphylococcus aureus*, a pathogen that suppresses host immune responses. *FASEB J.* **2011**, *25*, 3605–3612. [CrossRef]

# 13

# Hepatic Stellate Cells and Hepatocytes as Liver Antigen-Presenting Cells during *B. abortus* Infection

**Paula Constanza Arriola Benitez [1], Ayelén Ivana Pesce Viglietti [1], María Mercedes Elizalde [2], Guillermo Hernán Giambartolomei [1], Jorge Fabián Quarleri [2,*] and María Victoria Delpino [1,*]**

1 Instituto de Inmunología, Genética y Metabolismo (INIGEM), Universidad de Buenos Aires, CONICET, Buenos Aires 1120, Argentina; constanza-arriola@hotmail.com (P.C.A.B.); ayelenpv@gmail.com (A.I.P.V.); ggiambart@ffyb.uba.ar (G.H.G.)

2 Instituto de Investigaciones Biomédicas en Retrovirus y Sida (INBIRS), Universidad de Buenos Aires, CONICET, Buenos Aires 1121, Argentina; mecheeli@hotmail.com

* Correspondence: quarleri@fmed.uba.ar (J.F.Q.); mdelpino@ffyb.uba.ar (M.V.D.)

**Abstract:** In Brucellosis, the role of hepatic stellate cells (HSCs) in the induction of liver fibrosis has been elucidated recently. Here, we study how the infection modulates the antigen-presenting capacity of LX-2 cells. *Brucella abortus* infection induces the upregulation of class II transactivator protein (CIITA) with concomitant MHC-I and -II expression in LX-2 cells in a manner that is independent from the expression of the type 4 secretion system (T4SS). In concordance, *B. abortus* infection increases the phagocytic ability of LX-2 cells and induces MHC-II-restricted antigen processing and presentation. In view of the ability of *B. abortus*-infected LX-2 cells to produce monocyte-attracting factors, we tested the capacity of culture supernatants from *B. abortus*-infected monocytes on MHC-I and –II expression in LX-2 cells. Culture supernatants from *B. abortus*-infected monocytes do not induce MHC-I and -II expression. However, these supernatants inhibit MHC-II expression induced by IFN-$\gamma$ in an IL-10 dependent mechanism. Since hepatocytes constitute the most abundant epithelial cell in the liver, experiments were conducted to determine the contribution of these cells in antigen presentation in the context of *B. abortus* infection. Our results indicated that *B. abortus*-infected hepatocytes have an increased MHC-I expression, but MHC-II levels remain at basal levels. Overall, *B. abortus* infection induces MHC-I and -II expression in LX-2 cells, increasing the antigen presentation. Nevertheless, this response could be modulated by resident or infiltrating monocytes/macrophages.

**Keywords:** *Brucella*; HSC; MHC; IL-10

---

## 1. Introduction

*Brucella* spp. are Gram-negative intracellular bacteria that infect domestic and natural animals and produce an incapacitating chronic disease when transmitted to humans. In many countries, brucellosis remains endemic. The most frequent clinical characteristics are hepatomegaly, splenomegaly and peripheral lymphadenopathy, revealing the preference of *Brucella* for the reticuloendothelial system [1,2].

As a frequent niche of infections, the liver provides a tolerogenic environment. Such immunotolerant capacity is based on the presence of a resident immune cell repertoire in constant stimulation and the hepatic blood source that spread a unique growth factor and cytokine milieu [3].

However, the immune system of the liver is capable of inducing a prompt-response to tumor cells and pathogenic microorganisms [4]. Thus, the majority of the microorganisms that arrive in the liver are eradicated. Nonetheless, even though these several mechanisms can remove infectious agents, *Brucella* spp. can escape the immune response and persist in the liver. Accordingly, in humans infected with *Brucella*, the liver is frequently implicated, with a frequency between 5% to 52% or more [5]. Liver biopsies from *Brucella abortus*-infected patients revealed the presence of granulomas with single of multiple localizations in portal and parenchymal tissue, inflammatory infiltrations, and parenchymal necrosis [6,7].

Among the non-parenchymal cells, hepatic stellate cells (HSCs) are placed among hepatocyte and small blood vessels. They are characterized by their contents of intracellular lipid droplets and protuberances that spread nearby the blood vessels. During liver injury, HSCs are activated and realize collagen with the development of scar tissue, producing chronic fibrosis or cirrhosis [8]. Furthermore, they also have a role in liver fibrosis to heal restore inflammatory injury.

During *B. abortus* infection, the protagonism of HSCs during the generation of fibrosis has recently been revealed [9]. Besides their function during liver damage through the production of fibrosis, HSCs cans also participate as local antigen-presenting cells (APCs). HSCs express MHC class I and II molecules, as well as co-stimulatory molecules such as CD40 and CD80 [10]. Accordingly, HSCs can interact with $CD4^+$ T cells to induce effector responses [11]. In addition, HSCs direct naïve $CD4^+$ T-cell activation to $T_{reg}$ differentiation in the presence of Dendritic cells (DC) [12]. Thus, the main role of HSCs is the ability to induce a tolerogenic liver milieu that can favor the chronicity of *B. abortus* infection.

Nucleated cells express MHC class I molecules, but MHC class II molecule expression is restricted for cell types such as dendritic cells, macrophages and B lymphocytes. MHC class II expression is regulated in part by the class II transactivator protein (CIITA) at the transcription level. The $\alpha$- and $\beta$-chains of newly synthesized class II molecules are associated with the invariant change (Ii), giving rise to immature MHC-II. These molecules reach the cell surface, then recycle to the endosomal/lysosomal compartment, named MIIC. In this compartment, cathepsin S is one of the proteases responsible in Ii processing to Class II-associated invariant chain peptide (CLIP) in human antigen-presenting cells. Ii removal is an important step for the adequate export of the peptide-loaded class II molecule to the cell surface. Activation of HSCs by several agonists such as bacterial lipopolysaccharide (LPS) and IFN-$\gamma$ drive the increase of MHC class II expression and co-stimulatory molecules [11]. Immune responses to liver pathogens need to consider the possibility that unconventional Antigen presenting cells (APC) play an important function, and may account for the miscarriage of effective immunity. Thus, the aim of this study is to characterize the induction of surface MHC-I and -II expression during *B. abortus* infection.

## 2. Results

### *2.1. B. abortus Infection Induces MHC-I and -II Expression in LX-2 Cells via a T4SS-Independent Mechanism*

In this section, the capacity of *B. abortus* to induce the expression of MHC-I and -II molecules on LX-2 cells is determined. Cells were infected with *B. abortus* for 2 h, washed to eliminate the bacteria, and the infection was continued for additional 72 h. Our results indicate that *B. abortus* infection stimulated MHC-I and -II expression in LX-2 cells, yielding a level comparable to that of IFN-$\gamma$-stimulated cells used as a positive control (Figure 1).

The type 4 secretion system (T4SS) encoded by the *vir*B operon participates in the establishment of the intracellular replication niche of *Brucella* in different cell types [13], as well as contributing to the induction of a fibrotic phenotype in HSCs during *B. abortus* infection. We decided to test whether MHC-I and -II expression in infected LX-2 cells depends on a functional T4SS. No significant difference in MHC-I and -II expression was found between LX-2 cells infected with the wild-type strain and those infected with *vir*B10 isogenic mutants, indicating that the T4SS is not implicated in the modulation of MHC-I and -II expression (Figure 1). These results indicate that *B. abortus* infection induces MHC-I and -II upregulation in a T4SS-independent mechanism.

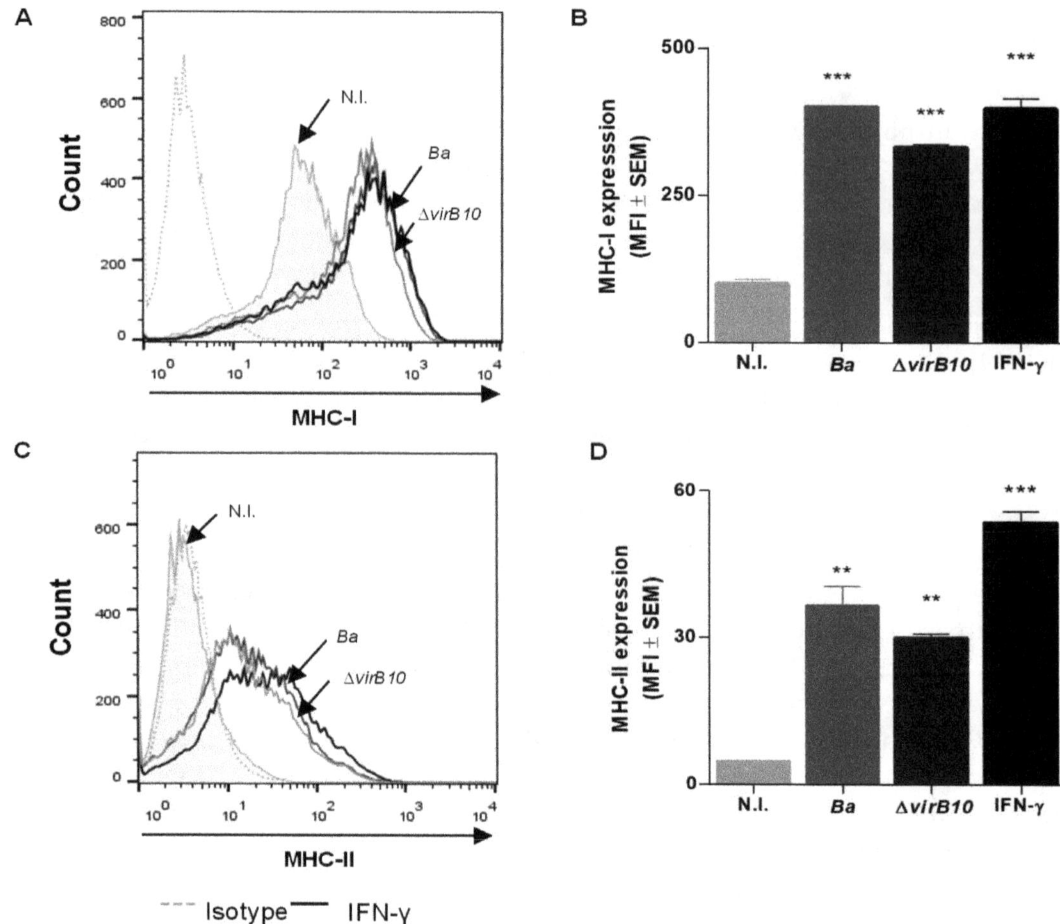


**Figure 1.** *Brucella abortus* infection induces MHC-I and -II expression in LX-2 cells. LX-2 cells were infected with *B. abortus* or *B. abortus virB10* mutant (Δ*virB10*) at a multiplicity of infection (MOI) of 1000 for 2 h, washed, and incubated for 72 h in complete media with antibiotics. IFN-γ (5oo U/mL) was used as a positive control. Non-infected cells (N.I.). MHC-I (**A**,**B**) and MHC-II (**C**,**D**) expression was assessed by flow cytometry. The histograms indicate the results of one representative of five independent experiments (**A**,**C**). The bars indicate the arithmetic means of five experiments, and the error bars indicate the standard errors of the means. MFI, mean fluorescence intensity (**B**,**D**). Non-specific binding was determined using a control isotype (Isotype). **, $p < 0.01$; ***, $p < 0.001$ versus non-infected cells (N.I.).

## 2.2. *B. abortus Infection Induce CIITA and Cathepsin S Transcription in LX-2 cells*

At transcription level, CIITA plays a key role in the MHC class II expression in professional antigen-presenting cells (APCs). Thus, experiments were conducted to evaluate whether increasing MHC-II expression correlated with increased transcription of CIITA. *B. abortus* infection induced up-regulation of CIITA mRNA after 72 h post-infection. IFN-γ was used as positive control (Figure 2A). The maturational processing of the MHC-II required the cleavage of li. The main protease involved in this process is cathepsin S [14]. Then, experiments were conducted to determine whether *B. abortus* infection was able to induce cathepsin S upregulation. *B. abortus* up-regulated cathepsin S mRNA at 72 h post-infection. IFN-γ was used as a positive control (Figure 2B). CIITA and cathepsin S primer specificity were determined by endpoint PCR (Figure 2C). These results indicate that *B. abortus* infection induces CIITA and cathepsin S transcription accordingly with the increase of MHC-II.

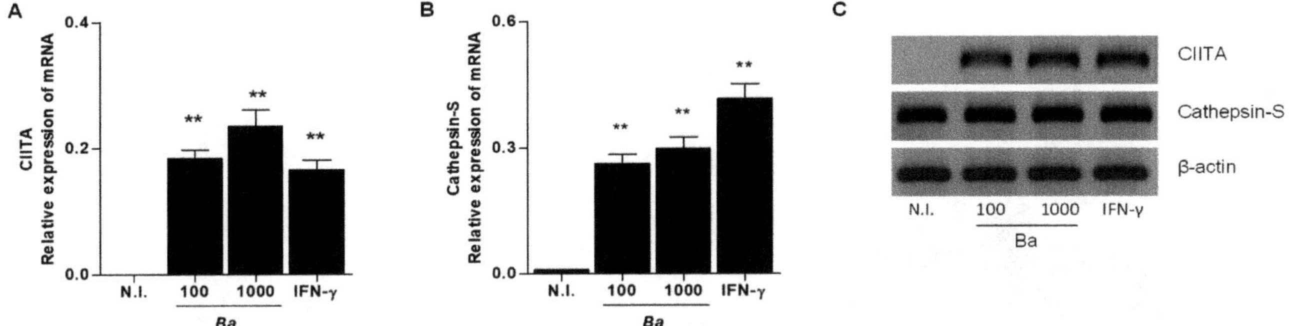


**Figure 2.** *B. abortus* infection induces Class II Major Histocompatibility Complex Transactivator (CIITA) and cathepsin-S expression in LX-2 cells. LX-2 cells were infected with *B. abortus* (*Ba*) at (MOIs) of 100 and 1000. At 72 h, post-infection levels of CIITA (**A**) and cathepsin-S (**B**) were determined by RT-qPCR. Agarose gel of PCR products from endpoint PCR products (**C**). Data are given as the means ± SD from three individual experiments. **, $p < 0.01$; ***, $p < 0.001$ versus non-infected cells (N.I.).

### *2.3. B. abortus Infection Does not Induce the Expression of Costimulatory Molecules CD80, CD86 and CD40*

For T cells, activation the recognition of antigen/MHC complex by the T cell receptor (TCR) must be complemented by a second signal that is provided by costimulatory molecules. Experiments were conducted to determine whether *B. abortus* infection could induce CD80, CD86 and CD40 expression on LX-2 cells. *B. abortus* was unable to induce costimulatory molecule expression measured at 72 h post-infection by using specific antibodies (not shown). These results indicated that even though *B. abortus* was able to induce MHC upregulation, the costimulatory molecules remained at basal levels.

### *2.4. B. abortus Increase the Phagocytic Capability of LX-2 Cells*

To determine whether *B. abortus* infection increase the phagocytic ability of LX-2 cells, cells were infected with *B. abortus* for 24 h then cultured with *Escherichia coli* for 30 min. Antibiotics were added to kill non-phagocyted *E. coli*. Counting of colony-forming unit (CFU) was performed to determine the phagocyted bacteria. *B. abortus*-infected LX-2 cells at an MOI of 1000 significantly increased phagocytosed *E. coli* in relation to uninfected cells (Figure 3A). We compared the phagocytic capacity of LX-2 cells with respect to the macrophage cell line J774.A1. Uninfected J774.A1 cells had an increased phagocytic capacity with respect to uninfected LX-2 cells. In addition, phagocytosed *E. coli* was significantly increased when J774.A1 cells were infected with *B. abortus* in an MOI-dependent fashion (Figure 3A). These results indicate that *B. abortus* infection increases the phagocytic capacity of LX-2 cells. However, its phagocytic capacity was lower than that observed in macrophages.

### *2.5. B. abortus Induces MHC-II-Restricted Antigen Processing and Presentation by LX-2 Cells*

To determine if the MHC-II upregulation promoted by *B. abortus* infection was related to changes in antigen processing and the presentation of soluble antigens for MHC-II-restricted T cells, LX-2 cells were infected for 72 h then incubated with Ag85B from *Mycobacterium tuberculosis* and DB1 T-cell hybridoma, which identify soluble Ag85B processed and presented by LX-2 cells (HLA-DR1). The infection with *B. abortus* significantly increased antigen processing and presentation at multiple Ag85B concentrations, as was revealed by the increased amount of IL-2 produced by T-cell hybridoma (Figure 3B). Thus, *B. abortus* infection induces processing and presentation of soluble antigens by LX-2 cells.

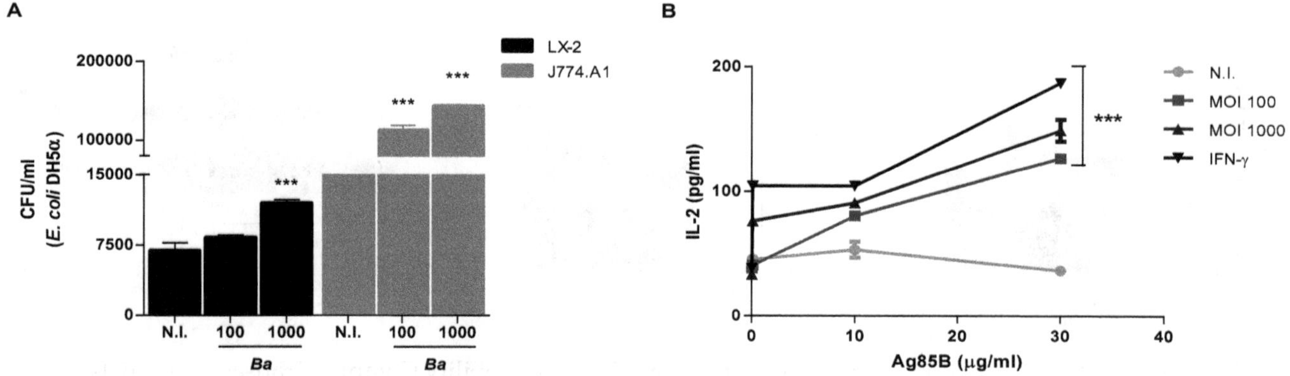


**Figure 3.** *B. abortus* increased the phagocytic capability and induced MHC-II-restricted processing and presentation in LX-2 cells. LX-2 cells were infected with *B. abortus* (*Ba*) at MOIs of 100 and 1000. After 72 h post-infection, *Escherichia coli* was added to the culture. Phagocytized *E. coli* were evaluated by intracellular colony-forming unit (CFU) counting (**A**); or after 72 post-infection, cells were pulsed with Ag85B for 6 h, followed by incubation with DB1 cells for 24 h. Supernatants were harvested and the amount of IL-2 was determined by ELISA (**B**). Data are given as the means ± SD from five individual experiments. ***, $p < 0.001$ versus non-infected cells (N.I.).

### *2.6. Culture Supernatants from B. abortus Infected THP-I Cells Do not Induce MHC-I and MHC-II Expression by LX2 Cells*

In view of the capacity of *B. abortus*-infected LX-2 cells to produce chemoattractant factors of a monocyte [9] that could attract monocytes to the site of infection, we evaluated whether supernatants from *B. abortus*-infected THP-1 cells were able to modulate MHC-I and -II in LX-2 cells. To this end, LX-2 cells were treated with a 1/2 dilution of supernatants from *B. abortus*-infected and uninfected monocytes over 72 h. IFN-γ was used as a positive control. Supernatants from *B. abortus*-infected monocytes did not alter the MHC-I and -II expression levels in LX-2 cells (Figure 4A–D).

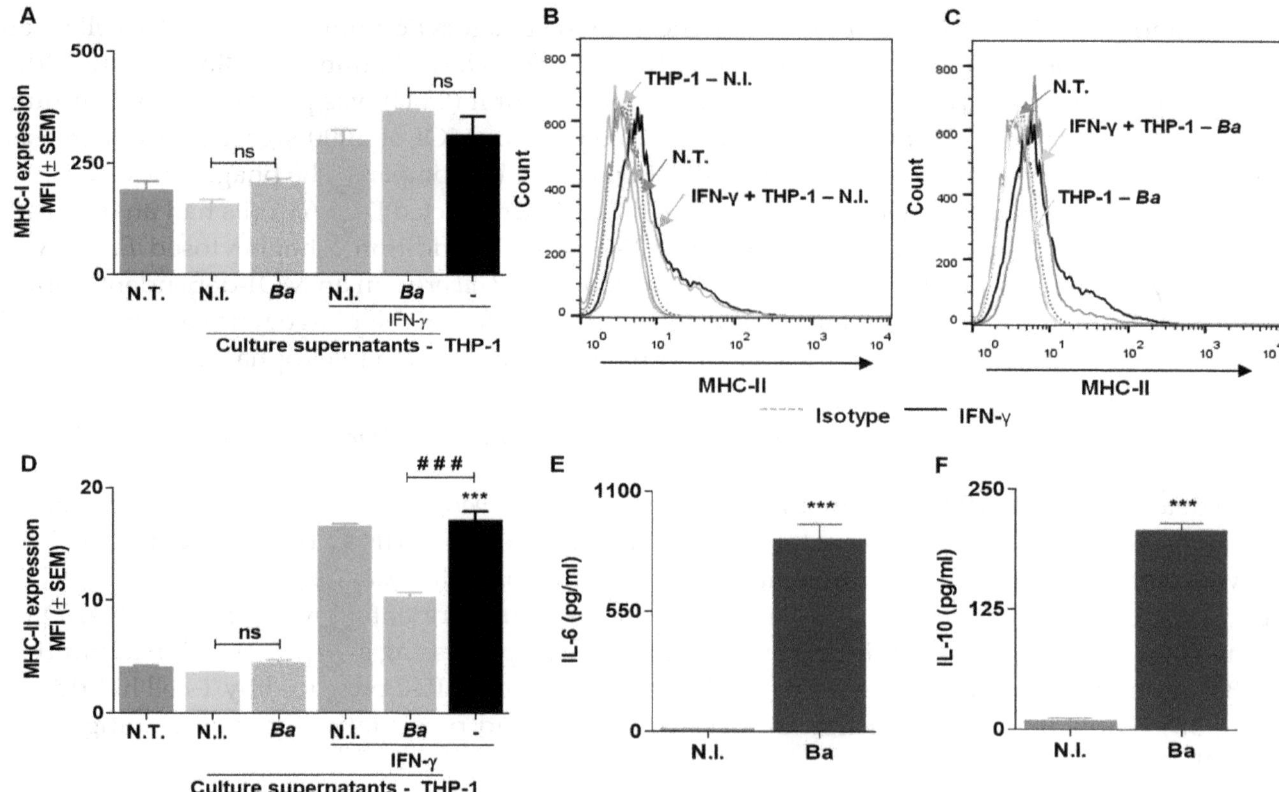


**Figure 4.** *Cont.*

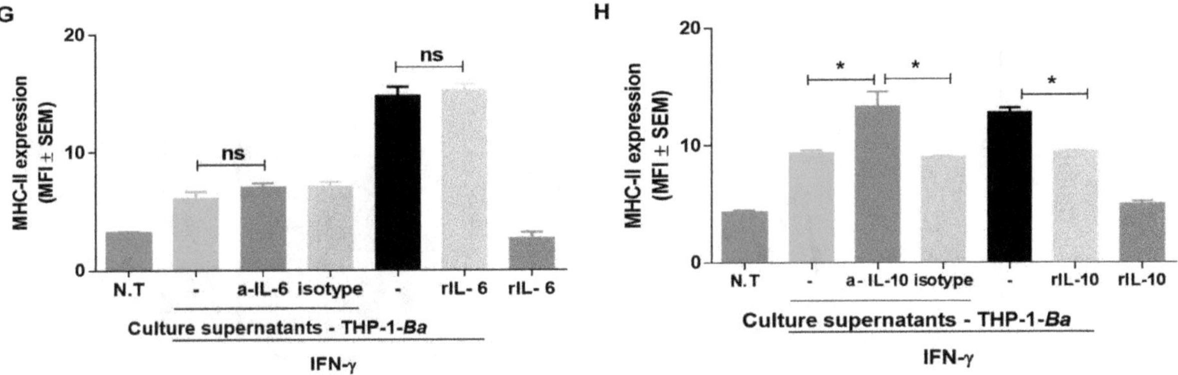


**Figure 4.** Modulation of MHC-I and -II expression in LX-2 cells by culture supernatants from *B. abortus*-infected monocytes. LX-2 cells were stimulated with culture supernatants from THP-1 cells infected at an MOI of 100 in the presence or not of IFN-$\gamma$ (500 U/mL) or culture supernatants from uninfected THP-1 cells as control at a 1/2 dilution. After 72 h post-stimulation, MHC-I (**A**), MHC-II (**B–D**), expression was assessed by flow cytometry. IL-6 and IL-10 were determined by ELISA in culture supernatants from *B. abortus*-infected THP-1 cells at a MOI of 100 (**E**,**F**). MHC-II expression in LX-2 treated with culture supernatants from infected THP-1 cells plus IFN-$\gamma$ was assayed in the presence of a neutralizing antibody anti-IL-6 (20 µg/mL), anti-IL-10 (20 ug/mL), or their isotype-matched control, with 10 ng/mL of recombinant human IL-6 (rIL-6) or 10 ng/mL of recombinant human IL-10 (rIL-10) alone or plus IFN-$\gamma$ used as a control (**G**,**H**). The histograms indicate the results of one representative of five independent experiments (**A–D**,**G**,**H**). The bars indicate the arithmetic means of five experiments, and the error bars indicate the standard errors of the means. MFI, mean fluorescence intensity (**A**,**D**). Non-specific binding was determined using a control isotype (Isotype). *, $p < 0.1$; **, $p < 0.01$; ***, $p < 0.001$ versus non-infected cells (N.I.) and cells stimulated with culture supernatants from uninfected THP-1 cells.

### *2.7. Culture Supernatants from B. abortus Infected THP-I Monocytes Inhibit MHC-II Expression Induced by IFN-$\gamma$ in an IL-10 Dependent Mechanism.*

After *B. abortus* infection of macrophages, their IFN-$\gamma$-induced expression of MHC-I and -II molecules is inhibited [15,16]. Here, we have demonstrated that the *B. abortus* infection of macrophages also hits the IFN-$\gamma$-induced MHC-II but not the MHC-I expression in HSC cells (Figure 4A–D). IL-6 and IL-10 were found to be involved in the inhibition of MHC-II in different cell types, including during *B. abortus* infection [15,17], and THP-1 cells were found to secrete IL-6 and IL-10 in response to *B. abortus* infection (Figure 4E,F). Moreover, when the IL-6 and IL-10 involvement was assessed using the neutralization of specific antibodies, IL-10 but not IL-6 participated in MHC-II downregulation in LX-2 cells. In addition, IL-10 present in culture supernatants from THP-1 cells was involved in the inhibition of MHC-II expression induced by IFN-$\gamma$, since a neutralizing antibody (anti IL-10) was able to reverse the inhibitory effect (Figure 4H). In contrast, recombinant IL-6 did not inhibit MHC-II expression induced by IFN-$\gamma$ in LX-2 cells (Figure 4G).

### *2.8. B. abortus Infection Induces MHC-I Expression in HepG2 Cells but Does not Alter MHC-II Levels*

Previously we have demonstrated that *B. abortus* infects and replicates in HepG2 hepatocytes [18]. These cells represent around 60% of liver mass, and do not express MHC-II molecules under physiological conditions. However, under inflammatory conditions, hepatocytes can express MHC-II molecules and also activate T cells [19]. Experiments were conducted to determine if *B. abortus* infection is capable of modulating MHC-II expression in HepG2 hepatocytes. To this end, HepG2 cells were infected with *B. abortus* and at 72 h post-infection, and the expression of MHC-I and -II molecules were measured. *B. abortus* infection was able to differentially induce MHC-I but not MHC-II expression (Figure 5). In addition, IFN-$\gamma$ was unable to induce MHC-II expression in HepG2 cells (Figure 5B).

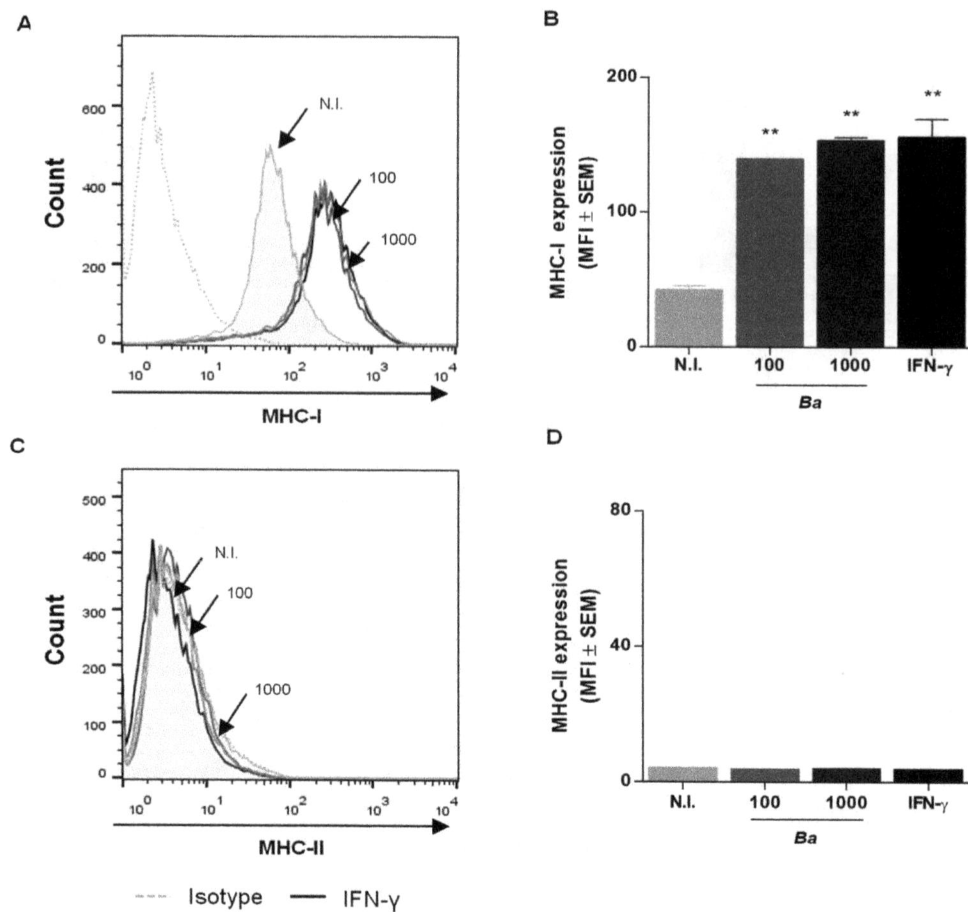


**Figure 5.** *B. abortus* infection induces MHC-I expression in HepG2 cells but does not alter MHC-II levels. HepG2 cells were infected with *B. abortus* at multiplicity of infections (MOIs) of 100 and 1000 for 2 h, washed, and incubated for 72 h in complete media with antibiotics. IFN-γ (500 U/mL) was used as a positive control. MHC-I (**A**,**B**) and MHC-II (**C**,**D**) expression were assessed by flow cytometry. The histograms indicate the results of one representative of five independent experiments (**A**,**C**). The bars indicate the arithmetic means of five experiments, and the error bars indicate the standard errors of the means. MFI, mean fluorescence intensity (**B**,**D**). Non-specific binding was determined using a control isotype (Isotype). **, $p < 0.01$ versus non-infected cells (N.I.).

## 3. Discussion

Most of the microorganisms that arrive in the liver are eliminated due to the balance between tolerance and inflammation of the hepatic microenvironment [3,20,21]. *Brucella* has a panoply of defensive strategies to evade immune response, including intracellular lifestyle and the prevention of the development of an appropriate adaptive immune response [22,23]. Thus, *Brucella* escapes from the immune response and persists in the liver, as demonstrated by the high frequency of liver pathology in human disease [5]. HSCs depict a pivotal function for wound healing of the liver [24]. Notwithstanding, the antigen-presenting capacity of HSCs has been previously reported in studies that revealed the expression of basal levels of costimulatory molecules and increases in MHC-II expression in response to IFN-γ [11,25,26]. In this study, we demonstrated the ability of *B. abortus* infection of HSCs (LX-2 cells) to upregulate MHC-I and -II expression, while the expression of the costimulatory molecules (CD80, CD86 and CD40) remained at basal levels.

The antigen presentation process involves recognition, uptake and processing by antigen-presenting cells. Previously, it has been described that the uptake of antigens by HSC is less effective than other professional APCs [27]. However, it is known that mature dendritic cells have a poor endocytic capacity, but effectively present antigens to T cells [28]. Nevertheless, *B. abortus* infection increased the efficiency of antigen uptake significantly via HSCs. Moreover, when these HSC-infected cells were cocultured with T cells, a higher level of IL-2 secretion was measured, thus inferring an increased antigen processing and further MHC-II-restricted T cell presentation after *B. abortus* infection. These results opposed other studies that have indicated that HSCs not only do not induce an effective T-cell response, but also induce the apoptosis of T cells through B7-H1 and B7-H4 signaling [26,29,30]. Such a discrepancy could be attributed to the fact that these studies eliminated the uptake, processing and presentation of antigens, since the T cell responses were performed by peptide pulsed-HSCs.

In B cells, thymus epithelial cells, and myeloid dendritic cells, CIITA is the master regulator of major histocompatibility complex (MHC) gene expression, which is constitutively expressed. However, in HSCs (among several cell types), the transcription of CIITA requires IFN-$\gamma$ among others factors for both MHC-II expression [31,32] and the modulation of the transcription of MHC-I genes [33–35]. Accordingly, when HSCs are infected by *B. abortus*, the expression of MHC-I and -II are upregulated.

Antigen processing and presentation require several lysosomal proteases, including cathepsin B, L, D, and S, which are involved in the maturation of MHC-II through the processing of Ii and the cleavage of antigen peptides that will be presented [36–39]. However, the most effective proteases involved in the last step of the Ii cleavage process are cathepsin S and L. Depending on the cell type, cathepsin L and S are involved in peptide degradation [39,40]. Recently, it has been shown that cathepsin S is expressed in HSCs, which can be induced by proinflammatory cytokines such as IFN-$\gamma$. This suggests a contribution to Ii processing. In contrast, cathepsin L expression has not been significantly increased at the transcription level upon stimulation with IFN-$\gamma$ [41], indicating that cathepsin S has a central role in antigen presentation in HSCs. In accordance with the increase in MHC-II expression, antigen processing, and presentation in MHC-II restricted T cells, *B. abortus* infection has also been able to induce cathepsin S mRNA transcription in HSCs.

The T4SS encoded by *virB* genes has been involved in the ability of *Brucella* to begin its intracellular replication niche [22]. In HSCs, we have previously reported that the T4SS is required to induce inflammasome activation and a fibrotic phenotype during *B. abortus* infection [42]. This system has been found to participate in the stimulation of inflammatory response during *B. abortus* infection both in vivo and in vitro [43]. However, our experiments using an isogenic a *B. abortus virB*10 polar mutant indicated that the T4SS was not involved in the induction of MHC-I and -II expression stimulated by *B. abortus* infection in HSCs.

The virulence of *B. abortus* relies on the ability of this bacteria to interact with macrophages as a central event for launching chronic *Brucella* infections [44,45]. In previous studies we have demonstrated that HSCs secrete MCP-1 in response to *B. abortus* infections [9], indicating that monocytes/macrophages could be attracted to the site of infection and, in conjunction with the resident macrophages, could modulate HSC responses. However, our results indicate that supernatants from *B. abortus*-infected macrophages were unable to induce MHC-I and -II expression.

*B. abortus* infection has been shown to potently activate a proinflammatory response that triggers the differentiation of T-cell responses to T-helper 1 (Th1) [46] with the simultaneous production of IFN-$\gamma$ [47]. This cytokine enhances not only microbicide activities of macrophages, but also antigen-presenting functions in cells [48]. However, *B. abortus* infection can stimulate not only inflammatory but also immunomodulatory mediators such as IL-10 and IL-6 through monocytes [49,50]. These cytokines have been reported as responsible for inhibiting IFN-$\gamma$-induced MHC-II expression in immune cells [51,52]. Our experiments demonstrate that during the *B. abortus* infection of HSCs, IL-10 but not IL-6 present in supernatants from *B. abortus*-infected monocytes was implicated, at least in part, in the inhibition of IFN-$\gamma$-induced MHC-II expression.

*B. abortus* infection can infect and replicate in hepatocytes, inducing an inflammatory response [18]. Here, we demonstrate that in the setting of *B. abortus* infection, the MHC-I but not the MHC-II expression was induced in hepatocytes, thus enabling the hepatocytes to be susceptible to CD8+ cytotoxic T cell action.

In conclusion, the *B. abortus* infection of hepatic stellate cells and hepatocytes is able to regulate differentially the MHC expression, thus stimulating the T-cell specific-immune response at the liver. However, due to a cellular interplay, such responses may also be modified by resident or infiltrating *B. abortus*-infected monocytes/macrophages. Such bacterial skills exerted on hepatic cells may promote the evasion of immune surveillance, thus favoring its chronicity in the liver.

## 4. Materials and Methods

### 4.1. Bacterial Culture

*Brucella abortus* S2308 or the isogenic *B. abortus vir*B10 polar mutant (kindly provided by Diego Comerci, UNSAM University, Argentina) were cultivated in 10 ml of tryptic soy broth (Merck, Buenos Aires, Argentina) for 18 h with constant agitation at 37 °C. Bacteria were harvested and the inoculum were prepared as described previously [53]. All experiments with live *Brucella* were carried out in biosafety level 3 facilities located at the Instituto de Investigaciones Biomédicas en Retrovirus y SIDA (INBIRS).

### 4.2. Cell Culture

The spontaneously immortalized human hepatic stellate cell line (LX-2) was kindly provided by Dr. Scott L. Friedman (Mount Sinai School of Medicine, New York, NY, USA). LX-2 cells were maintained in Dulbecco's Modified Eagle Medium (DMEM, Life Technologies, Grand Island, NY, USA) and supplemented with 5% fetal bovine serum (FBS; Life Technologies), L-glutamine (2 mM), sodium pyruvate (1 mM), 100 U/mL penicillin, and 100 μg/mL streptomycin. The human hepatoma cell line HepG2, the murine J774.A1 cell line, and the human monocytic cell line THP-1 were obtained from the ATCC (Manassas, VA, USA) and were cultured as previously described [18]. Monocyte differentiation from THP-1 cells was achieved through cultivation in the presence of 0.05 mmol/L 1, 25-dihydroxyvitamin D3 (Calbiochem-Nova Biochem International, La Jolla, CA, USA) for 72 h. DB1 T hybridoma cells (Ag85B specific) was kindly provided by W. H. Boom (Case Western Reserve University, Cleveland, OH, USA) and was maintained in DMEM supplemented as indicated above. All cultures were grown at 37 °C and 5% $CO_2$.

### 4.3. Cellular Infection

LX-2 cells were dispensed in 24-well plates and infected with *B. abortus* S2308 or *B. abortus virB10* polar mutant at a multiplicity of infection (MOI) of 100 or 1000. HepG2 cells were infected with *B. abortus* S2308 at an MOI of 100 or 1000, and THP-1 cells at an MOI of 100. After the bacterial suspension was dispensed, the plates were centrifuged for 10 min at 2000 rpm, then incubated for 2 h at 37 °C under a 5% $CO_2$ atmosphere. To remove extracellular bacteria, Cells were extensively washed with DMEM then incubated in medium supplemented with 100 μg/mL gentamicin and 50 μg/mL streptomycin to kill extracellular bacteria. LX-2 cells were harvested at 72 h to determine major histocompatibility complex class I (MHC-I), MHC-II, CD40, CD80, and CD86 surface expression and CIITA and cathepsin-S gene expression. Supernatants from THP-1 cells were harvested 24 h after infection to be used as conditioned medium.

### 4.4. Flow Cytometry

Infected LX-2 cells, cells treated with culture supernatants at a 1/2 dilution from THP-1 cells, or recombinant human IFN-$\gamma$-treated-LX-2 cells (500 U/mL; Endogen) were washed and incubated with fluorescein isothiocyanate-labeled (FITC) anti-human HLA-DR monoclonal antibody (MAb) (clone L243; BD Bioscience, San Diego, CA, USA), FITC-labeled anti-human HLA-ABC (clone G46-2.6; BD Bioscience), phycoerythrin (PE)-labeled anti-human CD40 (clone 5C3; BD Bioscience), PE-labeled anti-human CD86 (clone 2331(FUN-1); BD Bioscience) FITC-labeled anti-human CD80 (clone 2D10; BioLegend)m or isotype-matched control antibody (Ab) for 30 min on ice. Cells were then washed, stained with 7-Amino-Actinomycin D (7-AAD; BD Biosciences) for 10 min at 4 °C in darkness, and analyzed with a FACScan flow cytometer (Becton-Dickinson, Franklin Lakes, NJ, USA), gating on viable cells (7-AAD negative cells). Data were processed using CellQuest software (Becton Dickinson). Results were expressed as mean fluorescence intensities (arithmetic means ± standard errors of the means). MHC-II expression was also assayed in the presence of a neutralizing antibody anti-IL-6 (20 μg/mL, BD Bioscience), anti-IL-10 (20 g/mL, BD Bioscience), or their isotype matched control, with 10 ng/mL of recombinant human IL-6 (rIL-6, BD Bioscience) or 10 ng/mL of recombinant human IL-10 (rIL-10, BD Bioscience) alone or plus IFN-γ used as a control.

### 4.5. *Cytokine ELISA*

The IL-2, IL-6, and IL-10 level were measured in culture supernatants by ELISA according to the manufacturer's instructions (BD Biosciences).

### 4.6. *Phagocytosis Assays*

To study the phagocytosis capability of LX-2 cells, the phagocytic uptake of *E. coli* DH5α (Invitrogen) was measured as described [54]. Briefly, cells were infected with *B. abortus* at different MOIs, as described previously. Cells were washed twice and cultured in the presence of *E. coli* for 30 min at 37 °C in 5% $CO_2$. Extracellular bacteria were washed and killed with gentamicin (100 mg/mL) for 30 min. Cells were washed, lysed with 0.1% (v/v) Triton X-100, plated overnight on tripteine soy broth (TSB) agar, and colony forming units (CFU) were counted. As a positive control, the same bacteria phagocytic test was assessed using the murine macrophage cell line J774.A1.

### 4.7. *mRNA Preparation and RT-qPCR*

Total cellular RNA from LX-2 cells was extracted using Quick-RNA MiniPrep Kit (Zymo Research) and 1 μg of RNA was employed to perform the reverse transcription by means of Improm-II Reverse Transcriptase (Promega). Quantitative reverse-transcription polymerase chain reaction (qRT-PCR) analysis was achieved run on a StepOne real-time PCR detection system (Life Technology) using SYBR Green as a fluorescent DNA binding dye. The conditions of the amplification reaction were the following: 10 min 95 °C, 40 cycles for 15 s at 95 °C, 58 °C for 30 s, and 72 °C for 60 s. Primer sequences used for amplification were: β-actin, forward AACAGTCCGCCTAGAAGCAC, reverse 5′-CGTTGACATCCGTAAAGACC; cathepsin-S, forward 5′-TTATGGCAGAGAAGATGTCC, reverse 5′-AAGAGGGAAAGCTAGCAATC; CIITA, forward 5′-CCGACACAGACACCATCAAC, reverse 5′-TTTTCTGCCCAACTTCTGCT. All primer sets yielded a single product of the correct size. Relative transcript levels were calculated using the ΔΔCt method using as normalizer gene β-actin.

Endpoint PCR products were subjected to electrophoresis in 1% agarose gel, stained with ethidium bromide, visualized under UV light, and photographed. In order to normalize the qRT-PCR, the β-actin gene was included as housekeeping.

### 4.8. *Ag Processing and Presentation Assays*

LX-2 cells were cultured in 96-well flat-bottom plates ($10^5$ cells/well) and infected with *B. abortus* or stimulated with 500 U/mL of IFN-γ (Endogen) for 72 h. Following incubation and medium remotion, the cells were widely washed prior to Ag exposure. The cells then were pulsed with Ag85B (Abcam) 1, 10, and 30 μg/mL for 6 h, followed by incubation with DB1 T hybridoma cells ($10^5$ cells/well). After 2 to 24 h the supernatants were harvested and the amount of interleukin-2 (IL-2) secreted by T hybridoma cells was determined by ELISA.

### 4.9. Statistical Analysis

One-way ANOVA, followed by a Post Hoc Tukey Test using GraphPad Prism 4.0 software, was used to perform the statistical analysis of the results. The obtained data were represented as mean ± SEM.

**Author Contributions:** Conceptualization, J.F.Q., and M.V.D.; methodology, P.C.A.B., A.I.P.V., and M.M.E.; investigation, P.C.A.B., A.I.P.V., and M.M.E.; writing—original draft preparation, M.V.D.; writing—review and editing, G.H.G., J.F.Q., and M.V.D.; funding acquisition, J.F.Q., and M.V.D. All authors have read and agreed to the published version of the manuscript.

**Acknowledgments:** We thank Horacio Salomón and the staff of the Instituto de Investigaciones Biomédicas en Retrovirus y Sida (INBIRS) for their assistance with biosafety level 3 laboratory uses. We thank W. H. Boom (Case Western Reserve University, Cleveland, OH) for providing the DB1T-cell hybridoma. We also thank David H. Canaday (Case Western Reserve University) for his outstanding guidance concerning the culture of T-cell hybridomas.

## References

1. Pappas, G.; Akritidis, N.; Bosilkovski, M.; Tsianos, E. Brucellosis. *N. Engl. J. Med.* **2005**, *352*, 2325–2336. [CrossRef]
2. Hayoun, M.A.; Smith, M.E.; Shorman, M. *Brucellosis*; StatPearls Publishing: Treasure Island, FL, USA, 2020.
3. Kelly, A.M.; Golden-Mason, L.; Traynor, O.; Geoghegan, J.; McEntee, G.; Hegarty, J.E.; O'Farrelly, C. Changes in hepatic immunoregulatory cytokines in patients with metastatic colorectal carcinoma: Implications for hepatic anti-tumour immunity. *Cytokine* **2006**, *35*, 171–179. [CrossRef]
4. Robinson, M.W.; Harmon, C.; O'Farrelly, C. Liver immunology and its role in inflammation and homeostasis. *Cell. Mol. Immunol.* **2016**, *13*, 267–276. [CrossRef] [PubMed]
5. Madkour, M.M. Gastrointestinal brucellosis. In *Madkour's Brucellosis*, 2nd ed.; Madkour, M.M., Ed.; Springer: Berlin/Heidelberg, Germany, 2001; pp. 150–158.
6. Akritidis, N.; Tzivras, M.; Delladetsima, I.; Stefanaki, S.; Moutsopoulos, H.M.; Pappas, G. The liver in brucellosis. *Clin. Gastroenterol. Hepatol.* **2007**, *5*, 1109–1112. [CrossRef]
7. Heller, T.; Belard, S.; Wallrauch, C.; Carretto, E.; Lissandrin, R.; Filice, C.; Brunetti, E. Patterns of hepatosplenic brucella abscesses on cross-sectional imaging: A review of clinical and imaging features. *Am. J. Trop Med. Hyg.* **2015**, *93*, 761–766. [CrossRef] [PubMed]
8. Crispe, I.N. The liver as a lymphoid organ. *Annu. Rev. Immunol.* **2009**, *27*, 147–163. [CrossRef]
9. Arriola Benitez, P.C.; Scian, R.; Comerci, D.J.; Serantes, D.R.; Vanzulli, S.; Fossati, C.A.; Giambartolomei, G.H.; Delpino, M.V. Brucella abortus induces collagen deposition and MMP-9 down-modulation in hepatic stellate cells via TGF-beta1 production. *Am. J. Pathol.* **2013**, *183*, 1918–1927. [CrossRef]
10. Thomson, A.W.; Knolle, P.A. Antigen-presenting cell function in the tolerogenic liver environment. *Nat. Rev. Immunol.* **2010**, *10*, 753–766. [CrossRef] [PubMed]
11. Winau, F.; Hegasy, G.; Weiskirchen, R.; Weber, S.; Cassan, C.; Sieling, P.A.; Modlin, R.L.; Liblau, R.S.; Gressner, A.M.; Kaufmann, S.H. Ito cells are liver-resident antigen-presenting cells for activating T cell responses. *Immunity* **2007**, *26*, 117–129. [CrossRef] [PubMed]
12. Dunham, R.M.; Thapa, M.; Velazquez, V.M.; Elrod, E.J.; Denning, T.L.; Pulendran, B.; Grakoui, A. Hepatic stellate cells preferentially induce Foxp3$^+$ regulatory T cells by production of retinoic acid. *J. Immunol.* **2013**, *190*, 2009–2016. [CrossRef]
13. Roop, R.M., II; Bellaire, B.H.; Valderas, M.W.; Cardelli, J.A. Adaptation of the Brucellae to their intracellular niche. *Mol. Microbiol.* **2004**, *52*, 621–630. [CrossRef]
14. McCormick, P.J.; Martina, J.A.; Bonifacino, J.S. Involvement of clathrin and AP-2 in the trafficking of MHC class II molecules to antigen-processing compartments. *Proc. Natl. Acad. Sci. USA* **2005**, *102*, 7910–7915. [CrossRef] [PubMed]
15. Barrionuevo, P.; Cassataro, J.; Delpino, M.V.; Zwerdling, A.; Pasquevich, K.A.; Garcia Samartino, C.;

Wallach, J.C.; Fossati, C.A.; Giambartolomei, G.H. Brucella abortus inhibits major histocompatibility complex class II expression and antigen processing through interleukin-6 secretion via Toll-like receptor 2. *Infect. Immun.* **2008**, *76*, 250–262. [CrossRef]

16. Barrionuevo, P.; Delpino, M.V.; Pozner, R.G.; Velasquez, L.N.; Cassataro, J.; Giambartolomei, G.H. Brucella abortus induces intracellular retention of MHC-I molecules in human macrophages down-modulating cytotoxic CD8(+) T cell responses. *Cell. Microbiol.* **2013**, *15*, 487–502. [CrossRef] [PubMed]
17. Mittal, S.K.; Roche, P.A. Suppression of antigen presentation by IL-10. *Curr. Opin. Immunol.* **2015**, *34*, 22–27. [CrossRef] [PubMed]
18. Delpino, M.V.; Barrionuevo, P.; Scian, R.; Fossati, C.A.; Baldi, P.C. Brucella-infected hepatocytes mediate potentially tissue-damaging immune responses. *J. Hepatol.* **2010**, *53*, 145–154. [CrossRef]
19. Herkel, J.; Jagemann, B.; Wiegard, C.; Lazaro, J.F.; Lueth, S.; Kanzler, S.; Blessing, M.; Schmitt, E.; Lohse, A.W. MHC class II-expressing hepatocytes function as antigen-presenting cells and activate specific CD4 T lymphocyutes. *Hepatology* **2003**, *37*, 1079–1085. [CrossRef]
20. Golden-Mason, L.; Douek, D.C.; Koup, R.A.; Kelly, J.; Hegarty, J.E.; O'Farrelly, C. Adult human liver contains CD8pos T cells with naive phenotype, but is not a site for conventional alpha beta T cell development. *J. Immunol.* **2004**, *172*, 5980–5985. [CrossRef] [PubMed]
21. Giuffre, M.; Campigotto, M.; Campisciano, G.; Comar, M.; Croce, L.S. A story of liver and gut microbes: How does the intestinal flora affect liver disease? A review of the literature. *Am. J. Physiol. Gastrointest. Liver Physiol.* **2020**, *318*, G889–G906. [CrossRef] [PubMed]
22. Comerci, D.J.; Martinez-Lorenzo, M.J.; Sieira, R.; Gorvel, J.P.; Ugalde, R.A. Essential role of the VirB machinery in the maturation of the Brucella abortus-containing vacuole. *Cell. Microbiol.* **2001**, *3*, 159–168. [CrossRef] [PubMed]
23. De Figueiredo, P.; Ficht, T.A.; Rice-Ficht, A.; Rossetti, C.A.; Adams, L.G. Pathogenesis and immunobiology of brucellosis: Review of Brucella-host interactions. *Am. J. Pathol.* **2015**, *185*, 1505–1517. [CrossRef] [PubMed]
24. Friedman, S.L. Hepatic stellate cells: Protean, multifunctional, and enigmatic cells of the liver. *Physiol. Rev.* **2008**, *88*, 125–172. [CrossRef] [PubMed]
25. Vinas, O.; Bataller, R.; Sancho-Bru, P.; Gines, P.; Berenguer, C.; Enrich, C.; Nicolas, J.M.; Ercilla, G.; Gallart, T.; Vives, J.; et al. Human hepatic stellate cells show features of antigen-presenting cells and stimulate lymphocyte proliferation. *Hepatology* **2003**, *38*, 919–929. [CrossRef] [PubMed]
26. Yu, M.C.; Chen, C.H.; Liang, X.; Wang, L.; Gandhi, C.R.; Fung, J.J.; Lu, L.; Qian, S. Inhibition of T-cell responses by hepatic stellate cells via B7-H1-mediated T-cell apoptosis in mice. *Hepatology* **2004**, *40*, 1312–1321. [CrossRef]
27. Ebrahimkhani, M.R.; Mohar, I.; Crispe, I.N. Cross-presentation of antigen by diverse subsets of murine liver cells. *Hepatology* **2011**, *54*, 1379–1387. [CrossRef]
28. Garrett, W.S.; Chen, L.M.; Kroschewski, R.; Ebersold, M.; Turley, S.; Trombetta, S.; Galan, J.E.; Mellman, I. Developmental control of endocytosis in dendritic cells by Cdc42. *Cell* **2000**, *102*, 325–334. [CrossRef]
29. Chinnadurai, R.; Grakoui, A. B7-H4 mediates inhibition of T cell responses by activated murine hepatic stellate cells. *Hepatology* **2010**, *52*, 2177–2185. [CrossRef]
30. Charles, R.; Chou, H.S.; Wang, L.; Fung, J.J.; Lu, L.; Qian, S. Human hepatic stellate cells inhibit T-cell response through B7-H1 pathway. *Transplantation* **2013**, *96*, 17–24. [CrossRef]
31. Reith, W.; LeibundGut-Landmann, S.; Waldburger, J.M. Regulation of MHC class II gene expression by the class II transactivator. *Nat. Rev. Immunol.* **2005**, *5*, 793–806. [CrossRef]
32. Jabrane-Ferrat, N.; Nekrep, N.; Tosi, G.; Esserman, L.; Peterlin, B.M. MHC class II enhanceosome: How is the class II transactivator recruited to DNA-bound activators? *Int. Immunol.* **2003**, *15*, 467–475. [CrossRef]
33. Gobin, S.J.; Peijnenburg, A.; Keijsers, V.; van den Elsen, P.J. Site alpha is crucial for two routes of IFN gamma-induced MHC class I transactivation: The ISRE-mediated route and a novel pathway involving CIITA. *Immunity* **1997**, *6*, 601–611. [CrossRef]
34. Gobin, S.J.; Peijnenburg, A.; van Eggermond, M.; van Zutphen, M.; van den Berg, R.; van den Elsen, P.J. The RFX complex is crucial for the constitutive and CIITA-mediated transactivation of MHC class I and beta2-microglobulin genes. *Immunity* **1998**, *9*, 531–541. [CrossRef]

35. Gobin, S.J.; van Zutphen, M.; Westerheide, S.D.; Boss, J.M.; van den Elsen, P.J. The MHC-specific enhanceosome and its role in MHC class I and beta(2)-microglobulin gene transactivation. *J. Immunol.* **2001**, *167*, 5175–5184. [CrossRef] [PubMed]
36. Reyes, V.E.; Lu, S.; Humphreys, R.E. Cathepsin B cleavage of Ii from class II MHC alpha- and beta-chains. *J. Immunol.* **1991**, *146*, 3877–3880.
37. Bevec, T.; Stoka, V.; Pungercic, G.; Dolenc, I.; Turk, V. Major histocompatibility complex class II-associated p41 invariant chain fragment is a strong inhibitor of lysosomal cathepsin L. *J. Exp. Med.* **1996**, *183*, 1331–1338. [CrossRef]
38. Fineschi, B.; Sakaguchi, K.; Appella, E.; Miller, J. The proteolytic environment involved in MHC class II-restricted antigen presentation can be modulated by the p41 form of invariant chain. *J. Immunol.* **1996**, *157* 3211–3215.
39. Riese, R.J.; Mitchell, R.N.; Villadangos, J.A.; Shi, G.P.; Palmer, J.T.; Karp, E.R.; De Sanctis, G.T.; Ploegh, H.L.; Chapman, H.A. Cathepsin S activity regulates antigen presentation and immunity. *J. Clin. Investig.* **1998**, *101*, 2351–2363. [CrossRef]
40. Nakagawa, T.; Roth, W.; Wong, P.; Nelson, A.; Farr, A.; Deussing, J.; Villadangos, J.A.; Ploegh, H.; Peters, C.; Rudensky, A.Y. Cathepsin L: Critical role in Ii degradation and CD4 T cell selection in the thymus. *Science* **1998**, *280*, 450–453. [CrossRef] [PubMed]
41. Maubach, G.; Lim, M.C.; Kumar, S.; Zhuo, L. Expression and upregulation of cathepsin S and other early molecules required for antigen presentation in activated hepatic stellate cells upon IFN-gamma treatment. *Biochim. Biophys. Acta* **2007**, *1773*, 219–231. [CrossRef] [PubMed]
42. Arriola Benitez, P.C.; Pesce Viglietti, A.I.; Gomes, M.T.R.; Oliveira, S.C.; Quarleri, J.F.; Giambartolomei, G.H.; Delpino, M.V. Brucella abortus infection elicited hepatic stellate cell-mediated fibrosis through inflammasome-dependent IL-1beta production. *Front. Immunol.* **2019**, *10*, 3036. [CrossRef] [PubMed]
43. Gomes, M.T.; Campos, P.C.; Oliveira, F.S.; Corsetti, P.P.; Bortoluci, K.R.; Cunha, L.D.; Zamboni, D.S.; Oliveira, S.C. Critical role of ASC inflammasomes and bacterial type IV secretion system in caspase-1 activation and host innate resistance to Brucella abortus infection. *J. Immunol.* **2013**, *190*, 3629–3638. [CrossRef] [PubMed]
44. Gorvel, J.P.; Moreno, E. Brucella intracellular life: From invasion to intracellular replication. *Vet. Microbiol.* **2002**, *90*, 281–297. [CrossRef]
45. Kohler, S.; Michaux-Charachon, S.; Porte, F.; Ramuz, M.; Liautard, J.P. What is the nature of the replicative niche of a stealthy bug named Brucella? *Trends Microbiol.* **2003**, *11*, 215–219. [CrossRef]
46. Zhan, Y.; Cheers, C. Endogenous gamma interferon mediates resistance to Brucella abortus infection. *Infect. Immun.* **1993**, *61*, 4899–4901. [CrossRef]
47. Dornand, J.; Gross, A.; Lafont, V.; Liautard, J.; Oliaro, J.; Liautard, J.P. The innate immune response against Brucella in humans. *Vet. Microbiol.* **2002**, *90*, 383–394. [CrossRef]
48. Schroder, K.; Hertzog, P.J.; Ravasi, T.; Hume, D.A. Interferon-gamma: An overview of signals, mechanisms and functions. *J. Leukoc. Biol.* **2004**, *75*, 163–189. [CrossRef]
49. Hop, H.T.; Reyes, A.W.B.; Huy, T.X.N.; Arayan, L.T.; Min, W.; Lee, H.J.; Rhee, M.H.; Chang, H.H.; Kim, S. Interleukin 10 suppresses lysosome-mediated killing of Brucella abortus in cultured macrophages. *J. Biol. Chem.* **2018**, *293*, 3134–3144. [CrossRef]
50. Giambartolomei, G.H.; Scian, R.; Acosta-Rodriguez, E.; Fossati, C.A.; Delpino, M.V. Brucella abortus-infected macrophages modulate T lymphocytes to promote osteoclastogenesis via IL-17. *Am. J. Pathol.* **2012**, *181* 887–896. [CrossRef]
51. Kitamura, H.; Kamon, H.; Sawa, S.; Park, S.J.; Katunuma, N.; Ishihara, K.; Murakami, M.; Hirano, T. IL-6-STAT3 controls intracellular MHC class II alphabeta dimer level through cathepsin S activity in dendritic cells. *Immunity* **2005**, *23*, 491–502. [CrossRef]
52. Fumeaux, T.; Pugin, J. Role of interleukin-10 in the intracellular sequestration of human leukocyte antigen-DR in monocytes during septic shock. *Am. J. Respir. Crit. Care Med.* **2002**, *166*, 1475–1482. [CrossRef]

53. Scian, R.; Barrionuevo, P.; Giambartolomei, G.H.; De Simone, E.A.; Vanzulli, S.I.; Fossati, C.A.; Baldi, P.C.; Delpino, M.V. Potential role of fibroblast-like synoviocytes in joint damage induced by Brucella abortus infection through production and induction of matrix metalloproteinases. *Infect. Immun.* **2011**, *79*, 3619–3632. [CrossRef]
54. Gaikwad, S.; Agrawal-Rajput, R. Lipopolysaccharide from rhodobacter sphaeroides attenuates microglia-mediated inflammation and phagocytosis and directs regulatory T cell response. *Int. J. Inflam.* **2015**, *2015*, 361326. [CrossRef]

# 14

# Simultaneous Immunization with Omp25 and L7/L12 Provides Protection against Brucellosis in Mice

**Sonal Gupta [1], Surender Mohan [1], Vikas Kumar Somani [1,2], Somya Aggarwal [1,3] and Rakesh Bhatnagar [1,4,*]**

1 Laboratory of Molecular Biology and Genetic Engineering, School of Biotechnology, Jawaharlal Nehru University, New Delhi 110067, India; sonalmole@gmail.com (S.G.); mohan.surender@gmail.com (S.M.); vikass@wustl.edu (V.K.S.); s.aggarwal@wustl.edu (S.A.)
2 Department of Oncology, Washington University School of Medicine, St. Louis, MO 63110, USA
3 Department of Molecular Microbiology, Washington University School of Medicine, St. Louis, MO 63110, USA
4 Banaras Hindu University, Varanasi, Uttar Pradesh 221005, India
* Correspondence: rakeshbhatnagar@jnu.ac.in

**Abstract:** Currently used *Brucella* vaccines, *Brucella abortus* strain 19 and RB51, comprises of live attenuated *Brucella* strains and prevent infection in animals. However, these vaccines pose potential risks to recipient animals such as attenuation reversal and virulence in susceptible hosts on administration. In this context, recombinant subunit vaccines emerge as a safe and competent alternative in combating the disease. In this study, we formulated a divalent recombinant vaccine consisting of Omp25 and L7/L12 of *B. abortus* and evaluated vaccine potential individually as well as in combination. Sera obtained from divalent vaccine (Omp25+L7/L12) immunized mice group exhibited enhanced IgG titers against both components and indicated specificity upon immunoblotting reiterating its authenticity. Further, the IgG1/IgG2a ratio obtained against each antigen predicted a predominant Th2 immune response in the Omp25+L7/L12 immunized mice group. Upon infection with virulent *B. abortus* 544, Omp25+L7/L12 infected mice exhibited superior Log10 protection compared to individual vaccines. Consequently, this study recommends that simultaneous immunization of Omp25 and L7/L12 as a divalent vaccine complements and triggers a Th2 mediated immune response in mice competent of providing protection against brucellosis.

**Keywords:** Recombinant vaccine; divalent vaccine; brucellosis; Omp25; L7/L12; *Brucella abortus* 544

---

## 1. Introduction

Brucellosis, one of the major bacterial zoonoses across the globe, is caused by members of the *Brucella* genus, (*B. abortus*, *B. melitensis*, *B. suis* and *B. canis*). It is still considered one of the seven "neglected zoonoses" worldwide, in spite of a huge public health burden in many countries with low income [1]. The transmission of brucellosis in humans occurs through coincidental exposure to pathogenic bacterium from infected animals or animal products [2–6]. Domestic animals affected by brucellosis are more prone to abortions whereas human brucellosis leads to debilitating symptoms such as recurrent fevers, spondylitis, joint pains, and osteomyelitis [7–9]. The successful vaccines available commercially against brucellosis are Strain 19 and RB51 [10]. Strain 19 is an attenuated *B. abortus* strain with smooth morphology that has the capability to induce antibody responses and protect cattle against brucellosis. *B. abortus* RB51 is a rifampicin-resistant strain with rough morphology which provides protection against infection and abortion [11]. But there are several disadvantages associated with these vaccines such as occurrence of abortions in pregnant cows, restriction on age of vaccination, and reversion of vaccine strain back to pathogenic strain upon administration [12].

Further, antibodies are directed majorly against the lipopolysaccharide O-side chain during natural infection or S19 immunization which causes obstruction during a brucellosis diagnostic test. Therefore, development of an effective and safe vaccine is required.

In this context, recombinant protein-based subunit vaccines have an advantage over traditional live-attenuated vaccines as being protective and safe for human administration [13–15]. Recombinant subunit vaccines rely on specific parts of the pathogenic microorganism such as proteins or capsular polysaccharides containing protective epitopes to result in a protective immune response. Since these vaccines cannot replicate in the host, they are not pathogenic on administration [16]. Numerous intracellular components and surface proteins from bacteria have been researched as potential protective antigens against *Brucella* infection, such as L7/L12 [17], Omp19, Omp31 [18], BLS [19], BP26 [15] and Omp25 [20,21], and some have shown to be effective in providing significant protection against *Brucella* infection [15,17,20,21]. Although recombinant subunit vaccines offer no residual virulence but these require administration of multiple boosters along with providing lower levels of protection [22]. Further, when these *Brucella* protective antigens were administered in combination, the induced immune responses were superior in clearing intracellular *Brucella* as compared to their univalent counterpart [14,15,18,23–27]. Earlier studies have suggested that Omp25 and L7/L12 can serve as efficient protective antigens by inducing strong humoral and cellular immune response [17,20,28]. Outer membrane protein 25 (Omp25) of *Brucella* species has been identified as a potential antigen [20,28] capable of inhibiting TNF-$\alpha$ production [21]. Importantly, *omp25* gene is highly conserved in multiple *Brucella* species, strains, and biovars [8]. In addition, L7/L12 is a conserved protein which is found to be immunogenic and a stimulant of Th1 type $CD4^+$ cellular response in mice [17], making both these antigens relevant as components for divalent vaccine formulation against brucellosis. Further, many protective antigens from *Brucella* have been classified as specific and novel diagnostic target as compared to LPS based conventional tests [29,30].

Aluminum hydroxide (Alum) has been validated as an economical and safe adjuvant by U.S. Food and Drug administration for veterinary and human use. Aluminum salts form a short-term depot at the site of injection and slowly release antigen to the body's immune system [31,32]. In this study, we co-immunized Omp25 and L7/L12 of *Brucella abortus* using alum as an adjuvant. Elicited antibody titers and antibody subtype profile were analyzed when administered as a divalent vaccine candidate in BALB/c mice. Further, the protective efficacy of individual proteins and the divalent vaccine candidate against virulent *B. abortus* 544 challenge were determined.

## 2. Results

### 2.1. Expression and Purification of Recombinant Proteins rOmp25 and rL7/L12

The recombinant proteins rOmp25 and rL7/L12 were expressed in *E. coli* C43 cells and *E. coli* BL21 (DE3) cells respectively and purified using Ni-NTA chromatography. Further, the size of the expressed recombinant proteins was verified using 12% SDS-PAGE electrophoresis (Figure 1a). The immunoreactivity of purified proteins was confirmed by immunoblotting using (anti-Omp25+ L7/L12) mice serum signifying that serum from mice immunized with Omp25+L7/L12 consisted of antibodies specifically against its component proteins rOmp25 and rL7/L12 (Figure 1b).

Further, the virulence potential of rOmp25 and rL7/L12 was analyzed using bioinformatics analysis using VirulentPred and VaxiJen. VirulentPred is a tool used for prediction of virulent protein sequences in bacteria based on bi-layer cascade support vector machine (SVM) [33]. The SVM classifiers in this tool were trained and optimized using individual protein sequence features such as their amino acid and dipeptide composition along with position-specific iterated blast (PSI-BLAST). This tool distinguishes virulent proteins from non-virulent bacterial proteins with an accuracy of 81.8% [33]. On the basis of VirulentPred, rOmp25 and rL7/L12 were concluded to be virulent with predicted scores of 1.0411 and 0.2440, respectively (Figure 1c). In addition, we used the VaxiJen tool [34], which uses an alignment-independent approach for prediction of protective antigens. The antigen classification is purely based on the physiochemical properties of proteins without applying sequence alignment [34], and depends on auto cross covariance (ACC) transformation of protein sequences into uniform vectors of amino acid properties. VaxiJen results categorized Omp25 and L7/L12 as vaccine antigens with predicted scores of 0.7506 and 0.6442, respectively, at the threshold value of 0.4 (Figure 1d).

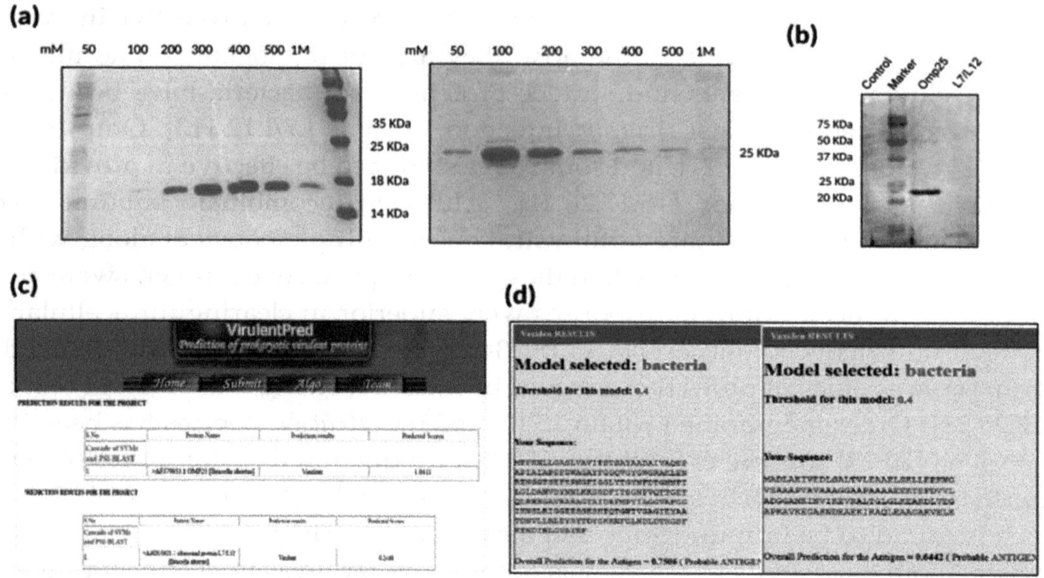


**Figure 1.** (**a**) Purification of rL7/L12 and rOmp25: SDS-PAGE gel stained with coomassie blue stain showing purification of rL7/L12 and rOmp25 recombinant proteins corresponding to 17 KDa and 25 KDa, respectively. (**b**) Immunoblotting with polyclonal sera of mice immunized with divalent vaccine (Omp25+L7/L12): The reactivity of purified proteins was confirmed by immunoblotting using anti-Omp25+L7/L12 mice serum. Negative control (lane 1; *E. coli* BL21 (DE3) cells with pET28a only), marker (lane 2) Precision Plus Protein™ Dual Color Standards, BIORAD #1610374, rOmp25 (lane 3), and rL7/L12 (lane 4). (**c**) Prediction of virulent proteins in a bacterium using VirulentPred: The sequences for *Brucella abortus* protein, Omp25 and L7/L12, have been submitted as input and their predicted scores have been calculated using VirulentPred software. (**d**) Prediction of vaccine antigens using VaxiJen: The sequences for *Brucella abortus* protein, Omp25 and L7/L12, have been submitted as input and the probability of these proteins being vaccine antigens has been predicted using VaxiJen.

### *2.2. Determination of IgG Antibody Titre upon Divalent Vaccine Immunization*

To assess the levels of IgG antibody titer generated in each of the immunized mice groups, sera was collected at day 28 and 42 post-priming and levels were estimated using Enzyme linked immunosorbent assay (ELISA). Our results revealed that immunization with Omp25+L7/L12 supported a robust anti-L7/L12 IgG response that was detectable at day 28 and remained stable until day 42. At day 28, anti-L7/L12 antibodies were observed to be higher in Omp25+L7/L12 mice compared to L7/L12 only immunized mice ($p < 0.05$; Figure 2a). At day 42, anti-L7/L12 levels were found to be similar in both L7/L12 and Omp25+L7/L12 immunized mice (approximately $6 \times 10^5$ in both). Immunization with divalent vaccine elicited a vigorous anti-Omp25 IgG response as well. Antibody levels were observed to be similar in Omp25+L7/L12 and Omp25 only immunized mice at day 28 and day 42 (Figure 2b). Therefore, the antigen alone vaccinated group generated antibodies only against a single antigen, whereas mice immunized with divalent vaccine (Omp25+L7/L12) produced antibodies against both components (rOmp25 and rL7/L12) in a cumulative manner, indicating that co-immunization of two proteins didn't hinder the immune response and supported generation of antibodies against its individual components.

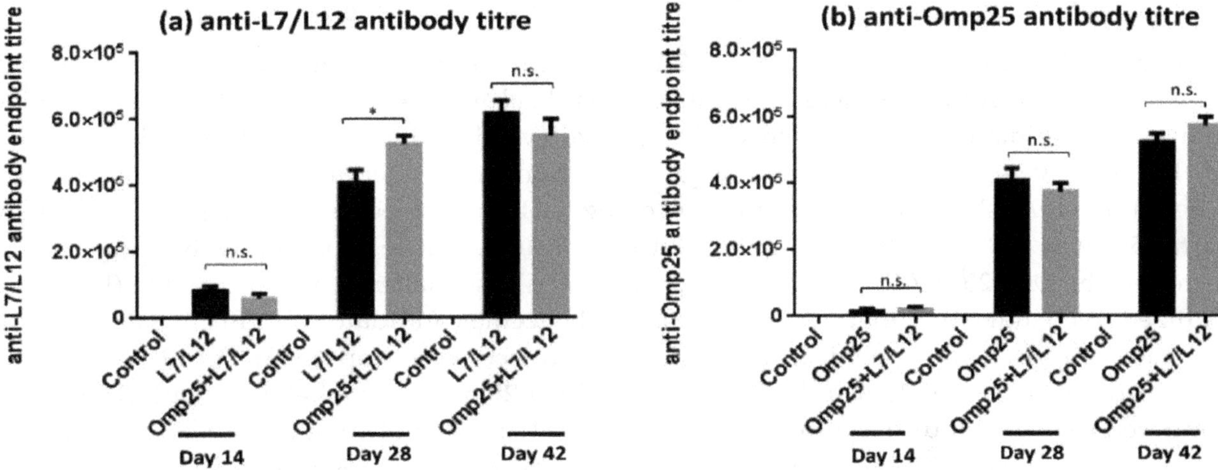


**Figure 2.** IgG antibody response elicited after immunization with L7/L12, Omp25, and divalent vaccine (Omp25+L7/L12): The mice were immunized with proteins rOmp25, rL7/L12 and rOmp25+rL7/L12 followed by isolation of serum samples from tail veins on day 14, 28, and 42. Estimation of IgG antibody end point titer was done through Enzyme linked immunosorbent assay (ELISA) and data is plotted as (mean ± SD).

### 2.3. *Evaluation of IgG Isotype Levels upon Divalent Vaccine Immunization*

In order to predict the Th1/Th2 bias of immune response, the relative IgG isotypes levels (IgG1, IgG2a and IgG2b) were analyzed in Omp25+L7/L12 immunized mice along with mice immunized solely with L7/L12 and Omp25. In the case of anti-L7/L12 antibodies, IgG1 levels were found to be significantly higher than IgG2a levels in the divalent vaccine as well as L7/L12 only immunized mice group, indicating a Th2 biased immune response in both (Figure 3a). Similarly, in the case of anti-Omp25 antibody levels, IgG1 levels were found to be higher than IgG2a (IgG1/IgG2a = 1.86), suggesting a Th2 biased immune response in divalent vaccine immunized mice. Interestingly, levels of IgG2a and IgG2b antibodies in Omp25+L7/L12 immunized mice were noteworthy (Figure 3b), predicting an elicitation of Th1 immune response in divalent vaccine immunized mice as well.

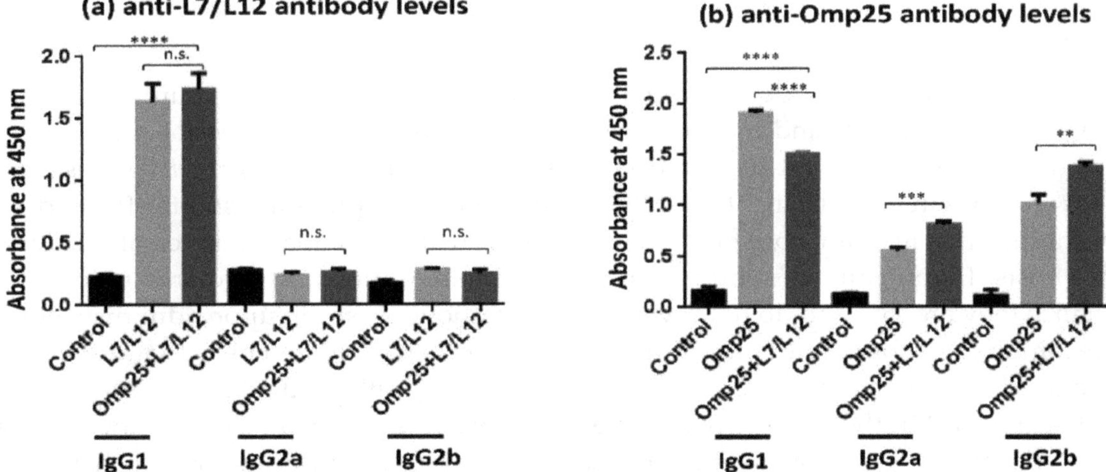


**Figure 3.** IgG antibody isotypes (IgG1, IgG2a, and IgG2b) elicited after immunization with rL7/L12, rOmp25, and divalent vaccine (Omp25+L7/L12): The recombinant *B. abortus* proteins rL7/L12, rOmp25, and Omp25+ L7/L12 were immunized in mice and isolation of serum samples was done from tail veins on day 42. Estimation of IgG isotype levels in serum of immunized mice was done through ELISA. The antibodies used for ELISA were horseradish peroxidase (HRP) conjugated anti-mouse IgG1, IgG2a, and IgG2b antibodies and data is plotted as (mean ($OD_{450}$ ± SD)).

### 2.4. Evaluation of Protective Efficacy Conferred by Divalent Vaccine Candidate against B. abortus 544 Challenge

The protective efficacy of the Omp25+L7/L12 vaccine candidate along with groups immunized solely with Omp25 and L7/L12 was analyzed against *B. abortus* 544 infection. Two weeks after last immunization, the immunized mice were challenged with virulent *B. abortus* 544 through intraperitoneal route. The mice were sacrificed four weeks post-infection, and bacterial colony forming units (CFU) were determined. As shown in Table 1, the level of $\log_{10}$ CFU per spleen at 28 days post-challenge with *B. abortus* 544 was (4.820 ± 0.18) in Omp25+L7/L12 immunized mice. Consecutively, $\log_{10}$ protection conferred by the Omp25+L7/L12 group was 1.98 at 28 day post-challenge as compared to (PBS + alum) immunized mice indicating that the Omp25+L7/L12 vaccine candidate was effective at eliminating pathogenic *B. abortus* 544 from a mice model. Mice immunized with Omp25 and L7/L12 alone exhibited $\log_{10}$ units of protection as 1.46 and 1.75, respectively at 28 days post-challenge with *B. abortus* 544 as compared to (PBS + alum) immunized mice. Overall, upon analyzing the levels of protection of the divalent vaccine candidate against *B. abortus* 544 challenge, it was found that Omp25+L7/L12 immunized mice exhibited efficacious $\log_{10}$ units of protection against *B. abortus* 544 challenge along with its individual components, however S19 exhibited the maximum.

**Table 1.** Bacterial proliferation in the spleen of mice immunized with rOmp25, rL7/L12, divalent vaccine candidate (Omp25+L7/L12) and control, using alum as adjuvant. The mice were infected with *B. abortus* 544 through intraperitoneal route and the splenic bacterial load was determined by plating dilutions of the splenocytes suspension on the tryptic soya agar plates followed by incubation at 37 °C in the presence of 5% $CO_2$ for 48 h. Data is represented as mean ± S.D.

| S. No | Vaccine Candidate | $\text{Log}_{10}$ Spleen Bacillary Load at Day 28 Post-Challenge ($\text{Log}_{10}$ CFU) | $\text{Log}_{10}$ Units of Protection at Day 28 Post-Challenge |
|---|---|---|---|
| 1. | rOmp25 | 5.338 ± 0.75 | 1.46 |
| 2. | rL7/L12 | 5.05 ± 0.27 | 1.75 |
| 3. | Omp25+L7/L12 | 4.820 ± 0.18 | 1.98 |
| 4. | PBS | 6.80 ± 0.58 | – |
| 5. | *B. abortus* S19 | 4.21 ± 0.31 | 2.59 |

## 3. Discussion

Brucellosis is still a major public health concern and endemic in many countries, mainly in the Mediterranean region, eastern and western Africa, and parts of South and Central America. There is a substantial requirement to control and eradicate this disease caused by the *Brucella* genus [4,5,7,35,36]. The current vaccines in use, Strain 19 and RB51 despite being popular are still far from ideal [12,37]. They prevent infection in animal but offer potential risks such as attenuation reversal and virulence in susceptible hosts. In this context, subunit vaccines are better options as compared to live attenuated vaccines, since they are safe and do not revert back to pathogenic strain upon administration [13–15]. Further, recombinant subunit vaccines protect against a given pathogen by activating humoral and cellular arms of immunity based upon a specific antigen along with used adjuvant, which makes them competent and useful in the vaccine field. There are certain proteins in *Brucella* species which can provide significant protection against the disease and are conserved throughout, such as L7/L12 [17], Omp19, Omp31 [18], BLS [19], BP26 [15], and Omp25 [20,21]. Among these, *omp25* and *l7/l12* genes in *Brucella* species encode for Omp25 and L7/L12 immunodominant proteins respectively, and have the potential to stimulate a strong humoral immune response along with providing protection against *Brucella* infections in mice models [15,17,20,28]. Earlier studies have suggested that outer membrane proteins from *Brucella* such as Omp25 [38], Omp10 [39], Omp19 [39], BP26 [40], Omp28 [41,42], and Omp31 [30] can distinguish between *Brucella*-infected animals and non-infected ones in an efficient and accurate way, withdrawing the false positive results in the field due to cross-reacting antibodies.

Further multivalent subunit vaccine formulations possess the capability to generate a wide range of immunogens that may result in better protection than their univalent counterpart [24–26]. In this study, we evaluated humoral immune response and protective efficacy of a divalent vaccine candidate consisting of rOmp25 and rL7/L12 against *Brucella* infection in mice. The foremost point to be explored in this study was whether two components combined together in a divalent vaccine have the capability to show a synergistic response and promote a heightened immune response, or if some kind of competitive inhibition occurs among them? Aluminum hydroxide (Alum) is beneficial since it is inexpensive and has been certified as the safest adjuvant for use by the United States Food and Drug Administration [31,32,43]. Alum creates a depot effect at the site of injection, resulting in a slow release of adsorbed antigens and an elevation in the immune response. The intraperitoneal route of administration was chosen because it helps to quickly absorb antigens into the vasculature, which leads to a rise of antigen drainage into the spleen and activation of immune cells circulating in the lymph nodes [44].

The results in this study exhibited that humoral immune response was elevated in mice immunized with the divalent vaccine Omp25+L7/L12 as compared to the control group (PBS + alum). The divalent vaccine (Omp25+L7/L12) was found immunogenic with high IgG levels against both of its components, rOmp25 and rL7/L12 (Figure 2), depicting that divalent vaccine has the potential to exhibit synergy among its individual components and elevate the immune response against virulent *Brucella abortus* 544 challenge. During bacterial infection, Th2-mediated immune response is characterized by synthesis and increase in the level of IgG1 antibodies [45,46] whereas Th1 immune response is represented by levels of IgG2a antibody along with IFN-$\gamma$ cytokine levels. The IgG subclass (IgG1, IgG2a and IgG2b) detection in Omp25+L7/L12 immunized mice exhibited that IgG1 levels were significantly higher as compared to IgG2a levels, predicting a more prominent Th2 immune response in case of anti-Omp25 and anti-L7/L12 antibodies (Figure 3). Although individual vaccinated group generated antibodies specifically to single antigen immunized whereas divalent vaccine candidate resulted in generation of antibodies against both antigens, rOmp25 and rL7/L12. Further, immunization with alum as an adjuvant has also been suggested to enhance antibody response in mice [43,47]. It helps in enhancement of antigen uptake and presentation to antigen-presenting cells, which results in promotion of Th2 immune responses [47,48]. Therefore, it is possible that alum has helped to increase antibody titer and elevate the humoral immune response in divalent vaccine immunized mice [49]. The analysis of protective efficacy in different mice groups after infection with virulent *Brucella abortus* 544 showed that Omp25+L7/L12 immunized mice exhibited a significant increase in $\log_{10}$ protection (1.98) as compared to the control (i.e. alum immunized mice) at 28 days after challenge ($p$ value $< 0.001$; Table 1). This specifies that Omp25+L7/L12 immunized mice were capable of eliminating pathogenic *B. abortus* 544 compared to the control group. On comparing $\log_{10}$ units of protection at 28 days after challenge in individual protein immunized mice, rOmp25 (1.46) and rL7/12 (1.75) with alum as the adjuvant, it was observed that Omp25+L7/L12 immunized mice showed a superior level of protection against *B. abortus* 544 infection, however S19 exhibited the maximum.

It is noteworthy that although *B. abortus* recombinant subunit vaccines show very promising results in mice models, the immune responses recognized in mice models may not reflect the protection achieved in natural hosts such as cattle after immunization [11]. Therefore, further studies determining protective efficacy in other animal models such as rats, guinea pigs, and monkeys are also encouraged before proceeding towards cattle administration [50]. Recombinant vaccines also need multiple booster administrations along with adjuvants and a combination of several antigens, which makes them economically unsuitable for cattle immunization [51]. Hence, there is a need to decrease the production cost, search for effective and affordable adjuvants, and reduce the expense of recombinant protein purification in order to make these vaccines economical for mass administration.

In a nutshell, this preliminary study shows that the combination of rOmp25 and rL7/L12 elicited steady immune responses against both antigens in mice. Further, when mice were immunized with the Omp25+L7/L12 vaccine candidate, a significant reduction in *B. abortus* 544 load in mice spleens was observed, implying the use of divalent vaccine (Omp25+L7/L12) as an improved vaccine candidate against brucellosis. Nevertheless, this study illustrates the potential of a divalent vaccine in providing host immunity and protection against *B. abortus* challenge, suggesting the use of a divalent recombinant vaccine candidate as an advanced approach in the future against brucellosis.

## 4. Materials and Methods

### *4.1. Plasmids and Bacterial Strains*

*E. coli* DH5α was used for propagation of recombinant plasmids. *E. coli* BL21 (DE3) and C43 strains were used for expression of rL7/L12 and rOmp25 proteins, respectively. *E. coli* strains were cultured using Luria–Bertani (LB) medium. Kanamycin was added to the medium at a final concentration of 50 μg/mL. *B. abortus* 544 and S19 strains were obtained from the Indian Veterinary Research Institute, Bareilly, India. *Brucella abortus* 544 was cultured in tryptic soy medium. Experiments involving *B. abortus* 544 and S19 strains were performed in a biosafety Level 3 laboratory at Jawaharlal Nehru University (JNU), Delhi, India.

### *4.2. Expression and Purification of Recombinant Proteins*

For formulation of the divalent vaccine, Omp25 and L7/L12 antigens of *Brucella abortus* were PCR amplified using gene specific primers and cloned in pET28(a) vector (Table 2). The expression of proteins was done in *E. coli*. To purify rOmp25, recombinants were grown in terrific broth until $OD_{600}$ ~ 0.5–0.6, and then induction was done using 1 mM IPTG for 5 h at 37 °C. Further purification of rOmp25 was done from the insoluble inclusion bodies fraction, using the urea-denaturing method and on-column refolding [20]. To purify rL7/L12, recombinants were grown in LB medium containing kanamycin to $OD_{600}$ ~ 0.7–0.8 followed by induction using 1 mM IPTG for 5 h at 37 °C. Both the proteins were affinity purified using immobilized nickel-nitrilotriacetic acid (Ni-NTA) agarose columns equilibrated in PB buffer (100 mM potassium phosphate buffer, pH 8.0) and eluted using 100–500 mM imidazole in PB. Purified proteins were analyzed by SDS-PAGE for content and purity. The dialysis of purified proteins was done against phosphate buffer saline (pH 7.4).

**Table 2.** Description of strains used in this study.

| S. No. | Protein Name | Strain Used for Purification of Protein | Reference |
|---|---|---|---|
| 1. | rOmp25 | *omp25* gene was cloned in pET28a at BamHI and SalI sites and expressed in *E. coli* C43 cells. | Goel et al. 2012 [20] |
| 2. | rL7/L12 | L7/L12 ribosomal gene was cloned in pET28a at NcoI and XhoI sites and expressed in *E. coli* BL21(DE3) cells. | Singh et al. 2015 [1] |

### *4.3. Immunoblotting*

For immunoblotting, the recombinant proteins were resolved by 12% SDS-PAGE followed by electroblotting onto nitrocellulose membrane. The membrane was further blocked using 3% BSA followed by incubation with anti-Omp25+L7/L12 antibody (1:5000 dilution, raised in mice) for 1 h. After subsequent washing, binding specificity was checked using AP-conjugated goat anti-mouse IgG antibody (Catalog no. sc-2047, Santa Cruz Biotechnology, USA) [52–54].

### *4.4. Immunization of Respective Proteins in Mice*

Four to six week old female BALB/c mice (inbred) were obtained from the National Centre for Laboratory Animal Sciences, Hyderabad, India. Recommendations from the Institutional Animal Ethics and Biosafety Committee were regularly followed during mouse experiments. In brief, mice were caged under sterile conditions in micro-isolators, fed with pathogen-free food and water *ad libitum* during consecutive immunizations. Once infected with *B. abortus* 544, mice were maintained at the BSL-3 animal facility of JNU for evaluation of protective efficacy.

For immunization of rL7/L12 and rOmp25, the optimized dose of each antigen was considered as mentioned in earlier reports [1,20]. Briefly, mice were grouped and immunized through the intraperitoneal route, either with Omp25 (30 μg) or L7/L12 (40 μg) alone or in combination as a divalent vaccine candidate with alum as an adjuvant. Two boosters were administered at regular intervals of 2 weeks, and 1X PBS with alum and *B. abortus* S19 immunized mice groups were taken as controls. For prime immunization and subsequent booster immunization, 100 μl emulsion of the required antigen and alum in 1X PBS was injected in each mouse. The blood was collected from each mouse on day 0, 14, 28, and 42 from tail veins and sera was extracted through centrifugation at 15,600 *g* for 20 min, followed by storage at −80 °C for further analysis.

### 4.5. Elucidation of End-Point Antibody Titer

An enzyme-linked immunosorbent assay (ELISA) was used to analyze serum antibody titer. In brief, 96-well microtiter plates (NuncMaxiSorp) were coated overnight with 500 ng/well of capture antigen (rOmp25 or rL7/L12) in PBS at 4 °C. The plates were washed three times using PBST (PBS with 0.1% tween 20) followed by blocking using 2% BSA in PBS for 2 h at 37 °C. The antibody titer in the sera of respective antigen immunized mice along with the divalent vaccine immunized mouse group was assessed by priming dilutions of the same, in triplicates, at 37 °C for 1 h. Washing of the plates was done using PBST followed by addition of horseradish peroxidase (HRP)-conjugated anti-mouse secondary antibodies (Catalog no. sc-2005, Santa Cruz Biotechnology, USA) at 1:10,000 dilution for 1 h at 37 °C [53,54]. The plates were further incubated with OptEIA TMB substrate (BD Biosciences, USA) for calorimetric assay and the reaction was stopped using 1N HCl. Absorbance of the plate was measured at 450 nm through Tecan's Sunrise absorbance microplate reader. End point titer was evaluated as the reciprocal of highest dilution giving absorbance greater than the threshold value. Threshold value was calculated as the mean of absorbance plus three times standard deviation of 1:1000 dilution of the control group (PBS + alum).

### 4.6. Analysis of IgG Isotypes in Immunized Mice

The IgG isotypes (IgG1, IgG2a and IgG2b) were detected in immunized mice using ELISA as described above. For secondary antibodies, anti-mouse IgG1-HRP (Catalog no. sc-2060), anti-mouse IgG2a-HRP (Catalog no. sc-2061) and anti-mouse IgG2b-HRP conjugated antibodies (Catalog no. sc-2062) (raised in goat; Santa Cruz Biotechnology, USA) were used and absorbance at 450 nm was measured [53].

### 4.7. Evaluation of Protective Efficacy of Vaccine Candidate

Two weeks after the final booster immunization (day 42), mice groups immunized with PBS, rOmp25, rL7/L12, and divalent vaccine candidate (rOmp25+rL7/L12) were challenged with $2 \times 10^5$ cells of *B. abortus* 544 through the intraperitoneal route. *B.abortus* S19 was injected on day 0 in respective group, and challenge was done after 21 days with $2 \times 10^5$ cells of virulent *B. abortus* 544. After 4 weeks of infection, mice from each group were euthanized through cervical dislocation. Their spleen was extracted under sterile conditions and finally homogenized in PBS using probe homogenizer. For CFU count, various dilutions of the spleen homogenate were prepared and plated on tryptic soya agar followed by incubation at 37 °C for 48 h in the presence of 5% $CO_2$. Total splenic load was calculated and represented as $\text{Log}_{10}$ CFU mean ± standard deviation (SD). $\text{Log}_{10}$ units of protection were determined by calculating the difference between the $\log_{10}$ CFU of PBS injected group (control) and vaccinated group.

### 4.8. Statistical Analysis

The results are represented as mean ± SD and are reported as data of three different sets of experiments. The statistical significance in antibody titer was calculated using two-tailed Student's t-test. (* represents $P < 0.05$; ** represents $P < 0.01$; *** represents $P < 0.001$, **** represents $P < 0.0001$).

**Ethical statement:** All mice experiments were performed while abiding by the rules of Institutional Animal Ethics Committee (IAEC), Jawaharlal Nehru University, New Delhi, India guidelines. All experiments involving virulent *Brucella abortus* 544 and *Brucella abortus* S19 strain have been performed in Biosafety level-3 (BSL-3) facility.

**Author Contributions:** Each author has made valuable contributions to this work. Conceptualization, S.G. and R.B.; methodology, S.G., RB; validation, S.G., S.M., V.K.S., S.A. and R.B.; formal analysis, S.G., V.K.S., S.A.; investigation, R.B.; resources, R.B.; data curation, S.G.; writing—original draft preparation, S.G.; writing—review and editing, S.G., S.M., V.K.S., S.A., R.B.; visualization, S.G., S.M., V.K.S. and S.A.; supervision, R.B.; project administration, R.B.; funding acquisition, R.B. All authors have read and agreed to the published version of the manuscript.

## References

1. Singh, D.; Goel, D.; Bhatnagar, R. Recombinant L7/L12 protein entrapping PLGA (poly lactide-*co*-glycolide) micro particles protect BALB/c mice against the virulent B. abortus 544 infection. *Vaccine* **2015**, *33*, 2786–2792. [CrossRef] [PubMed]
2. Moreno, E.; Stackebrandt, E.; Dorsch, M.; Wolters, J.; Busch, M.; Mayer, H. Brucella abortus 16S rRNA and lipid A reveal a phylogenetic relationship with members of the alpha-2 subdivision of the class Proteobacteria. *J. Bacteriol.* **1990**, *172*, 3569–3576. [CrossRef] [PubMed]
3. Renukaradhya, G.J.; Isloor, S.; Rajasekhar, M. Epidemiology, zoonotic aspects, vaccination and control/eradication of brucellosis in India. *Vet. Microbiol.* **2002**, *90*, 183–195. [CrossRef]
4. Godfroid, J.; Cloeckaert, A.; Liautard, J.P.; Kohler, S.; Fretin, D.; Walravens, K.; Garin-Bastuji, B.; Letesson, J.J. From the discovery of the Malta fever's agent to the discovery of a marine mammal reservoir, brucellosis has continuously been a re-emerging zoonosis. *Vet. Res.* **2005**, *36*, 313–326. [CrossRef] [PubMed]
5. Tan, S.Y.; Davis, C. David Bruce (1855–1931): Discoverer of brucellosis. *Singap. Med. J.* **2011**, *52*, 138–139.
6. Babaoglu, U.T.; Ogutcu, H.; Demir, G.; Sanli, D.; Babaoglu, A.B.; Oymak, S. Prevalence of Brucella in raw milk: An example from Turkey. *Niger. J. Clin. Pract.* **2018**, *21*, 907–911.
7. Glynn, M.K.; Lynn, T.V. Brucellosis. *J. Am. Vet. Med. Assoc.* **2008**, *233*, 900–908. [CrossRef]
8. Cloeckaert, A.; Verger, J.M.; Grayon, M.; Grepinet, O. Restriction site polymorphism of the genes encoding the major 25 kDa and 36 kDa outer-membrane proteins of Brucella. *Microbiology* **1995**, *141*, 2111–2121. [CrossRef
9. Pappas, G.; Bosilkovski, M.; Akritidis, N.; Mastora, M.; Krteva, L.; Tsianos, E. Brucellosis and the respiratory system. *Clin. Infect. Dis.* **2003**, *37*, e95–e99. [CrossRef]
10. Olsen, S.C.; Stoffregen, W.S. Essential role of vaccines in brucellosis control and eradication programs for livestock. *Expert Rev. Vaccines* **2005**, *4*, 915–928. [CrossRef]
11. Dorneles, E.M.; Sriranganathan, N.; Lage, A.P. Recent advances in Brucella abortus vaccines. *Vet. Res.* **2015**, *46*, 76. [CrossRef] [PubMed]
12. Nicoletti, P. Prevalence and persistence of Brucella abortus strain 19 infections and prevalence of other biotypes in vaccinated adult dairy cattle. *J. Am. Vet. Med. Assoc.* **1981**, *178*, 143–145. [PubMed]
13. Nascimento, I.P.; Leite, L.C.C. Recombinant vaccines and the development of new vaccine strategies. *Braz. J. Med Biol. Res. Rev. Bras. De Pesqui. Med. E Biol.* **2012**, *45*, 1102–1111. [CrossRef] [PubMed]
14. Hop, H.T.; Arayan, L.T.; Huy, T.X.N.; Reyes, A.W.B.; Min, W.; Lee, H.J.; Park, S.J.; Chang, H.H.; Kim, S. Immunization of BALB/c mice with a combination of four recombinant Brucella abortus proteins, AspC, Dps, InpB and Ndk, confers a marked protection against a virulent strain of Brucella abortus. *Vaccine* **2018**, *36*, 3027–3033. [CrossRef]

15. Yang, X.; Walters, N.; Robison, A.; Trunkle, T.; Pascual, D.W. Nasal immunization with recombinant Brucella melitensis bp26 and trigger factor with cholera toxin reduces B. melitensis colonization. *Vaccine* **2007**, *25*, 2261–2268. [CrossRef]
16. Hansson, M.; Nygren, P.A.; Stahl, S. Design and production of recombinant subunit vaccines. *Biotechnol. Appl. Biochem.* **2000**, *32*, 95–107. [CrossRef]
17. Oliveira, S.C.; Splitter, G.A. Immunization of mice with recombinant L7/L12 ribosomal protein confers protection against Brucella abortus infection. *Vaccine* **1996**, *14*, 959–962. [CrossRef]
18. Cassataro, J.; Estein, S.M.; Pasquevich, K.A.; Velikovsky, C.A.; de la Barrera, S.; Bowden, R.; Fossati, C.A.; Giambartolomei, G.H. Vaccination with the recombinant Brucella outer membrane protein 31 or a derived 27-amino-acid synthetic peptide elicits a $CD^{4+}$ T helper 1 response that protects against Brucella melitensis infection. *Infect. Immun.* **2005**, *73*, 8079–8088. [CrossRef]
19. Velikovsky, C.A.; Goldbaum, F.A.; Cassataro, J.; Estein, S.; Bowden, R.A.; Bruno, L.; Fossati, C.A.; Giambartolomei, G.H. Brucella lumazine synthase elicits a mixed $Th_1$-$Th_2$ immune response and reduces infection in mice challenged with Brucella abortus 544 independently of the adjuvant formulation used. *Infect. Immun.* **2003**, *71*, 5750–5755. [CrossRef]
20. Goel, D.; Bhatnagar, R. Intradermal immunization with outer membrane protein 25 protects Balb/c mice from virulent B. abortus 544. *Mol. Immunol.* **2012**, *51*, 159–168. [CrossRef]
21. Paul, S.; Peddayelachagiri, B.V.; Nagaraj, S.; Kingston, J.J.; Batra, H.V. Recombinant outer membrane protein 25c from Brucella abortus induces $Th_1$ and $Th_2$ mediated protection against Brucella abortus infection in mouse model. *Mol. Immunol.* **2018**, *99*, 9–18. [CrossRef] [PubMed]
22. Lalsiamthara, J.; Lee, J.H. Development and trial of vaccines against Brucella. *J. Vet. Sci.* **2017**, *18*, 281–290. [CrossRef] [PubMed]
23. Al-Mariri, A.; Tibor, A.; Mertens, P.; De Bolle, X.; Michel, P.; Godefroid, J.; Walravens, K.; Letesson, J.J. Protection of BALB/c mice against Brucella abortus 544 challenge by vaccination with bacterioferritin or P39 recombinant proteins with CpG oligodeoxynucleotides as adjuvant. *Infect. Immun.* **2001**, *69*, 4816–4822. [CrossRef] [PubMed]
24. Delpino, M.V.; Estein, S.M.; Fossati, C.A.; Baldi, P.C.; Cassataro, J. Vaccination with Brucella recombinant DnaK and SurA proteins induces protection against Brucella abortus infection in BALB/c mice. *Vaccine* **2007**, *25*, 6721–6729. [CrossRef]
25. Ghasemi, A.; Jeddi-Tehrani, M.; Mautner, J.; Salari, M.H.; Zarnani, A.H. Simultaneous immunization of mice with Omp31 and TF provides protection against Brucella melitensis infection. *Vaccine* **2015**, *33*, 5532–5538. [CrossRef]
26. Tabatabai, L.B.; Pugh, G.W. Modulation of immune responses in Balb/c mice vaccinated with Brucella abortus Cu Zn superoxide dismutase synthetic peptide vaccine. *Vaccine* **1994**, *12*, 919–924. [CrossRef]
27. Tadepalli, G.; Singh, A.K.; Balakrishna, K.; Murali, H.S.; Batra, H.V. Immunogenicity and protective efficacy of Brucella abortus recombinant protein cocktail (rOmp19+rP39) against B. abortus 544 and B. melitensis 16M infection in murine model. *Mol. Immunol.* **2016**, *71*, 34–41. [CrossRef]
28. Goel, D.; Rajendran, V.; Ghosh, P.C.; Bhatnagar, R. Cell mediated immune response after challenge in Omp25 liposome immunized mice contributes to protection against virulent Brucella abortus 544. *Vaccine* **2013**, *31*, 1231–1237. [CrossRef]
29. Vatankhah, M.; Beheshti, N.; Mirkalantari, S.; Khoramabadi, N.; Aghababa, H.; Mahdavi, M. Recombinant Omp2b antigen-based ELISA is an efficient tool for specific serodiagnosis of animal brucellosis. *Braz. J. Microbiol. Publ. Braz. Soc. Microbiol.* **2019**, *50*, 979–984. [CrossRef]
30. Ahmed, I.M.; Khairani-Bejo, S.; Hassan, L.; Bahaman, A.R.; Omar, A.R. Serological diagnostic potential of recombinant outer membrane proteins (rOMPs) from Brucella melitensis in mouse model using indirect enzyme-linked immunosorbent assay. *BMC Vet. Res.* **2015**, *11*, 275. [CrossRef]
31. Gupta, R.K.; Rost, B.E.; Relyveld, E.; Siber, G.R. Adjuvant properties of aluminum and calcium compounds. *Pharm. Biotechnol.* **1995**, *6*, 229–248. [CrossRef] [PubMed]

32. Sivakumar, S.M.; Safhi, M.M.; Kannadasan, M.; Sukumaran, N. Vaccine adjuvants—Current status and prospects on controlled release adjuvancity. *Saudi. Pharm. J.* **2011**, *19*, 197–206. [CrossRef] [PubMed]
33. Garg, A.; Gupta, D. VirulentPred: A SVM based prediction method for virulent proteins in bacterial pathogens. *BMC Bioinform.* **2008**, *9*, 62. [CrossRef] [PubMed]
34. Doytchinova, I.A.; Flower, D.R. VaxiJen: A server for prediction of protective antigens, tumour antigens and subunit vaccines. *BMC Bioinform.* **2007**, *8*, 4. [CrossRef] [PubMed]
35. Fugier, E.; Pappas, G.; Gorvel, J.P. Virulence factors in brucellosis: Implications for aetiopathogenesis and treatment. *Expert Rev. Mol. Med.* **2007**, *9*, 1–10. [CrossRef]
36. Martirosyan, A.; Moreno, E.; Gorvel, J.P. An evolutionary strategy for a stealthy intracellular Brucella pathogen. *Immunol. Rev.* **2011**, *240*, 211–234. [CrossRef]
37. Moriyon, I.; Grillo, M.J.; Monreal, D.; Gonzalez, D.; Marin, C.; Lopez-Goni, I.; Mainar-Jaime, R.C.; Moreno, E.; Blasco, J.M. Rough vaccines in animal brucellosis: Structural and genetic basis and present status. *Vet. Res.* **2004**, *35*, 1–38. [CrossRef]
38. Cloeckaert, A.; Zygmunt, M.S.; Bezard, G.; Dubray, G. Purification and antigenic analysis of the major 25-kilodalton outer membrane protein of Brucella abortus. *Res. Microbiol.* **1996**, *147*, 225–235. [CrossRef]
39. Simborio, H.L.; Lee, J.J.; Bernardo Reyes, A.W.; Hop, H.T.; Arayan, L.T.; Min, W.; Lee, H.J.; Yoo, H.S.; Kim, S. Evaluation of the combined use of the recombinant Brucella abortus Omp10, Omp19 and Omp28 proteins for the clinical diagnosis of bovine brucellosis. *Microb. Pathog.* **2015**, *83–84*, 41–46. [CrossRef]
40. Cloeckaert, A.; Baucheron, S.; Vizcaino, N.; Zygmunt, M.S. Use of recombinant BP26 protein in serological diagnosis of Brucella melitensis infection in sheep. *Clin. Diagn. Lab. Immunol.* **2001**, *8*, 772–775. [CrossRef]
41. Salih-Alj Debbarh, H.; Cloeckaert, A.; Bezard, G.; Dubray, G.; Zygmunt, M.S. Enzyme-linked immunosorbent assay with partially purified cytosoluble 28-kilodalton protein for serological differentiation between Brucella melitensis-infected and B. melitensis Rev.1-vaccinated sheep. *Clin. Diagn. Lab. Immunol.* **1996**, *3*, 305–308. [CrossRef] [PubMed]
42. Chaudhuri, P.; Prasad, R.; Kumar, V.; Gangaplara, A. Recombinant OMP28 antigen-based indirect ELISA for serodiagnosis of bovine brucellosis. *Mol. Cell. Probes* **2010**, *24*, 142–145. [CrossRef] [PubMed]
43. Awate, S.; Babiuk, L.A.; Mutwiri, G. Mechanisms of action of adjuvants. *Front. Immunol.* **2013**, *4*, 114. [CrossRef] [PubMed]
44. Olin, T.; Saldeen, T.J.C.R. The lymphatic pathways from the peritoneal cavity: A lymphangiographic study in the rat. *Cancer Res.* **1964**, *24*, 1700–1711. [PubMed]
45. Maecker, H.T.; Do, M.S.; Levy, S. CD81 on B cells promotes interleukin 4 secretion and antibody production during T helper type 2 immune responses. *Proc. Natl. Acad. Sci. USA* **1998**, *95*, 2458–2462. [CrossRef] [PubMed]
46. Pasquevich, K.A.; Estein, S.M.; Garcia Samartino, C.; Zwerdling, A.; Coria, L.M.; Barrionuevo, P.; Fossati, C.A.; Giambartolomei, G.H.; Cassataro, J. Immunization with recombinant Brucella species outer membrane protein Omp16 or Omp19 in adjuvant induces specific $CD^{4+}$ and $CD^{8+}$ T cells as well as systemic and oral protection against Brucella abortus infection. *Infect. Immun.* **2009**, *77*, 436–445. [CrossRef]
47. Ghimire, T.R. The mechanisms of action of vaccines containing aluminum adjuvants: An in vitro vs in vivo paradigm. *SpringerPlus* **2015**, *4*, 181. [CrossRef]
48. Mannhalter, J.W.; Neychev, H.O.; Zlabinger, G.J.; Ahmad, R.; Eibl, M.M. Modulation of the human immune response by the non-toxic and non-pyrogenic adjuvant aluminium hydroxide: Effect on antigen uptake and antigen presentation. *Clin. Exp. Immunol.* **1985**, *61*, 143–151.
49. Audibert, F.M.; Lise, L.D. Adjuvants: Current status, clinical perspectives and future prospects. *Immunol. Today* **1993**, *14*, 281–284. [CrossRef]
50. Silva, T.M.; Costa, E.A.; Paixao, T.A.; Tsolis, R.M.; Santos, R.L. Laboratory animal models for brucellosis research. *J. Biomed. Biotechnol.* **2011**, *2011*, 518323. [CrossRef]
51. Schurig, G.G.; Sriranganathan, N.; Corbel, M.J. Brucellosis vaccines: Past, present and future. *Vet. Microbiol.* **2002**, *90*, 479–496. [CrossRef]
52. Ghosh, S.; Sulistyoningrum, D.C.; Glier, M.B.; Verchere, C.B.; Devlin, A.M. Altered glutathione homeostasis in heart augments cardiac lipotoxicity associated with diet-induced obesity in mice. *J. Biol. Chem.* **2011**, *286*, 42483–42493. [CrossRef] [PubMed]

53. Gupta, S.; Singh, D.; Gupta, M.; Bhatnagar, R. A combined subunit vaccine comprising BP26, Omp25 and L7/L12 against brucellosis. *Pathog. Dis.* **2019**, *77*. [CrossRef] [PubMed]
54. Aggarwal, S.; Somani, V.K.; Gupta, S.; Garg, R.; Bhatnagar, R. Development of a novel multiepitope chimeric vaccine against anthrax. *Med. Microbiol. Immunol.* **2019**, *208*, 185–195. [CrossRef]

# 15

# *Brucella abortus* Proliferates in Decidualized and Non-Decidualized Human Endometrial Cells Inducing a Proinflammatory Response

**Lucía Zavattieri [1,2,†], Mariana C. Ferrero [1,2,†], Iván M. Alonso Paiva [1,2], Agustina D. Sotelo [1,2], Andrea M. Canellada [1,2] and Pablo C. Baldi [1,2,*]**

1 Facultad de Farmacia y Bioquímica, Cátedra de Inmunología, Universidad de Buenos Aires, Buenos Aires 1113, Argentina; mv.lzavattieri@gmail.com (L.Z.); ferrerom@ffyb.uba.ar (M.C.F.); ivan_alonsopaiva@yahoo.com.ar (I.M.A.P.); agustinadsotelo@gmail.com (A.D.S.); acanell@ffyb.uba.ar (A.M.C.)

2 CONICET-Universidad de Buenos Aires, Instituto de Estudios de la Inmunidad Humoral (IDEHU), Buenos Aires 1033, Argentina

* Correspondence: pablobal@ffyb.uba.ar

† These authors contributed equally to this work.

**Abstract:** *Brucella* spp. have been associated with abortion in humans and animals. Although the mechanisms involved are not well established, it is known that placental *Brucella* infection is accompanied by inflammatory phenomena. The ability of *Brucella abortus* to infect and survive in human endometrial stromal cells (T-HESC cell line) and the cytokine response elicited were evaluated. *B. abortus* was able to infect and proliferate in both non-decidualized and decidualized T-HESC cells. Intracellular proliferation depended on the expression of a functional *virB* operon in the pathogen. *B. abortus* internalization was inhibited by cytochalasin D and to a lower extent by colchicine, but was not affected by monodansylcadaverine. The infection did not induce cytotoxicity and did not alter the decidualization status of cells. *B. abortus* infection elicited the secretion of IL-8 and MCP-1 in either decidualized or non-decidualized T-HESC, a response also induced by heat-killed *B. abortus* and outer membrane vesicles derived from this bacterium. The stimulation of T-HESC with conditioned media from *Brucella*-infected macrophages induced the production of IL-6, MCP-1 and IL-8 in a dose-dependent manner, and this effect was shown to depend on IL-1β and TNF-α. The proinflammatory responses of T-HESC to *B. abortus* and to factors produced by infected macrophages may contribute to the gestational complications of brucellosis.

**Keywords:** *Brucella abortus*; human endometrial cells; internalization; intracellular replication; decidualization; chemokines; macrophages

---

## 1. Introduction

Human brucellosis, a zoonotic disease mostly caused by *Brucella melitensis*, *B. suis* and *B. abortus*, affects over 500,000 people each year around the world [1]. The infection can be found in several domestic animals (cattle, sheep, goats, pigs, and dogs) and in some wild species. Transmission to humans usually occurs by contact with infected animal tissues and consumption of dairy products.

The clinical manifestations of human brucellosis are usually linked to inflammatory phenomena in the affected organs [2]. Involvement with the reproductive organs is common in animals, which frequently present problems such as abortion and perinatal death. Studies performed in animals have shown that placental *Brucella* infection is accompanied by the infiltration of inflammatory cells [3,4]. The fact that placental inflammatory responses are involved in infection-triggered abortion by several pathogens [5–7] suggests that placental inflammation may also have a role in *Brucella*-induced abortion.

Abortion due to *Brucella* infection has been also reported in humans, with an incidence that ranges from 7% to 40% according to different studies [8–10]. Among pregnant women who presented with acute brucellosis at a Saudi Arabian hospital, 43% had spontaneous abortion during the first and second trimester, and 2% in the third trimester [11]. In spite of the importance of *Brucella*-related abortion, the pathophysiology of this complication in humans is largely unknown. Recent studies have shown that *Brucella* spp. can infect and replicate in human trophoblasts, and that the infection elicits a proinflammatory response [12,13]. These trophoblastic inflammatory responses may be relevant to the pathogenesis of abortion in human brucellosis. However, the potential of other placental cell populations to contribute to an inflammatory environment during *Brucella* infection has not been explored.

For several microorganisms that reach the placenta by the hematogenous route, including *Brucella abortus*, in vivo studies in animal models have indicated that the maternal decidua is the initial site of placental colonization [14,15]. Decidualization of the endometrium, a process essential for successful implantation and maintenance of pregnancy, involves progesterone-driven morphological and biochemical changes of fibroblast-like endometrial stromal cells (ESCs) to differentiate into decidual stromal cells (DSCs). These DSCs are characterized by the secretion of prolactin, insulin growth factor-binding protein and several cytokines that act as regulators of the innate immunity [16].

Given the relevance of the decidua as the initial site of placental colonization for several hematogenously spread infections, the ability of decidual cells to respond to pathogens is especially relevant. Primary DSC and ESC cell lines have been shown to express several Toll-like receptors (TLRs) and Nod-like receptors (NLRs), and respond to pathogen-associated molecular patterns (PAMPs) with an enhanced production of matrix metalloproteinases (MMPs) and proinflammatory cytokines including MCP-1, IL-6, IL-8, IL-1β, and CCL5 (RANTES) [17]. At least for group B streptococcal infection the cytokine response of endometrial stromal cells is modulated by decidualization, so that decidualized cells produce IL-6, TNF-α, IL-10, and TGF-β while non-decidualized cells do not [18].

In addition to decidual stromal cells, the decidua also contains significant proportions of immune cells, including macrophages, natural killer cells, dendritic cells, and T cells [19]. Early pregnancy is considered to resemble an open wound which requires a strong inflammatory response, thus the first trimester is considered a proinflammatory phase, which turns to an anti-inflammatory phase in the second trimester [20,21]. Although decidual macrophages exhibit an M2 phenotype and exert an immunosuppressive effect on local lymphocyte populations, in the context of local infection they may increase their production of proinflammatory cytokines and contribute to pregnancy disorders [19]. Of note, DSC or ESC have been shown to interact with macrophages in several ways [22,23]. In response to stimulation with lipopolysaccharide (LPS) from *Escherichia coli*, a coculture of ESC and PMA-differentiated THP-1 cells (human monocytes) produced enhanced levels of many cytokines (IL-1β, RANTES, MCP-1, IL-10, TGF-β, MIC-1, G-CSF) as compared to the respective monocultures [24]. Importantly, *B. abortus* is known to survive and replicate in macrophages from several animal species, inducing the secretion of proinflammatory cytokines [25–27].

The T-HESC cell line, derived from normal primary human ESC by telomerase immortalization, has been widely used to study several aspects of human ESC biology, including infection and cytokine production [23,24,28–31]. T-HESC are karyotypically, morphologically, and phenotypically similar to the primary parent cells, and after treatment with estradiol and medroxyprogesterone acetate (MPA) display the morphological and biochemical pattern of decidualization [32]. In the present study we evaluated the ability of *Brucella* spp. to infect and survive in decidualized T-HESC, and also assessed the cytokine production induced in these cells by the infection or by their interaction with infected macrophages.

## 2. Results

### 2.1. Brucella abortus Infects and Replicates in Both Decidualized and Non-Decidualized T-HESC Cells

Both decidualized and non-decidualized T-HESC endometrial cells were infected with *B. abortus* at a multiplicity of infection (MOI) of 250 bacteria/cell, and colony-forming units (CFU) of intracellular bacteria were determined at different times post-infection (p.i.). As shown in Figure 1, *B. abortus* was able to infect T-HESC cells in both conditions, although the initial number of intracellular bacteria (2 h p.i.) was slightly higher for non-decidualized cells (1125 ± 250 vs. 345 ± 32 CFU/well, mean ± SD). Besides wild type *B. abortus*, two additional strains carrying mutations in genes relevant for virulence were also tested for their capacity to infect and survive in T-HESC cells. These included a mutant lacking the *virB10* gene, widely reported as essential for the intracellular survival and replication of *Brucella* [33,34], and a double mutant lacking *btpA* and *btpB* genes which encode proteins able to interfere with TLR signaling [35,36]. As shown in Figure 1A, both mutant strains were able to infect decidualized and non-decidualized T-HESC at levels similar to the wild type strain. However, the ability to survive and replicate intracellularly differed between the *virB10* mutant and the other two strains. While CFU of intracellular bacteria increased along time for wild type *B. abortus* and the *btpAbtpB* mutant, showing intracellular replication, the CFU of the *virB10* mutant declined at the same time and no viable bacteria were detected in either condition at 48 h p.i. This later result confirmed in endometrial cells the essential role of *virB10* for the intracellular survival of *Brucella*.

Infection experiments were also carried out in the presence of specific inhibitors to examine whether *B. abortus* internalization by T-HESC cells depends on actin polymerization (cytochalasin D), microtubules (colchicine), or clathrin-mediated endocytosis (monodansylcadaverine, MDC). As shown in Figure 1B, *B. abortus* internalization was highly inhibited by cytochalasin D and to a lower extent by colchicine, but was not affected by MDC.

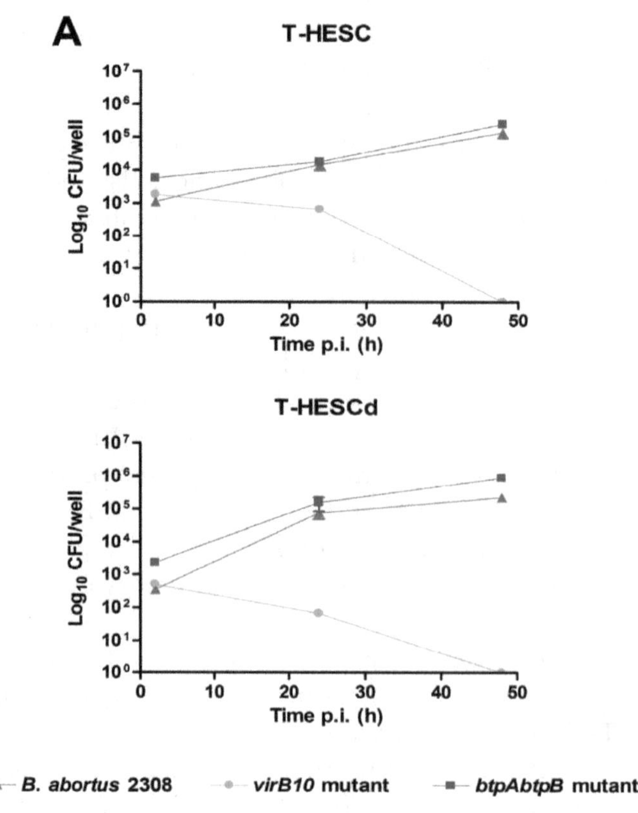


**Figure 1.** *Cont.*

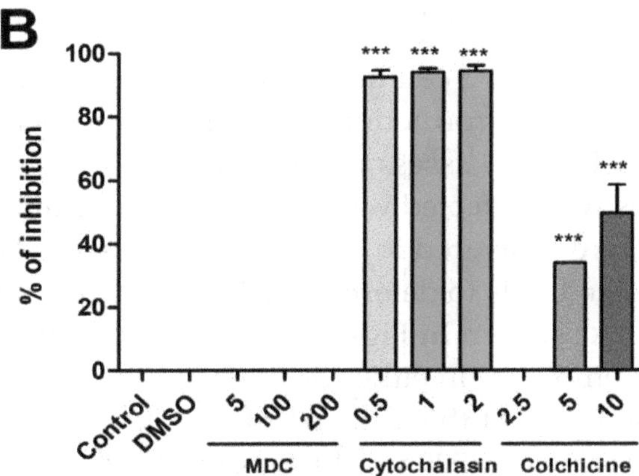


**Figure 1.** *Brucella abortus* invades and replicates in T-HESC cells. (**A**) Non-decidualized (T-HESC) and decidualized (T-HESCd) endometrial cells were infected with wild type *B. abortus* and two isogenic mutants (*virB10* and *btpAbtpB*), and colony forming unit (CFU) numbers of intracellular bacteria were determined at different times post-infection (p.i.). (**B**). Decidualized T-HESC were pretreated for 1 h with different doses of Colchicine (10, 5, 2.5 μM), Monodansylcadaverine (MDC; 200, 100, 5 μM), Cytochalasin D (2, 1, 0.5 μg/mL), or DMSO (vehicle) before infection with wild type *B. abortus*. Intracellular CFU were determined at 1 h p.i. Results are expressed as mean ± SD from three independent experiments run in duplicates. *** $p < 0.001$ versus control.

To assess whether the infection affected the viability of T-HESC cells or their decidualization status, the levels of lactate dehydrogenase (LDH) and prolactin were measured in culture supernatants of infected cells at 24 and 48 h p.i. and also in non-infected cells cultured in parallel. As shown in Figure 2, the infection with either wild type *B. abortus* or the *btpAbtpB* mutant did not modify the levels of LDH or prolactin as compared to non-infected cells at any time point, showing that it does not induce cytotoxicity or affect the decidualization of cells.

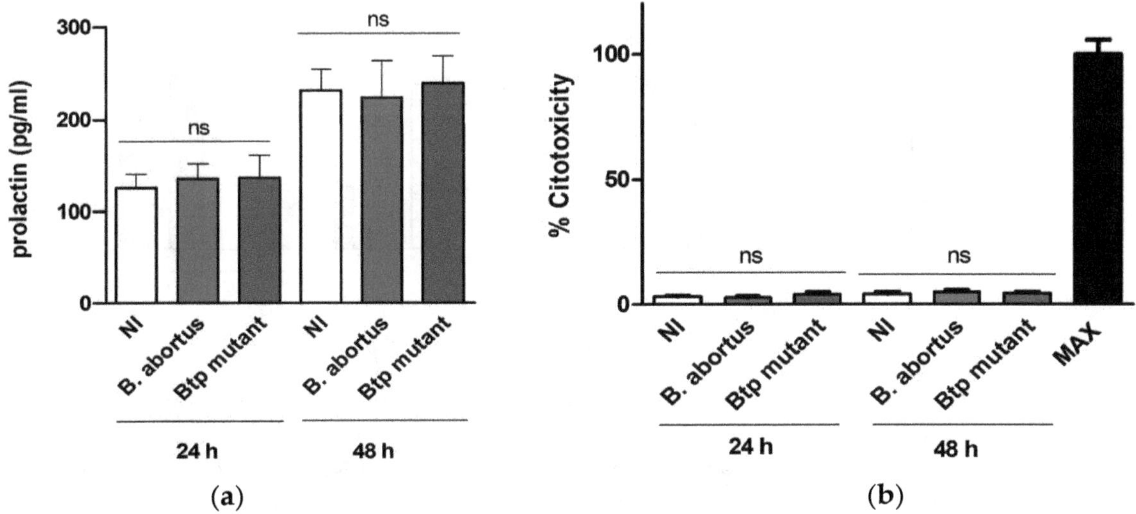


**Figure 2.** *B. abortus* infection does not induce cytotoxicity or alterations in decidualization in T-HESC cells. Decidualized T-HESC cells were infected or not (NI) with *B. abortus* wild type or its isogenic *btpAbtpB* mutant, and culture supernatants were harvested at 24 and 48 h p.i. to measure the levels of prolactin by ELISA (**a**) and the activity of lactate dehydrogenase (LDH) using a commercial non-radioactive cytotoxicity assay (**b**). In the latter assay, a control of 100% cell lysis (Max) was obtained by hypotonic lysis of the same number of non-infected cells. Results are expressed as mean ± SD from three independent experiments run in duplicates. ns: non-significant versus NI.

*2.2. B. abortus Infection Induces the Secretion of Proinflammatory Chemokines in T-HESC Cells*

As mentioned above, DSC and ESC cell lines express several TLRs and NLRs, and respond to microbial PAMPs with an enhanced production of proinflammatory cytokines, including MCP-1, IL-6, IL-8, IL-1β, and RANTES [17]. To assess the ability of *B. abortus* to induce a proinflammatory response in T-HESC, these cells were infected with the wild type strain and the *btpAbtpB* mutant, and the levels of IL-8 and MCP-1 were measured in culture supernatants. The studies were performed on decidualized and non-decidualized cells to determine whether the proinflammatory response depends on the decidualization status. As shown in Figure 3, the infection with any of the *B. abortus* strains elicited the secretion of both chemokines in either decidualized or non-decidualized T-HESC, and this effect was mostly evident at 48 h p.i. At this time point, IL-8 levels were higher in non-decidualized cells than in decidualized ones (mean, 8539 vs. 4948 pg/mL), whereas no significant difference was found for MCP-1 (mean, 3197 vs. 3621 pg/mL).

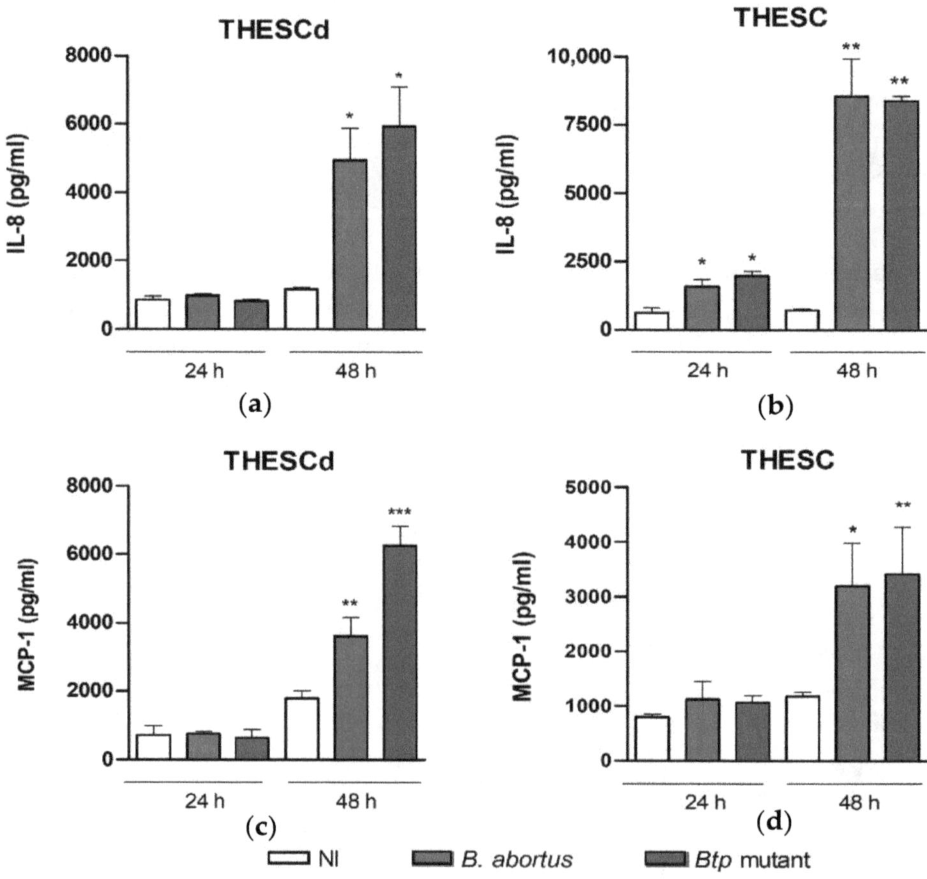


**Figure 3.** *B. abortus* infection elicits chemokine secretion in T-HESC cells. Decidualized (T-HESCd) (**a**,**c**) and non-decidualized (T-HESC) (**b**,**d**) endometrial cells were infected or not (NI) with wild type *B. abortus* and the *btpAbtpB* mutant, and the levels of IL-8 (**a**,**b**) and MCP-1 (**c**,**d**) were measured by ELISA in culture supernatants harvested at 24 or 48 h p.i. Results are expressed as mean ± SD from three independent experiments run in duplicates. * $p < 0.05$, ** $p < 0.01$, *** $p < 0.001$ versus NI.

To determine which signaling pathways may be involved in the induction of chemokine secretion, decidualized T-HESC cells were treated with SB203580 (p38 MAPK inhibitor), SP600125 (Jnk1/2 inhibitor), BAY 11-7082 (NF-κB inhibitor), or vehicle (dimethyl sulfoxide, DMSO) before and during the infection with *B. abortus*, and IL-8 and MCP-1 were measured as above. As shown in Figure 4, the secretion of both cytokines was not affected significantly by DMSO but was reduced to basal levels by all the inhibitors tested, suggesting that all the signaling pathways (p38, Jnk1/2, and NF-kB) are involved.

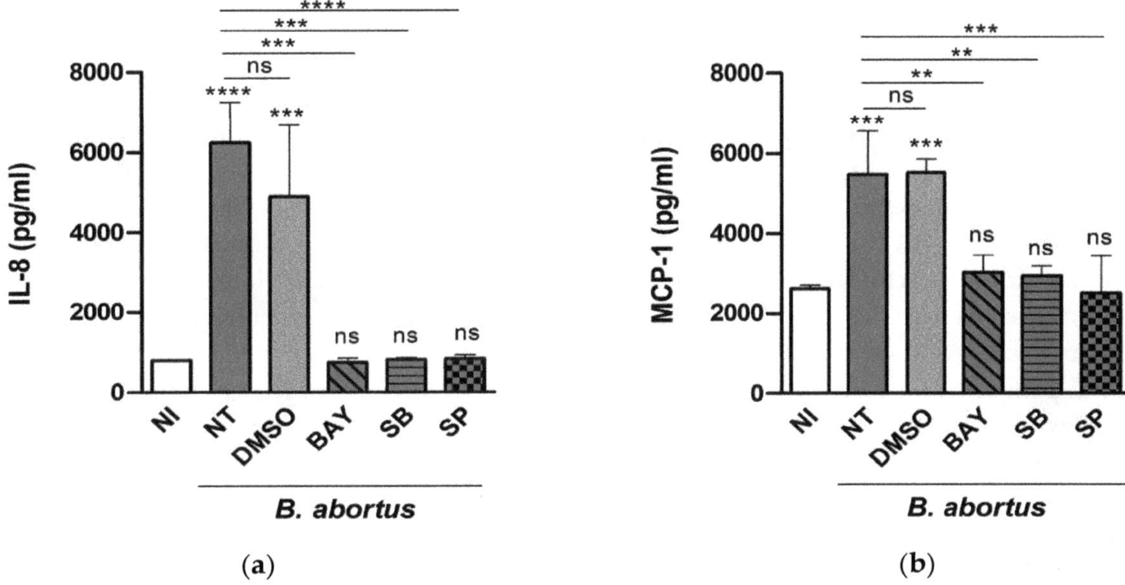


**Figure 4.** Several signaling pathways are involved in chemokine secretion in *Brucella*-infected T-HESC cells. Decidualized T-HESC cells were treated or not (NT) with SB203580 (SB, p38 MAPK inhibitor), SP600125 (SP, Jnk1/2 inhibitor), BAY 11-7082 (BAY, NF-κB inhibitor), or vehicle (DMSO) for 1 h, and were infected with wild type *B. abortus*. The inhibitors were kept throughout the experiment. At 48 h p.i. culture supernatants were harvested for measuring IL-8 (**a**) and MCP-1 (**b**) by ELISA. Non-treated non-infected cells (NI) served as controls. Results are expressed as mean ± SD from three independent experiments run in duplicates. Asterisks over bars indicate *** $p < 0.001$ or **** $p < 0.0001$ versus NI. Asterisks over lines indicate ** $p < 0.01$, *** $p < 0.001$ or **** $p < 0.0001$ versus NT. ns: non-significant.

Given the ability of *B. abortus* infection to induce the secretion of IL-8 and MCP-1 in T-HESC cells, experiments were carried out to determine whether such responses can be also elicited by stimulation with *B. abortus* antigens or, conversely, depend on bacterial viability. For this purpose, cells were stimulated with either heat-killed *B. abortus* (HKBA), or LPS or outer membrane vesicles (OMVs) from this bacterium, and chemokine levels were measured at 48 h p.i. As shown in Figure 5, HKBA (at $10^9$ CFU/mL) elicited IL-8 and MCP-1 secretion by T-HESC cells, albeit at lower levels than those attained by the infection. In addition, IL-8 secretion was significantly induced by *B. abortus* OMVs. These results show that the induction of chemokines in these cells does not depend on *Brucella* viability.

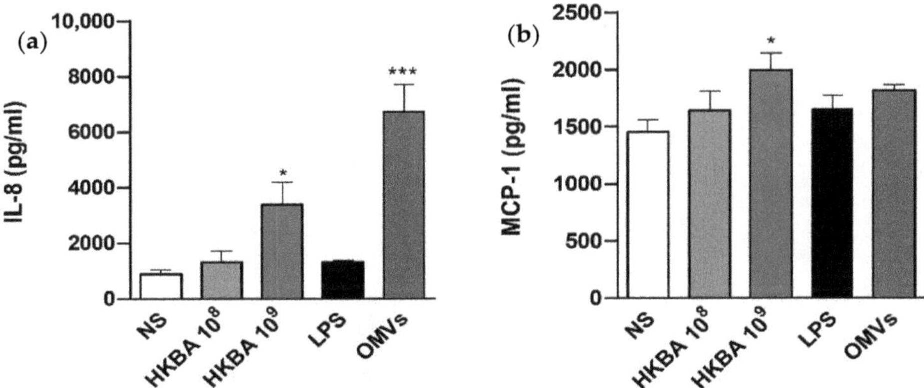


**Figure 5.** The chemokine response of T-HESC to *B. abortus* does not require bacterial viability. Decidualized T-HESC cells were stimulated or not (NS) with two doses ($10^8$ and $10^9$ CFU/mL) of heat-killed *B. abortus* (HKBA), or with lipopolysaccharide (LPS) or outer membrane vesicles (OMVs) from this bacterium, and IL-8 (**a**) and MCP-1 (**b**) levels were measured in culture supernatants at 48 h post-stimulation. Results are expressed as mean ± SD from three independent experiments run in duplicates. * $p < 0.05$, *** $p < 0.001$ versus NS.

*2.3. Factors Produced by Brucella-Infected Macrophages Stimulate Proinflammatory Responses in Decidualized T-HESC Cells*

The results shown above demonstrate that decidualized T-HESC produce proinflammatory mediators in response to infection with *B. abortus* or stimulation with its antigens. In the context of infection in the pregnant uterus, however, endometrial cells may also receive stimulation by factors produced by adjacent infected macrophages [22,23]. To model this scenario in vitro, decidualized T-HESC cells were stimulated with conditioned media (CM) from *B. abortus*-infected macrophages and the levels of proinflammatory cytokines were measured in culture supernatants 24 h later. The preexisting levels of these cytokines in the CM were subtracted in order to calculate the secretion specifically induced by the stimulation. As shown in Figure 6, stimulation with CM from *Brucella*-infected macrophages induced the production of IL-6, MCP-1, and IL-8 in a dose-dependent manner (higher secretion for stimulation with CM diluted at 1/2). No significant secretion of any of these cytokines was induced by stimulation with CM from non-infected monocytes. Previous similar studies on the stimulation of other non-phagocytic cells have shown that IL-1β and TNF-α are involved in the inducing effect of CM from *Brucella*-infected macrophages. To test whether these cytokines are also involved in the stimulation of IL-6, IL-8, and MCP-1 in decidualized T-HESC cells, experiments were performed in which CM were preincubated with a TNF-neutralizing antibody or T-HESC were preincubated with the natural antagonist of the IL-1 receptor (IL-1Ra). As shown in Figure 6, the stimulating effect of the CM on the secretion of IL-6 was significantly reduced by both pretreatments, implying that both TNF-α and IL-1β are involved. For MCP-1 and IL-8, in contrast, only the preincubation with the anti-TNF antibody produced a significant reduction. Although the isotype control also produced a significant reduction of MCP-1 levels, the reducing effect of the specific anti-TNF antibody was much more pronounced. In summary, TNF-α and/or IL-1β are involved in the ability of CM from *Brucella*-infected macrophages to stimulate the production of proinflammatory cytokines by decidualized T-HESC.

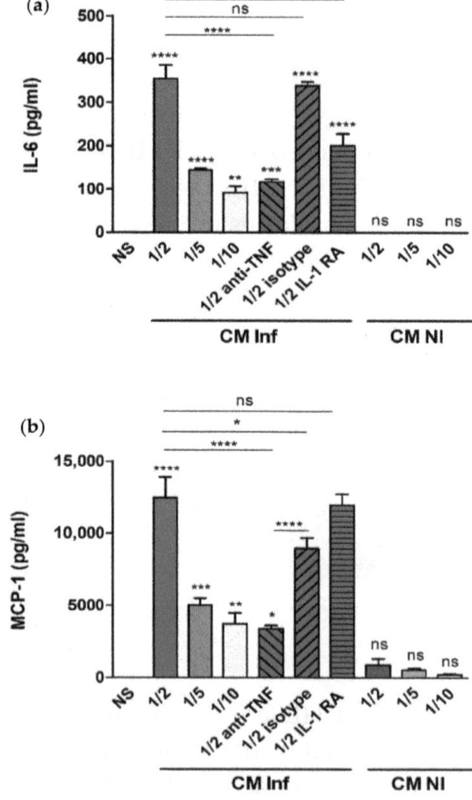


**Figure 6.** *Cont.*

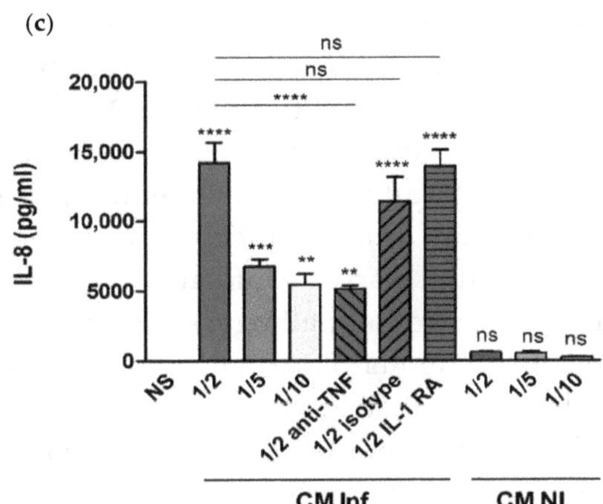


**Figure 6.** Factors produced by *B. abortus*–infected macrophages stimulate cytokine production by endometrial cells. Decidualized T-HESC cells were stimulated or not (NS) with conditioned media from *B. abortus*-infected macrophages (CM Inf) or from uninfected macrophages (CM NI) at different dilutions (1/2, 1/5 or 1/10), and 24 h later culture supernatants were harvested to measure IL-6 (**a**), MCP-1 (**b**) and IL-8 (**c**) levels. In parallel experiments, CM Inf was preincubated with a TNF-α neutralizing antibody or an isotype control before addition to cells, or T-HESC were preincubated with the natural antagonist of the IL-1 receptor (IL-1Ra) before stimulation with CM Inf, and cytokine levels were measured as described. Results are expressed as mean ± SD from three independent experiments run in duplicates. * $p < 0.05$, ** $p < 0.01$, *** $p < 0.001$, **** $p < 0.0001$, ns: non-significant. Asterisks over bars indicate differences versus NS.

## 3. Discussion

*Brucella* infections have been associated with abortion in both humans and animals. Although the pathophysiology of this complication has not been completely elucidated, the inflammatory phenomena observed in the affected placenta [3,4] suggest that, as with other pathogens causing abortion, placental inflammation may have a role in *Brucella*-induced abortion. As *Brucella* can reach the placenta by the hematogenous route, the maternal decidua is probably the initial site of placental colonization [14,15]. Given the known ability of decidual cells to respond to microbial PAMPs with an enhanced production of proinflammatory cytokines, and the known deleterious effect of placental inflammation on gestation, we decided to assess the ability of *Brucella* spp. to colonize decidualized stromal endometrial cells (T-HESC) and to induce the production of proinflammatory cytokines.

As shown here, *B. abortus* was able to infect both decidualized and non-decidualized T-HESC cells, although the initial number of intracellular bacteria was slightly higher for non-decidualized cells. This may relate to the fact that decidualized cells form an organized layer thus exposing less membrane surface to the environment. In addition, the pathogen was able to survive and replicate inside these cells. These findings are in line with the reported ability of *B. abortus* for intracellular replication in several phagocytic and non-phagocytic cells, including macrophages, epithelial cells and trophoblasts [13,25,37]. It has been widely demonstrated that this ability for intracellular survival in different cell types depends on the expression of a type IV secretion system encoded by the *virB* operon, which allows *Brucella* to modulate phagosome-lysosome fusion [33,34]. In line with this, we found that a *B. abortus* mutant lacking the *virB10* gene was unable to survive and replicate inside decidualized and non-decidualized T-HESC despite a similar ability of invasion compared to the wild type strain. In contrast, a mutant lacking the genes for the BtpA and BtpB proteins that interfere with TLR signaling exhibited invasion and replication abilities similar to the wild type strain. Importantly, *B. abortus* infection did not induce cytotoxicity, nor did it affect the decidualization status of cells, suggesting that the decidua might sustain the infection in affected individuals.

The mechanisms for *Brucella* invasion of non-phagocytic cells may vary according to the cell type considered. Whereas actin polymerization and microtubules have been involved in many cells [37], internalization in Vero cells does not depend on microtubules but depends on clathrin-mediated endocytosis [38]. The requirements for invasion of endometrial cells have not been reported. We found that *B. abortus* internalization was inhibited by cytochalasin D and to a lower extent by colchicine, which inhibit actin polymerization and microtubule formation, respectively. In contrast, internalization was not affected by MDC, an inhibitor of clathrin-mediated endocytosis.

As mentioned previously, placental *Brucella* infection is accompanied by the infiltration of inflammatory cells [3,4], which suggests that placental inflammation may have a role in *Brucella*-induced abortion as it does in abortion triggered by other pathogens. Our results show that *B. abortus* infection elicits the secretion of IL-8 and MCP-1 in either decidualized or non-decidualized T-HESC cells. IL-8 levels were higher in non-decidualized cells than in decidualized ones, whereas no significant difference was found for MCP-1. The higher production of IL-8 in non-decidualized cells may relate to the higher number of intracellular bacteria found in this condition as compared to decidualized cells, a downmodulating effect of decidualization on IL-8 production [39], or both. Nonetheless, these results suggest that, although the decidualization status may influence the levels of some proinflammatory mediators, decidualized endometrial cells are capable of mediating a proinflammatory response to *B. abortus*. A few previous studies have shown that *Brucella* BtpA and BtpB proteins, which contain TIR motifs and can thus modulate TLR signaling, can reduce cytokine production in dendritic cells in vitro (IL-12, TNF-$\alpha$) and in lung tissues in vivo (IL-12, CXCL-1, MCP-1) [35,36]. However, the potential modulating role of these proteins in *Brucella*-infected endometrial cells was unknown. At variance with those previous studies, we did not detect significant differences in the levels of the two chemokines here evaluated (IL-8 and MCP-1) between T-HESC infected with the wild type *B. abortus* strain or the *btpAbtpB* mutant strain. These results agree with those reported for the same chemokines in *Brucella*-infected human trophoblasts [12], and add support to the hypothesis that the immune responses of professional phagocytes are more influenced by the action of Btp proteins than those of non-phagocytic cells.

The secretion of both cytokines was reduced to basal levels by all the inhibitors tested, suggesting that all the signaling pathways (p38, Jnk1/2, and NF-kB) are involved. In line with these findings, previous studies have shown that several signaling pathways are involved in cytokine production by different cell types in response to *B. abortus*. For example, CCL20 secretion by human bronchial epithelial cells depends on p38, Jnk1/2, Erk1/2, and NF-kB [40], whereas in murine astrocytes TNF-$\alpha$ secretion depends on p38 and Erk1/2 signaling pathways [41].

Previous studies in several non-phagocytic cells have shown that not only live *B. abortus* but also some of its antigens can elicit the production of proinflammatory cytokines [42–44]. In line with these reports, we found that HKBA and OMVs from *B. abortus* elicit IL-8 and/or MCP-1 secretion in T-HESC cells. Obviously, these findings imply that the induction of chemokines in these cells does not depend on *Brucella* viability. At variance with HKBA and OMVs, *B. abortus* LPS did not elicit the production of the chemokines analyzed. This result is in line with previous studies in other cell types, which demonstrated that *B. abortus* LPS is a poor inducer of proinflammatory responses [42–45]. In contrast, most inflammatory responses are triggered by outer membrane lipoproteins, which induce TLR2 signaling [45].

As shown in this study, decidualized T-HESC produce proinflammatory mediators, including MCP-1, in response to infection with *B. abortus* or stimulation with its antigens. However, the decidua contains not only DSC but also a significant proportion of macrophages [19], with which DSC can establish reciprocal interactions [22,23]. In addition, the number of decidual macrophages could eventually augment in the context of locally increased MCP-1 levels induced by an infectious process. Therefore, it can be speculated that, during *B. abortus* infection in the pregnant uterus, endometrial cells may respond not only to the stimulus of bacterial antigens but also to stimulation by factors produced by adjacent *Brucella*-infected macrophages. In support of this hypothesis, we found that the stimulation of decidualized T-HESC with CM from *B. abortus*-infected macrophages induced the production of IL-6, MCP-1, and IL-8 in a dose-dependent manner, a phenomenon not produced by stimulation with CM from non-infected monocytes. Additional studies using specific blocking agents revealed that IL-6 induction by CM is mediated by TNF-$\alpha$ and IL-1$\beta$, whereas the induction of MCP-1 and IL-8 is mediated by TNF-$\alpha$. These findings are similar to those reported for the interaction between *Brucella*-infected macrophages and human trophoblasts [12].

Overall, these results suggest a possible scenario in which DSC produce IL-6 and chemoattractants for monocytes/macrophages in response to *B. abortus* infection and/or in response to cytokines produced by *Brucella*-infected placental macrophages. Reciprocal stimulations between DSC and phagocytes may amplify these phenomena. These interactions may be long-lasting due to the ability of *Brucella* to survive and replicate within macrophages and DSC. Altogether, these proinflammatory responses may contribute to the gestational complications of brucellosis.

## 4. Materials and Methods

### *4.1. Reagents*

LPS from *Brucella abortus* 2308 was provided by Ignacio Moriyón (University of Navarra, Pamplona, Spain). The purity and the characteristics of this preparation have been published previously [46].

### *4.2. Cell lines*

A human endometrial stromal cell line (T-HESC) was kindly provided Dr. Andrea Randi (Human Biochemistry Department, School of Medicine, University of Buenos Aires). This cell line was derived from normal stromal cells obtained from an adult patient subjected to hysterectomy, and conserved the characteristics of the regular endometrial cells [32]. The line was obtained by immortalization by transfection of telomerase (hTERT) using a retroviral system, and expressed puromycin resistance genes. Cells were maintained in DMEM-F12 supplemented with 10% FCS, 50 U/mL penicillin, 50 μg/mL streptomycin, 2 mM glutamine and 500 ng/mL puromycin. Decidualization was achieved following published procedures [47]. Briefly, T-HESC ($5 \times 10^4$ cells/well) were treated with medroxyprogesterone acetate (MPA, $10^{-7}$ M) and dibutyryl cAMP (0.5 mM) for 8 days, changing the culture media every 48 h. Decidualization was evaluated by morphology and by prolactin levels measured by sandwich ELISA (R&D Systems). For infection assays, cells were cultured for 24 h in antibiotic-free culture medium.

### *4.3. Monocyte Isolation and Macrophage Differentiation*

Peripheral blood samples were obtained from healthy volunteers after approval by the Ethics Committee of the School of Pharmacy and Biochemistry (Approval 2194/17). Written informed consent was obtained from all volunteers. Human monocytes were isolated by standard procedures. Briefly, whole blood diluted with sterile phosphate-buffered saline (PBS) was carefully layered on Ficoll-Paque (density: 1.077 g/mL) and centrifuged at 400× *g* for 30 min. The layer containing peripheral blood mononuclear cells was carefully removed by pipetting and washed with PBS by centrifugation at 250× *g* for 10 min. The pellet was resuspended in RPMI 1640 medium supplemented with 1 mM glutamine and was incubated for 2 h in 24-well plates. After washing with sterile PBS to eliminate nonadherent cells, RPMI medium supplemented with 10% sera from the same donors and antibiotics (100 U/mL penicillin and 100 μg/mL streptomycin) was added to the adherent cells. Cells were incubated at 37 °C in a 5% $CO_2$ atmosphere for 7 days for in vitro macrophagic differentiation [48]. Antibiotics were removed 24 h prior to infection.

### *4.4. Bacterial Strains and Growth Conditions*

*B. abortus* 2308 (wild type strain), its isogenic *btpAbtpB* double mutant and *virB10* polar mutant were obtained from our collection. The strains were grown in tryptic soy broth at 37 °C with agitation. After two washes with sterile PBS, bacterial inocula were adjusted to the desired concentration in sterile PBS based on optical density readings. An aliquot of each suspension was plated on tryptic soy agar (TSA) and incubated at 37 °C to determine the actual concentration of colony-forming units

(CFU) in the inocula. Cells were inoculated with *B. abortus* 2308 at an MOI of 200 and the plates were centrifuged for 10 min at 1200 rpm at room temperature. After 2 h, culture medium was removed and replaced with medium containing gentamicin and streptomycin. All live *Brucella* manipulations were performed in biosafety level 3 facilities. To prepare HKBA, bacteria were washed in sterile PBS, heat killed at 70 °C for 30 min, aliquoted, and stored at −80 °C until use. The absence of bacterial viability was checked by plating on TSA.

### *4.5. Isolation of Outer Membrane Vesicles*

OMVs from *B. abortus* 2308 were obtained essentially as described previously [49]. Briefly, bacteria were grown as described above, harvested by centrifugation and washed twice in sterile PBS. The pellet was resuspended in Gerhardt-Wilson minimal medium at an $OD_{600\ nm}$ of 0.1 and cultured for 72 h (early stationary phase of growth). The culture was centrifuged, and the cell-free supernatant was filter-sterilized. The filtrate was centrifuged at 15,000× *g* for 5 h at 4 °C. The pellets containing the OMVs were resuspended in PBS, and protein concentration was measured by the bicinchoninic acid assay (Pierce). The presence of OMVs was corroborated by electron microscopy. OMVs were stored at −20 °C until use.

### *4.6. Stimulation of T-HESC Cells with Brucella antigens*

Decidualized T-HESC cells ($5 \times 10^4$ cells/well) were stimulated with LPS from *B. abortus* (1 µg/mL), OMVs (1 µg/mL of protein), or HKBA ($10^9$ or $10^8$ CFU/mL). Cells were cultured at 37 °C in a 5% $CO_2$ atmosphere, and supernatants were collected 48 h after stimulation for chemokine measurement.

### *4.7. Cellular Infections*

Decidualized and non-decidualized T-HESC cells were infected with *B. abortus* 2308 at MOI of 250 bacteria/cell. Monocyte-derived macrophages were infected at MOI 100 bacteria/cell in culture medium containing no antibiotics. After dispensing the bacterial suspension, the plates were centrifuged (10 min at 400× *g*) and then incubated for 2 h at 37 °C in a 5% $CO_2$ atmosphere. Non-internalized bacteria were eliminated by several washes with medium alone followed by incubation in medium supplemented with 100 µg/mL gentamicin and 50 µg/mL streptomycin. After that, cells were washed and then incubated with culture medium without antibiotics. At different times post-infection (2, 24 or 48 h) culture supernatants were harvested for cytokine measurement, while the cells were washed with sterile PBS and lysed with 0.2% Triton X-100. Serial dilutions of the lysates were plated on TSA to enumerate intracellular CFU. In addition, the levels of prolactin were measured in culture supernatants as described above to assess the impact of infection on the decidualization status of the cells, and the levels of LDH were measured to assess cytotoxicity.

### *4.8. Evaluation of Cytotoxicity*

To analyze the effect of infection on cell integrity, the release of lactate dehydrogenase (LDH) from infected T-HESC cells was determined. LDH activity was determined using the CytotTox 96 Non-Radiactive Cytotoxicity Assay (Promega, USA) in culture supernatants obtained at 24 and 48 h p.i. Results were expressed as the ratio between LDH levels measured in the samples (infected or non-infected cultures) and those corresponding to a 100% cell lysis (obtained by hypotonic lysis of the same number of cells).

### *4.9. Internalization Pathways*

To assess the role of microtubules, actin or clathrin in *B. abortus* internalization, decidualized T-HESC were pretreated for 1 h with different doses of colchicine (10, 5, 2.5 µM, Sigma), monodansylcadaverine (MDC; 200, 100, 5 µM) or cytochalasin D (2, 1, 0.5 µg/mL, Sigma) and were later infected as described above but in the presence of these inhibitors. MDC and cytochalasin were solubilized in dimethyl sulfoxide (DMSO), and in all the experiments control cells were incubated without inhibitor or with DMSO for the same period as treated cells. Intracellular CFU were determined at 2 h p.i. as described above.

### 4.10. *Stimulation of T-HESC with Conditioned Media (CM) from Brucella-Infected Macrophages*

CM from macrophages infected with *B. abortus* 2308 (MOI 100) were harvested at 24 h p.i., filter-sterilized and used to stimulate noninfected decidualized T-HESC cells. After 24 and 48 h, supernatants from stimulated cultures were harvested to measure cytokines. The preexisting levels of cytokines in the CM were subtracted in order to calculate the secretion specifically induced by the stimulation. To determine if TNF-α might be involved in the stimulating effects of CM, in some experiments CM were preincubated for 1 h at 37 °C with a neutralizing monoclonal antibody against TNF-α or its isotype control (both from BD Pharmingen) before being transferred to T-HESC cultures. Alternatively, to determine the role of IL-1 β in the stimulating effect, decidualized T-HESC cells were preincubated with the IL-1β receptor antagonist IL-1Ra (R&D Systems) for 1 h at 37 °C before stimulation with CM from infected macrophages.

### 4.11. *Inhibition of Signaling Pathways*

To examine the signaling pathways involved in cytokine secretion, decidualized T-HESC cells were pretreated with 10μM SB203580 (p38 MAPK inhibitor, Gibco), 10 μM SP600125 (Jnk1/2 inhibitor, Sigma), 2.5 μM BAY11-7082 (NF-κB inhibitor, Sigma) or vehicle (DMSO). These reagents were added one hour before infection with *B. abortus* and were kept throughout the experiment (48 h). Cell viability after incubation with these inhibitors was higher than 90%, as assessed by staining with trypan blue.

### 4.12. *Measurement of Cytokines and Chemokines*

Levels of human IL-6, IL-8, MCP-1, and TNF-α were measured in culture supernatants by sandwich ELISA, using paired cytokine-specific monoclonal antibodies according to the manufacturer's instructions (BD Pharmingen).

### 4.13. *Statistical Analysis*

Each experiment was performed in duplicates on three independent occasions. The values obtained are presented as the mean ± SD. Statistical analysis was performed with one-way ANOVA, followed by Post Hoc Tukey's Test or Dunnett's Test using GraphPad Prism 6.0 software.

**Author Contributions:** Conceptualization, M.C.F., A.M.C., and P.C.B.; Methodology, L.Z., M.C.F., I.M.A.P., and A.D.S.; Formal analysis, L.Z., M.C.F., A.M.C., and P.C.B.; Writing—original draft preparation, L.Z., M.C.F., I.M.A.P., A.D.S., A.M.C., and P.C.B.; Supervision, A.M.C. and P.C.B. All authors have read and agreed to the published version of the manuscript.

**Acknowledgments:** We thank Andrea Randi for providing the T-HESC cells. We are grateful to the staff of the BSL3 facility of INBIRS for expert assistance.

## References

1. Pappas, G.; Papadimitriou, P.; Akritidis, N.; Christou, L.; Tsianos, E.V. The new global map of human brucellosis. *Lancet Infect. Dis.* **2006**, *6*, 91–99. [CrossRef]
2. Pappas, G.; Akritidis, N.; Bosilkovski, M.; Tsianos, E. Brucellosis. *N. Engl. J. Med.* **2005**, *352*, 2325–2336. [CrossRef] [PubMed]

3. Carmichael, L.E.; Kenney, R.M. Canine abortion caused by Brucella canis. *J. Am. Vet. Med. Assoc.* **1968**, *152*, 605–616.
4. Carvalho, A.V.; Mol, J.P.S.; Xavier, M.N.; Paixão, T.A.; Lage, A.P.; Santos, R.L. Pathogenesis of bovine brucellosis. *Vet. J.* **2010**, *184*, 146–155.
5. Bildfell, R.J.; Thomson, G.W.; Haines, D.M.; McEwen, B.J.; Smart, N. Coxiella burnetii infection is associated with placentitis in cases of bovine abortion. *J. Vet. Diagn. Invest.* **2000**, *12*, 419–425. [CrossRef]
6. Buxton, D.; Anderson, I.E.; Longbottom, D.; Livingstone, M.; Wattegedera, S.; Entrican, G. Ovine chlamydial abortion: Characterization of the inflammatory immune response in placental tissues. *J. Comp. Pathol.* **2002**, *127*, 133–141. [CrossRef]
7. Chattopadhyay, A.; Robinson, N.; Sandhu, J.K.; Finlay, B.B.; Sad, S.; Krishnan, L. Salmonella enterica Serovar Typhimurium-Induced Placental Inflammation and Not Bacterial Burden Correlates with Pathology and Fatal Maternal Disease. *Infect. Immun.* **2010**, *78*, 2292–2301. [CrossRef]
8. Lulu, A.R.; Araj, G.F.; Khateeb, M.I.; Mustafa, M.Y. Human Brucellosis in Kuwait: A Prospective Study of 400 Cases. *Q J. Med.* **1988**, *66*, 39–54.
9. Sarram, M.; Feiz, J.; Foruzandeh, M.; Gazanfarpour, P. Intrauterine fetal infection with Brucella melitensis as a possible cause of second-trimester abortion. *Am. J. Obstet. Gynecol.* **1974**, *119*, 657–660. [CrossRef]
10. Makhseed, M.; Harouny, A.; Araj, G.; Moussa, M.A.; Sharma, P. Obstetric and gynecologic implication of brucellosis in Kuwait. *J. Perinatol.* **1998**, *18*, 196–199.
11. Khan, M.Y.; Mah, M.W.; Memish, Z.A. Brucellosis in Pregnant Women. *Clin. Infect. Dis.* **2001**, *32*, 1172–1177. [CrossRef] [PubMed]
12. Fernández, A.G.; Ferrero, M.C.; Hielpos, M.S.; Fossati, C.A.; Baldi, P.C. Proinflammatory Response of Human Trophoblastic Cells to Brucella abortus Infection and upon Interactions with Infected Phagocytes. *Biol. Reprod.* **2016**, *94*, 48. [CrossRef] [PubMed]
13. Salcedo, S.P.; Chevrier, N.; Lacerda, T.L.S.; Ben Amara, A.; Gerart, S.; Gorvel, V.A.; De Chastellier, C.; Blasco, J.M.; Mege, J.L.; Gorvel, J.P. Pathogenic brucellae replicate in human trophoblasts. *J. Infect. Dis.* **2013**, *207*, 1075–1083. [CrossRef] [PubMed]
14. Robbins, J.R.; Bakardjiev, A.I. Pathogens and the placental fortress. *Curr. Opin. Microbiol.* **2012**, *15*, 36–43. [CrossRef]
15. Vigliani, M.B.; Bakardjiev, A.I. Intracellular organisms as placental invaders. *Fetal Matern. Med. Rev.* **2014**, *25*, 332–338. [CrossRef]
16. Dunn, C.L.; Kelly, R.W.; Critchley, H.O.D. Decidualization of the human endometrial stromal cell: An enigmatic transformation. *Reprod. Biomed. Online* **2003**, *7*, 151–161. [CrossRef]
17. Anders, A.P.; Gaddy, J.A.; Doster, R.S.; Aronoff, D.M. Current concepts in maternal-fetal immunology: Recognition and response to microbial pathogens by decidual stromal cells. *Am. J. Reprod. Immunol.* **2017**, *77*, e12623. [CrossRef]
18. Castro-Leyva, V.; Zaga-Clavellina, V.; Espejel-Nuñez, A.; Vega-Sanchez, R.; Flores-Pliego, A.; Reyes-Muñoz, E.; Giono-Cerezo, S.; Nava-Salazar, S.; Espino y Sosa, S.; Estrada-Gutierrez, G. Decidualization Mediated by Steroid Hormones Modulates the Innate Immunity in Response to Group B Streptococcal Infection in vitro. *Gynecol. Obstet. Invest.* **2017**, *82*, 592–600. [CrossRef]
19. Nagamatsu, T.; Schust, D.J. Review: The Immunomodulatory Roles of Macrophages at the Maternal-Fetal Interface. *Reprod. Sci.* **2010**, *17*, 209–218. [CrossRef]
20. Mor, G.; Cardenas, I.; Abrahams, V.; Guller, S. Inflammation and pregnancy: The role of the immune system at the implantation site. *Ann. N. Y. Acad. Sci.* **2011**, *1221*, 80–87. [CrossRef]
21. Mor, G.; Cardenas, I. The Immune System in Pregnancy: A Unique Complexity. *Am. J. Reprod. Immunol.* **2010**, *63*, 425–433. [CrossRef] [PubMed]
22. Oreshkova, T.; Dimitrov, R.; Mourdjeva, M. A Cross-Talk of Decidual Stromal Cells, Trophoblast, and Immune Cells: A Prerequisite for the Success of Pregnancy. *Am. J. Reprod. Immunol.* **2012**, *68*, 366–373. [CrossRef] [PubMed]
23. Eyster, K.M.; Hansen, K.A.; Winterton, E.; Klinkova, O.; Drappeau, D.; Mark-Kappeler, C.J. Reciprocal Communication Between Endometrial Stromal Cells and Macrophages. *Reprod. Sci.* **2010**, *17*, 809–822. [CrossRef] [PubMed]

24. Rogers, L.M.; Anders, A.P.; Doster, R.S.; Gill, E.A.; Gnecco, J.S.; Holley, J.M.; Randis, T.M.; Ratner, A.J.; Gaddy, J.A.; Osteen, K.; et al. Decidual stromal cell-derived PGE $_2$ regulates macrophage responses to microbial threat. *Am. J. Reprod. Immunol.* **2018**, *80*, e13032. [CrossRef] [PubMed]
25. Celli, J. Surviving inside a macrophage: The many ways of Brucella. *Res. Microbiol.* **2006**, *157*, 93–98. [CrossRef]
26. Zhan, Y.; Cheers, C. Differential induction of macrophage-derived cytokines by live and dead intracellular bacteria in vitro. *Infect. Immun.* **1995**, *63*, 720–723. [CrossRef]
27. Covert, J.; Mathison, A.J.; Eskra, L.; Banai, M.; Splitter, G. Brucella melitensis, B. neotomae and B. ovis elicit common and distinctive macrophage defense transcriptional responses. *Exp. Biol. Med.* **2009**, *234*, 1450–1467. [CrossRef]
28. Pagani, I.; Ghezzi, S.; Ulisse, A.; Rubio, A.; Turrini, F.; Garavaglia, E.; Candiani, M.; Castilletti, C.; Ippolito, G.; Poli, G.; et al. Human Endometrial Stromal Cells Are Highly Permissive To Productive Infection by Zika Virus. *Sci. Rep.* **2017**, *7*, 44286. [CrossRef]
29. Lu, Y.; Bocca, S.; Anderson, S.; Wang, H.; Manhua, C.; Beydoun, H.; Oehninger, S. Modulation of the Expression of the Transcription Factors T-Bet and GATA-3 in Immortalized Human Endometrial Stromal Cells (HESCs) by Sex Steroid Hormones and cAMP. *Reprod. Sci.* **2013**, *20*, 699–709. [CrossRef]
30. Gellersen, B.; Wolf, A.; Kruse, M.; Schwenke, M.; Bamberger, A.-M. Human Endometrial Stromal Cell-Trophoblast Interactions: Mutual Stimulation of Chemotactic Migration and Promigratory Roles of Cell Surface Molecules CD82 and CEACAM11. *Biol. Reprod.* **2013**, *88*, 80. [CrossRef]
31. Grasso, E.; Gori, S.; Soczewski, E.; Fernández, L.; Gallino, L.; Vota, D.; Martínez, G.; Irigoyen, M.; Ruhlmann, C.; Lobo, T.F.; et al. Impact of the Reticular Stress and Unfolded Protein Response on the inflammatory response in endometrial stromal cells. *Sci. Rep.* **2018**, *8*, 12274. [CrossRef] [PubMed]
32. Krikun, G.; Mor, G.; Alvero, A.; Guller, S.; Schatz, F.; Sapi, E.; Rahman, M.; Caze, R.; Qumsiyeh, M.; Lockwood, C.J. A Novel Immortalized Human Endometrial Stromal Cell Line with Normal Progestational Response. *Endocrinology* **2004**, *145*, 2291–2296. [CrossRef] [PubMed]
33. Sieira, R.; Comerci, D.J.; Sanchez, D.O.; Ugalde, R.A. A homologue of an operon required for DNA transfer in Agrobacterium is required in Brucella abortus for virulence and intracellular multiplication. *J. Bacteriol.* **2000**, *182*, 4849–4855. [CrossRef] [PubMed]
34. Comerci, D.J.; Martínez-Lorenzo, M.J.; Sieira, R.; Gorvel, J.P.; Ugalde, R.A. Essential role of the virB machinery in the maturation of the Brucella abortus-containing vacuole. *Cell. Microbiol.* **2001**, *3*, 159–168. [CrossRef]
35. Salcedo, S.P.; Marchesini, M.I.; Degos, C.; Terwagne, M.; Von Bargen, K.; Lepidi, H.; Herrmann, C.K.; Santos Lacerda, T.L.; Imbert, P.R.C.; Pierre, P.; et al. BtpB, a novel Brucella TIR-containing effector protein with immune modulatory functions. *Front. Cell. Infect. Microbiol.* **2013**, *3*, 28. [CrossRef]
36. Hielpos, M.S.; Ferrero, M.C.; Fernández, A.G.; Falivene, J.; Vanzulli, S.; Comerci, D.J.; Baldi, P.C. Btp Proteins from Brucella abortus modulate the lung innate immune response to infection by the respiratory route. *Front. Immunol.* **2017**, *8*, 1011. [CrossRef]
37. Pizarro-Cerdá, J.; Moreno, E.; Gorvel, J. Invasion and intracellular trafficking of Brucella abortus in nonphagocytic cells. *Microbes Infect.* **2000**, *2*, 829–835. [CrossRef]
38. Detilleux, P.G.; Deyoe, B.L.; Cheville, N.F. Effect of endocytic and metabolic inhibitors on the internalization and intracellular growth of Brucella abortus in Vero cells. *Am. J. Vet. Res.* **1991**, *52*, 1658–1664.
39. Lockwood, C.J.; Kumar, P.; Krikun, G.; Kadner, S.; Dubon, P.; Critchley, H.; Schatz, F. Effects of thrombin, hypoxia, and steroids on interleukin-8 expression in decidualized human endometrial stromal cells: Implications for long-term progestin-only contraceptive-induced bleeding. *J. Clin. Endocrinol. Metab.* **2004**, *89*, 1467–1475. [CrossRef]
40. Hielpos, M.S.; Ferrero, M.C.; Fernández, A.G.; Bonetto, J.; Giambartolomei, G.H.; Fossati, C.A.; Baldi, P.C. CCL20 and beta-defensin 2 production by human lung epithelial cells and macrophages in response to Brucella abortus infection. *PLoS ONE* **2015**, *10*, e0140408. [CrossRef]
41. Miraglia, M.C.; Scian, R.; Samartino, C.G.; Barrionuevo, P.; Rodriguez, A.M.; Ibañez, A.E.; Coria, L.M.; Velásquez, L.N.; Baldi, P.C.; Cassataro, J.; et al. Brucella abortus induces TNF-$\alpha$-dependent astroglial MMP-9 secretion through mitogen-activated protein kinases. *J. Neuroinflamm.* **2013**, *10*, 47. [CrossRef] [PubMed]
42. Ferrero, M.C.; Bregante, J.; Delpino, M.V.; Barrionuevo, P.; Fossati, C.A.; Giambartolomei, G.H.; Baldi, P.C. Proinflammatory response of human endothelial cells to Brucella infection. *Microbes Infect.* **2011**, *13*, 852–861. [CrossRef] [PubMed]

43. Scian, R.; Barrionuevo, P.; Giambartolomei, G.H.; De Simone, E.A.; Vanzulli, S.I.; Fossati, C.A.; Baldi, P.C.; Delpino, M.V. Potential role of fibroblast-like synoviocytes in joint damage induced by Brucella abortus infection through production and induction of matrix metalloproteinases. *Infect. Immun.* **2011**, *79*, 3619–3632. [CrossRef] [PubMed]
44. Ferrero, M.C.; Fossati, C.A.; Baldi, P.C. Direct and monocyte-induced innate immune response of human lung epithelial cells to Brucella abortus infection. *Microbes Infect.* **2010**, *12*, 736–747. [CrossRef]
45. Giambartolomei, G.H.; Zwerdling, A.; Cassataro, J.; Bruno, L.; Fossati, C.A.; Philipp, M.T. Lipoproteins, not lipopolysaccharide, are the key mediators of the proinflammatory response elicited by heat-killed Brucella abortus. *J. Immunol.* **2004**, *173*, 4635–4642. [CrossRef]
46. Velasco, J.; Bengoechea, J.A.; Brandenburg, K.; Lindner, B.; Seydel, U.; González, D.; Zähringer, U.; Moreno, E.; Moriyón, I. Brucella abortus and its closest phylogenetic relative, Ochrobactrum spp., differ in outer membrane permeability and cationic peptide resistance. *Infect. Immun.* **2000**, *68*, 3210–3218. [CrossRef]
47. Grasso, E.; Gori, S.; Paparini, D.; Soczewski, E.; Fernández, L.; Gallino, L.; Salamone, G.; Martinez, G.; Irigoyen, M.; Ruhlmann, C.; et al. VIP induces the decidualization program and conditions the immunoregulation of the implantation process. *Mol. Cell. Endocrinol.* **2018**, *460*, 63–72. [CrossRef]
48. Daigneault, M.; Preston, J.A.; Marriott, H.M.; Whyte, M.K.B.; Dockrell, D.H. The identification of markers of macrophage differentiation in PMA-stimulated THP-1 cells and monocyte-derived macrophages. *PLoS ONE* **2010**, *5*, e8668. [CrossRef]
49. Pollak, C.N.; Delpino, M.V.; Fossati, C.A.; Baldi, P.C. Outer Membrane Vesicles from Brucella abortus Promote Bacterial Internalization by Human Monocytes and Modulate Their Innate Immune Response. *PLoS ONE* **2012**, *7*, e50214. [CrossRef]

# Permissions

All chapters in this book were first published by MDPI; hereby published with permission under the Creative Commons Attribution License or equivalent. Every chapter published in this book has been scrutinized by our experts. Their significance has been extensively debated. The topics covered herein carry significant findings which will fuel the growth of the discipline. They may even be implemented as practical applications or may be referred to as a beginning point for another development.

The contributors of this book come from diverse backgrounds, making this book a truly international effort. This book will bring forth new frontiers with its revolutionizing research information and detailed analysis of the nascent developments around the world.

We would like to thank all the contributing authors for lending their expertise to make the book truly unique. They have played a crucial role in the development of this book. Without their invaluable contributions this book wouldn't have been possible. They have made vital efforts to compile up to date information on the varied aspects of this subject to make this book a valuable addition to the collection of many professionals and students.

This book was conceptualized with the vision of imparting up-to-date information and advanced data in this field. To ensure the same, a matchless editorial board was set up. Every individual on the board went through rigorous rounds of assessment to prove their worth. After which they invested a large part of their time researching and compiling the most relevant data for our readers.

The editorial board has been involved in producing this book since its inception. They have spent rigorous hours researching and exploring the diverse topics which have resulted in the successful publishing of this book. They have passed on their knowledge of decades through this book. To expedite this challenging task, the publisher supported the team at every step. A small team of assistant editors was also appointed to further simplify the editing procedure and attain best results for the readers.

Apart from the editorial board, the designing team has also invested a significant amount of their time in understanding the subject and creating the most relevant covers. They scrutinized every image to scout for the most suitable representation of the subject and create an appropriate cover for the book.

The publishing team has been an ardent support to the editorial, designing and production team. Their endless efforts to recruit the best for this project, has resulted in the accomplishment of this book. They are a veteran in the field of academics and their pool of knowledge is as vast as their experience in printing. Their expertise and guidance has proved useful at every step. Their uncompromising quality standards have made this book an exceptional effort. Their encouragement from time to time has been an inspiration for everyone.

The publisher and the editorial board hope that this book will prove to be a valuable piece of knowledge for researchers, students, practitioners and scholars across the globe.

# List of Contributors

**Angelino T. Tromp, Joris P. Jansen, Lisette M. Scheepmaker, Anneroos Velthuizen, Carla J.C. De Haas, Kok P.M. Van Kessel, Bart W. Bardoel, Jos A.G. Van Strijp, Robert Jan Lebbink and Pieter-Jan A. Haas**
Department of Medical Microbiology, University Medical Center Utrecht, 3584 CX Utrecht, The Netherlands

**Michiel Van Gent**
Department of Medical Microbiology, University Medical Center Utrecht, 3584 CX Utrecht, The Netherlands
Department of Microbiology, University of Chicago, Chicago, IL 60637, USA

**Michael T. McManus**
Department of Microbiology and Immunology, UCSF Diabetes Center, Keck Center for Noncoding RNA, University of California, San Francisco, San Francisco, CA 94143, USA

**Michael Boettcher**
Department of Microbiology and Immunology, UCSF Diabetes Center, Keck Center for Noncoding RNA, University of California, San Francisco, San Francisco, CA 94143, USA
Medical Faculty, Martin Luther University Halle-Wittenberg, 06120 Halle (Saale), Germany

**András N. Spaan**
Department of Medical Microbiology, University Medical Center Utrecht, 3584 CX Utrecht, The Netherlands
St. Giles Laboratory of Human Genetics of Infectious Diseases, Rockefeller Branch, The Rockefeller University, New York, NY 10065, USA

**Mathilde Van der Henst, Elodie Carlier and Xavier De Bolle**
Unité de Recherche en Biologie des Microorganismes (URBM), University of Namur, 61 rue de Bruxelles, 5000 Namur, Belgium

**Manuel Wolters, Martin Christner, Anna Both and Holger Rohde**
Institute of Medical Microbiology, Virology and Hygiene, University Medical Center Hamburg-Eppendorf (UKE), 20251 Hamburg, Germany

**Hagen Frickmann**
Department of Microbiology and Hospital Hygiene, Bundeswehr Hospital Hamburg, 20359 Hamburg, Germany
Institute for Medical Microbiology, Virology and Hygiene, University Medicine Rostock, 18057 Rostock, Germany

**Kwabena Oppong and Charity Wiafe Akenten**
Kumasi Centre for Collaborative Research in Tropical Medicine (KCCR), Kumasi, Ghana

**Jürgen May**
Tropical Medicine II, University Medical Center Hamburg-Eppendorf (UKE), 20251 Hamburg, Germany
Department of Infectious Disease Epidemiology, Bernhard Nocht Institute for Tropical Medicine (BNITM), 20359 Hamburg, Germany

**Denise Dekker**
Department of Infectious Disease Epidemiology, Bernhard Nocht Institute for Tropical Medicine (BNITM), 20359 Hamburg, Germany
German Centre for Infection Research (DZIF), Hamburg-Lübeck-Borstel-Riems, 38124 Braunschweig, Germany

**Ana María Rodríguez, Agustina P. Melnyczajko, M. Cruz Miraglia, M. Victoria Delpino and Guillermo Hernán Giambartolomei**
Instituto de Inmunología, Genética y Metabolismo (INIGEM), CONICET, Facultad de Farmacia y Bioquímica, Universidad de Buenos Aires, Buenos Aires C1120AAD, Argentina

**Paula Barrionuevo and Aldana Trotta**
Instituto de Medicina Experimental (IMEX) (CONICET-Academia Nacional de Medicina), Buenos Aires C1425ASU, Argentina

**Kwang Sik Kim**
Division of Pediatric Infectious Diseases, Department of Pediatrics, Johns Hopkins University School of Medicine, Baltimore, MD 21287, USA

**Janine J. Wilden, Eike R. Hrincius and Yvonne Boergeling**
Institute of Virology Muenster (IVM), Westfaelische Wilhelms-University Muenster, 48149 Muenster, Germany

**Silke Niemann**
Institute of Medical Microbiology, Westfaelische Wilhelms-University Muenster, 48149 Muenster, Germany

**Bettina Löffler**
Institute of Medical Microbiology, Jena University Hospital, 07747 Jena, Germany
Cluster of Excellence EXC 2051 "Balance of the Microverse", FSU Jena, 07743 Jena, Germany

**Christina Ehrhardt**
Section of Experimental Virology, Institute of Medical Microbiology, Jena University Hospital, 07745 Jena, Germany

**Stephan Ludwig**
Institute of Virology Muenster (IVM), Westfaelische Wilhelms-University Muenster, 48149 Muenster, Germany
Cluster of Excellence EXC 1003 "Cells in Motion", WWU Muenster, 48149 Muenster, Germany

**Raiany Santos**
Department of Genetics, Institute of Biological Sciences, Federal University of Minas Gerais—Belo Horizonte, Minas Gerais 31270-901, Brazil

**Priscila C. Campos, Marcella Rungue, Victor Rocha, David Santos, Viviani Mendes, Fabio V. Marinho, Angelica T. Vieira and Sergio C. Oliveira**
Department of Biochemistry and Immunology, Institute of Biological Sciences, Federal University of Minas Gerais—Belo Horizonte, Minas Gerais 31270-901, Brazil

**Flaviano Martins**
Department of Microbiology, Institute of Biological Sciences, Federal University of Minas Gerais—Belo Horizonte, Minas Gerais 31270-901, Brazil

**Mayra F. Ricci, Diego C. dos Reis and Geovanni D. Cassali**
Department of General Pathology, Institute of Biological Sciences, Federal University of Minas Gerais—Belo Horizonte, Minas Gerais 31270-901, Brazil

**José Carlos Alves-Filho**
Department of Pharmacology, Ribeirao Preto Medical School, University of Sao Paulo, Ribeirao Preto 14049-900, Brazil

**Juselyn D. Tupik, Sheryl L. Coutermarsh-Ott, Angela H. Benton, Kellie A. King, Hanna D. Kiryluk and Clayton C. Caswell**
Department of Biomedical Sciences and Pathobiology, Virginia-Maryland College of Veterinary Medicine, Virginia Tech, Blacksburg, VA 24061, USA

**Irving C. Allen**
Department of Biomedical Sciences and Pathobiology, Virginia-Maryland College of Veterinary Medicine, Virginia Tech, Blacksburg, VA 24061, USA
Department of Basic Science Education, Virginia Tech Carilion School of Medicine, Roanoke, VA 24016, USA

**Motoyasu Miyazaki**
Department of Pharmacy, Fukuoka University Chikushi Hospital, Fukuoka 818-8502, Japan

**Carmen Lozano, Rosa Fernández-Fernández, Laura Ruiz-Ripa, Paula Gómez, Myriam Zarazaga and Carmen Torres**
Area of Biochemistry and Molecular Biology, University of La Rioja, 26006 Logroño, Spain

**Inés Reigada, Kirsi Savijoki and Adyary Fallarero**
Drug Research Program, Division of Pharmaceutical Biosciences, Faculty of Pharmacy, University of Helsinki, FI-00014 Helsinki, Finland

**Clara Guarch-Pérez, Martijn Riool and Sebastian A. J. Zaat**
Department of Medical Microbiology and Infection Prevention, Amsterdam institute for Infection and Immunity, Amsterdam UMC, University of Amsterdam, 1105 AZ Amsterdam, The Netherlands

**Jayendra Z. Patel and Jari Yli-Kauhaluoma**
Drug Research Program, Division of Pharmaceutical Chemistry and Technology, Faculty of Pharmacy, University of Helsinki, FI-00014 Helsinki, Finland

**Magalí G. Bialer and Gabriela Sycz**
Fundación Instituto Leloir (FIL), IIBBA (CONICET-FIL), Buenos Aires 1405, Argentina

**Florencia Muñoz González, Mariana C. Ferrero and Pablo C. Baldi**
Cátedra de Inmunología, Facultad de Farmacia y Bioquímica, Universidad de Buenos Aires, Buenos Aires 1113, Argentina
Instituto de Estudios de la Inmunidad Humoral (IDEHU), CONICET-Universidad de Buenos Aires, Buenos Aires 1113, Argentina

**Angeles Zorreguieta**
Fundación Instituto Leloir (FIL), IIBBA (CONICET-FIL), Buenos Aires 1405, Argentina
Departamento de Química Biológica, Facultad de Ciencias Exactas y Naturales, Universidad de Buenos Aires, Buenos Aires 1428, Argentina

**Yutaka Ueda, Kota Mashima and Hidetoshi Kamimura**
Department of Pharmacy, Fukuoka University Hospital, Fukuoka 814-0180, Japan

**Satoshi Takagi**
Departments of Plastic, Reconstructive and Aesthetic Surgery, Faculty of Medicine, Fukuoka University, Fukuoka 814-0180, Japan

**Shuuji Hara**
Department of Drug Informatics, Faculty of Pharmaceutical Sciences, Fukuoka University, Fukuoka 814-0180, Japan

**Shiro Jimi**
Central Lab for Pathology and Morphology, Faculty of Medicine, Fukuoka University, Fukuoka 814-0180, Japan

**Muzaffar Hussain**
Institute of Medical Microbiology, University Hospital Münster, 48149 Münster, Germany

**Christian Kohler**
Friedrich Loeffler-Institute of Medical Microbiology, University Medicine Greifswald, 17475 Greifswald, Germany

**Karsten Becker**
Institute of Medical Microbiology, University Hospital Münster, 48149 Münster, Germany
Friedrich Loeffler-Institute of Medical Microbiology, University Medicine Greifswald, 17475 Greifswald, Germany

**Paula Constanza Arriola Benitez, Ayelén Ivana Pesce Viglietti, Guillermo Hernán Giambartolomei and María Victoria Delpino**
Instituto de Inmunología, Genética y Metabolismo (INIGEM), Universidad de Buenos Aires, CONICET, Buenos Aires 1120, Argentina

**María Mercedes Elizalde and Jorge Fabián Quarleri**
Instituto de Investigaciones Biomédicas en Retrovirus y Sida (INBIRS), Universidad de Buenos Aires, CONICET, Buenos Aires 1121, Argentina

**Sonal Gupta and Surender Mohan**
Laboratory of Molecular Biology and Genetic Engineering, School of Biotechnology, Jawaharlal Nehru University, New Delhi 110067, India

**Vikas Kumar Somani**
Laboratory of Molecular Biology and Genetic Engineering, School of Biotechnology, Jawaharlal Nehru University, New Delhi 110067, India
Department of Oncology, Washington University School of Medicine, St. Louis, MO 63110, USA

**Somya Aggarwal**
Laboratory of Molecular Biology and Genetic Engineering, School of Biotechnology, Jawaharlal Nehru University, New Delhi 110067, India
Department of Molecular Microbiology, Washington University School of Medicine, St. Louis, MO 63110, USA

**Rakesh Bhatnagar**
Laboratory of Molecular Biology and Genetic Engineering, School of Biotechnology, Jawaharlal Nehru University, New Delhi 110067, India
Banaras Hindu University, Varanasi, Uttar Pradesh 221005, India

**Lucía Zavattieri, Mariana C. Ferrero, Iván M. Alonso Paiva, Agustina D. Sotelo, Andrea M. Canellada and Pablo C. Baldi**
Facultad de Farmacia y Bioquímica, Cátedra de Inmunología, Universidad de Buenos Aires, Buenos Aires 1113, Argentina
CONICET-Universidad de Buenos Aires, Instituto de Estudios de la Inmunidad Humoral (IDEHU), Buenos Aires 1033, Argentina

# Index

## N



## O



## P



## R



## S



## T



## V



## W



## Z

Printed in the USA
CPSIA information can be obtained
at www.ICGtesting.com
JSHW051405091023
49903JS00006B/286